Axon Growth and Guidance

ADVANCES IN EXPERIMENTAL MEDICINE AND BIOLOGY

Recent Volumes in this Series

Volume 614
OXYGEN TRANSPORT TO TISSUE XXIX
Edited by Kyung A. Kang, David K. Harrison, and Duane F. Bruley

Volume 615
PROGRAMMED CELL DEATH IN CANCER PROGRESSION AND THERAPY
Edited by Roya Khosravi-Far, and Eileen White

Volume 616
TRANSGENIC MICROALGAE AS GREEN CELL FACTORIES
Edited by Rosa León, Aurora Graván, and Emilio Fernández

Volume 617
HORMONAL CARCINOGENESIS V
Edited by Jonathan J. Li

Volume 618
HYPOXIA AND THE CIRCULATION
Edited by Robert H. Roach, Peter Hackett, and Peter D. Wagner

Volume 619
PROCEEDINGS OF THE INTERAGENCY, INTERNATIONAL SYMPOSIUM ON CYANOBACTERIAL HARMFUL ALGAL BLOOMS (ISOC-HAB)
Edited by H. Kenneth Hudnell

Volume 620
BIO-APPLICATIONS OF NANOPARTICLES
Edited by Warren C.W. Chan

Volume 621
AXON GROWTH AND GUIDANCE
Edited by Dominique Bagnard

A Continuation Order Plan is available for this series. A continuation order will bring delivery of each new volume immediately upon publication. Volumes are billed only upon actual shipment. For further information please contact the publisher.

Axon Growth and Guidance

Edited by

Dominique Bagnard, Ph.D.

Laboratoire de Neurobiologie du Developpement et de la Regeneration (LNDR), Centre de Neurochimie de CNRS, Strasbourg, France

Springer Science+Business Media, LLC

Landes Bioscience

Springer Science+Business Media, LLC
Landes Bioscience

Printed in the U.S.A.

Springer Science+Business Media, LLC, 233 Spring Street, New York, New York 10013, U.S.A.
http://www.springer.com

Please address all inquiries to the Publishers:
Landes Bioscience, 1002 West Avenue, 2nd Floor, Austin, Texas 78701, U.S.A.
Phone: 512/ 637 6050; FAX: 512/ 637 6079
http://www.landesbioscience.com

Axon Growth and Guidance, edited by Dominique Bagnard, Landes Bioscience / Springer Science+Business Media, LLC dual imprint / Springer series: Advances in Experimental Medicine and Biology.

ISBN: 978-0-387-76714-7

Library of Congress Cataloging-in-Publication Data

Axon growth and guidance / edited by Dominique Bagnard.
p. ; cm. -- (Advances in experimental medicine and biology ; v. 621)
Includes bibliographical references and index.
ISBN 978-0-387-76714-7
1. Axons--Physiology. 2. Nervous system--Regeneration. 3. Nervous system--Growth. I. Bagnard, Dominique. II. Series.
[DNLM: 1. Axons--physiology. 2. Nerve Regeneration. 3. Nervous System--growth & development. W1 AD559 v.621 2007 / WL 102.5 A9703 2007]
QP363.A92 2007
612.8--dc22

2007042192

About the Editor...

DOMINIQUE BAGNARD, Ph.D., is an assistant professor directing the program of cellular and molecular neurobiology at the Université Louis Pasteur in Strasbourg. Dr. Bagnard's research focuses on the role of guidance molecules of the semaphorin family during cortical development. His research covers the signalling pathways triggered by semaphorins and their potential clinical significance during nerve regeneration. His lab recently demonstrated the existence of a functional link between matrix metalloproteinases and semaphorins to control axon guidance. Dominique Bagnard serves as an executive editor for *Current Pharmaceutical Design* and is the Editor-in-Chief of *Cell Adhesion and Migration*.

DEDICATION

To Antoine et Nathan, my favourite Tom Thumb story …

PREFACE

The complex architecture of neuronal networks together with the extraordinary associated functions make the nervous system a fascinating biological structure. The considerable work performed to explore this cellular machinery is nowadays successful because the mystery of nervous system development is being unravelled. As described in their outstanding review published 10 years ago in *Science*,[1] Marc Tessier-Lavigne and Corey Goodman—the pioneers of the molecular era of axon guidance—summarized the assembly of nervous system connections as a subtle game of attraction and repulsion of neuronal growth cones. The cellular ballet ensuring the formation of billions of synapses, which ultimately gives rise to the highest cognitive functions, is primarily orchestrated by a step-by-step mechanism of growth driven by multiple molecular cues. While our general concept of axon guidance remains identical, a profound evolution of our knowledge of the molecular identity of the guidance cues together with their interactions and signalling pathways occurred over the past ten years.

This book proposes an exhaustive and updated view of the current knowledge of the molecular and cellular mechanisms ensuring axon growth and guidance. An introductory chapter by C. Bouquet and F. Nothias will remind the readers of all the features of a growth cone and the mechanisms controlling its growth. From there, one enters a fabulous journey with a growth cone, a Tom Thumb story filled with molecular encounters and complex interactions leading to one of the most fantastic developmental achievements: the nervous system wiring.

The journey starts with a description of the classical guidance signals such as the netrins (by S.W. Moore, and colleagues), the semaphorins (by E. Koncina and colleagues), the ephrins (by M. Reber and colleagues) or the Slit family (by A. Chédotal). The question of the exact definition of a guidance signal is addressed in an exquisite chapter by S. Guthrie discussing whether neurotrophic factors can be considered as guidance cues. This complex question is also considered in a chapter by H. Kamiguchi presenting the role of adhesion molecules during axon guidance and T. Ruediger and J. Bolz reporting compelling evidence for a guidance effect of neurotransmitters. The diversity of the molecular cues used by growth cones to reach their targets is finally illustrated by F. Charron and M. Tessier-Lavigne in a chapter describing the unexpected role of morphogenes as direct guidance signals.

Much work has been done to elucidate the signalling pathways triggered by guidance signals. The complexity of the required intracellular pathways is detailed in all chapters, but we decided to include a special focus by M. Piper and colleagues on the role of cyclic nucleotides which appear to be the key regulators of the biological outcome of guidance signals in terms of attraction or repulsion. The identification of an additional level of regulation of the guidance cues is also illustrated in the chapter by C. Guirland and J.Q. Zheng presenting how lipid rafts interfere with guidance mechanisms. Hence, because 80% of the cells in the nervous system are glial cells, the story ends up with the role of glial cells in mediating axon guidance (by A.R. Learte and A. Hidalgo).

I wish to thank the authors whose contributions in this book provide excellent a goldmine of information for students and researchers interested by this challenging question of the development of axonal connections.

Dominique Bagnard, Ph.D.

[1] Tessier-Lavigne M, Goodman CS. The molecular biology of axon guidance. Science 1996; 274(5290):1123-33.

PARTICIPANTS

Dominique Bagnard
Laboratoire de Neurobiologie du Developpement et de la Regeneration (LNDR)
Centre de Neurochimie de CNRS
Strasbourg
France
Email: Dominique.Bagnard@inserm.u-strasbg.fr

Juergen Bolz
Institute für Allgemeine Zoologie und Tierphysiologie
Universitat Jena
Jena
Germany
Email: bolz@pan.zoo.uni-jena.de

Céline Bouquet
Université Pierre et Marie Curie
Paris
France

Frédéric Charron
Molecular Biology of Neural Development
Institut de Recherches Cliniques de Montréal (IRCM)
and
Canada Department of Medicine
University of Montreal
Montreal, Quebec
Canada

Alain Chédotal
Université Paris
Paris
France
Email: chedotal@infobiogen.fr

Bertand Gonthier
Physiopathologie du Systéme Nerveux
Strasbourg
France

Carmine Guirland
Department of Neuroscience and Cell Biology
Robert Wood Johnson Medical School
University of Medicine and Dentistry of New Jersey
Piscataway, New Jersey
U.S.A.

Sarah Guthrie
MRC Centre for Developmental Neurobiology
King's College London
London
U.K.

Robert Hindges
MRC Centre for Developmental Neurobiology
King's College London
London
U.K.

Alicia Hidalgo
NeuroDevelopment Group
School of Biosciences
University of Birmingham
Birmingham
U.K.

Christine Holt
Department of Anatomy
University of Cambridge
Cambridge
U.K.
Email: ceh@mole.bio.cam.ac.uk

Hiroyuki Kamiguchi
Laboratory for Neuronal Growth
Mechanisms
RIKEN Brain Science Institute
Wako, Saitama
Japan
Email: kamiguchi@brain.riken.jp

Timothy E. Kennedy
Montreal Neurological Institute
McGill University
Montreal, Quebec
Canada

Eric Koncina
Physiopathologie du Systéme Nerveux
Strasbourg
France

Anabel R. Learte
NeuroDevelopment Group
School of Biosciences
University of Birmingham
Birmingham
U.K.
Email: anabelearte@gmail.com

Greg Lemke
Molecular Neurobiology Laboratory
The Salk Institute
La Jolla, California
U.S.A.

Simon W. Moore
Montreal Neurological Institute
McGill University
Montreal, Quebec
Canada
Email: simon.moore@mail.mcgill.ca

Fatiha Nothias
Université Pierre et Marie Curie
Paris
France
Email: fatiha.nothias@snv.jussieu.fr

Michael Piper
Department of Anatomy
University of Cambridge
Cambridge
U.K.

Michael Reber
Centre de Neurochimie
Strasbourg
France
Email: michael.reber@inserm.u-strasbg.fr

Lise Roth
Physiopathologie du Systéme Nerveux
Strasbourg
France

Tina Ruediger
Institute für Allgemeine Zoologie
und Tierphysiologie
Universitat Jena
Jena
Germany

Marc Tessier-Lavigne
Genentech, Inc.
South San Francisco, California
U.S.A
Email: tessier-lavigne.marc@gene.com

Francis van Horck
Department of Anatomy
University of Cambridge
Cambridge
U.K.

James Q. Zheng
Department of Neuroscience
and Cell Biology
Robert Wood Johnson Medical School
University of Medicine and Dentistry
of New Jersey
Piscataway, New Jersey
U.S.A.

CONTENTS

CHAPTER 1

Molecular Mechanisms of Axonal Growth

Céline Bouquet and Fatiha Nothias*

Abstract

Outgrowth of axons during neuronal development, as well as their regeneration after injury, of the adult nervous system is controlled by specific extracellular cues which are diffusible, or bound to cell membranes or extracellular matrix. The exact molecular mechanisms through which these extracellular signals are integrated by the growing axon, are not yet well defined. However, it is widely accepted that most, if not all, signaling cascades triggered by guidance cues eventually converge onto the cytoskeleton. The action of extracellular guidance factors is thus modulated not only by specific membrane receptors, but also by cytoskeletal and cytoskeleton-associated molecules within the axon. In fact, the cytoskeleton represents a point of convergence and integration of both neuron-intrinsic and extrinsic factors. Moreover, in recent years, there has been increasing evidence for the involvement of a coordinated cross-talk between actin filaments and microtubules, the two main components of the growth cone cytoskeleton. Their reorganization is complex and involves numerous cytoskeleton-associated proteins whose function is regulated via activation or inhibition of particular signaling pathways.[1-4]

Introduction

The growth cone, highly motile distal tip of the axon, shares many properties with other motile structures, such as the leading edge of migrating cells. This is reflected in a similar cytoskeletal organization of these subcellular compartments, and the use of common signaling pathways, such as the one involving Rho-GTPases (see below). Despite these similarities, the behavior of neurons appears more complex than that of other cell types, in that they extend very long processes, and exhibit quite "sophisticated" responses when confronted to extracellular cues. Expression of cytoskeleton-associated molecules specific for the neuronal growth cone may, at least in part, explain some unique features of this motile structure (for review refs. 5-8).

Here, we will describe in some detail the cytoskeletal network within the neuronal growth cone, and how its organization is regulated in response to extracellular factors by integration of signaling pathways.

The Neuronal Growth Cone and Its Cytoskeletal Organization

Neurites should be thought of as exceptionally differentiated cellular processes. The growth cone tipping an axon (or dendrite) is an extremely motile and dynamic structure that explores the environment. To guide an axon towards the appropriate target, the growth cone fulfills different functions: it acts as sensor of environmental cues, signal transducer, and motility device. Growth cone advance is mediated by the polymerization/depolymerization of cytoskeletal

*Corresponding Author: Fatiha Nothias—UMR7101/IFR83, CNRS-UPMC, Bat. A/3ème Etage, Case-2, Université P&M Curie, 7 Quai Saint Bernard, 75005-Paris, France. Email: fatiha.nothias@snv.jussieu.fr

Axon Growth and Guidance, edited by Dominique Bagnard. ©2007 Landes Bioscience and Springer Science+Business Media.

elements, and their specific interactions. The axonal cytoskeleton is composed of three main filamentous polymers: neurofilaments, microtubules and actin microfilaments. Within the growth cone, microtubules and actin filaments are actually the major cytoskeletal components, and have been the focus of most studies. Their spatial organization and relative position in the growth cone define different, functionally specific zones, described in Figure 1.

The Peripheral Domain

The peripheral domain (P-domain) is the most distal part of the growth cone, a highly dynamic, actin-rich structure. This domain bears lamellipodia, membranous flat veil-like protrusions, from which extend many filopodia. These very thin, finger-like structures contain mainly actin filament bundles, and undergo permanent elongation and retraction cycles as they organize their content in response to the environment.[9,10] In the P-domain, equilibrium between actin polymerization and depolymerization (actin "treadmilling") constantly generates protrusion forces, and retrograde flow of actin (see below).

The Transition Zone

The transition zone (T-zone) is situated at the interface between the actin-rich P-domain and the MT-rich central domain. The molecular motor myosin, concentrated in the T-zone, can serve to contract the actin network, thereby inducing the formation of an actin-filament arc.[11] Movements of this arc, in association with retrograde actin flow, limit the penetration of MTs into the P-domain.

The Central Domain

The central domain (C-domain) represents the main site of MT polymerization. Neurofilaments, which transport vesicles and organelles along with the MTs, are also present. The size of the C-domain varies in correlation to the growth mode of the axon: Relatively large when the growth cone is pausing, whereas the C-domain exhibits a thinner shape during fast advance mode.

Actin Filaments and Associated Proteins

Actin filaments (AFs) are helical polymers formed by addition of ATP-actin monomers. AFs are polarized structures characterized by a "pointed" and a "barbed" end. Dissociation of ADP-actin is favored at the pointed end, suggesting that polymerization occurs at the barbed end, and depolymerization at the pointed end.[7,12]

Actin is present in both the P-domain and the T-zone, where it is organized in two different types of networks:[10,13,14] filopodia are composed of thick actin bundles, while in lamellipodia, AFs are organized in a loose meshwork. Similarly, contraction of the actin meshwork in the T-zone by myosin action results in formation of a thick actin arc, oriented perpendicularly to the axon.[11]

In the P-domain, AFs polymerize close to the distal membrane, and polymers are retrogradely transported to the T-zone by a myosin-dependent mechanism.[15] Increased contractile forces in the T-zone then induce severing and depolymerization of AFs. This permanent actin treadmilling accounts for the high dynamics of the P-domain. Moreover, the retrograde actin flow generates a backward force suspected to limit MT invasion into the P-domain.

Dozens of actin-associated proteins have been described in the neuronal growth cone, and were classified according to their function (for review see ref. 16). There are two main groups regulating actin polymerization/ depolymerization:

Actin Nucleation/Polymerization

Actin nucleation/polymerization factors increasing the number of free barbed ends incidentally increase actin polymerization. The Arp2/3 complex, thus, not only favors de novo actin polymerization, but also by binding sideways to preexisting filaments, creating a new branch and hence a new barbed end (for review see ref. 17). Members of the formin protein family

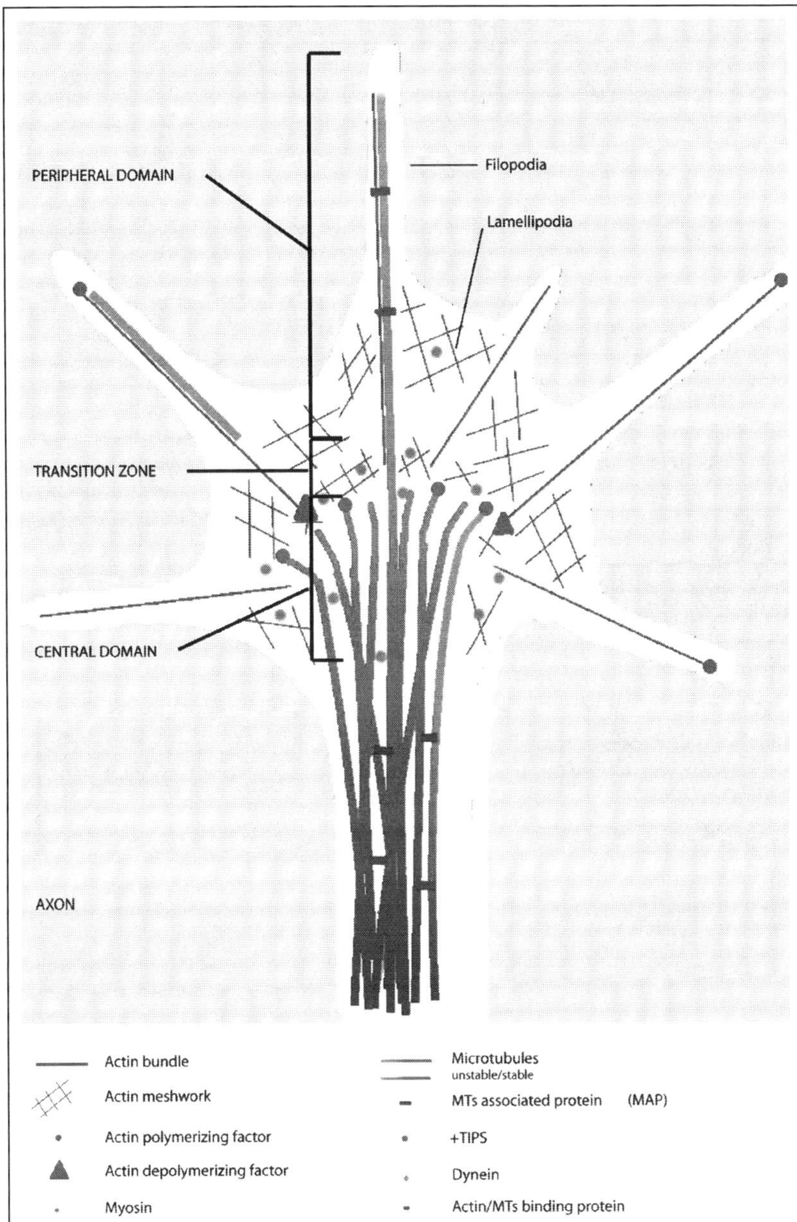

Figure 1. Cytoskeleton growth cone organization. The growth cone, tipping an axon, is divided in three different, functional zones. The C-domain is the polymerization site of MTs, which are thereafter stabilized in the axon shaft. The T-zone limits MTs penetration in the C-domain, and contains a high density actin meshwork associated with myosin. The P-domain is very dynamic and mainly contains actin, organized in bundles in filopodia, and meshwork in lamellipodia. Transient interactions between actin filaments and MTs are observed in the P-domain that are mediated by still unknown factors. MTs- and actin- associated proteins regulate their transport, polymerization and stabilization.

bind to the barbed end of elongating actin polymers, enhance filament elongation, and prevent binding of actin capping molecules (for review see ref. 18).

Actin Depolymerization/Severing

Actin depolymerizing factors (ADFs), as well as cofilins, are important regulators of actin dynamics in the growth cone. Although encoded by different genes, ADFs and cofilin have very similar effects, are regulated by reversible phosphorylation, and colocalize in the cells (for review see ref. 7). When phosphorylated, both bind to the rear end of actin filaments, generating actin fragments. Interestingly, actin severing leads to generation of new free barbed ends, thereby promoting actin polymerization, a mechanism perpetuating the retrograde actin flow (for review see refs. 7,19). Other proteins, such as gelsolin, stop actin polymerization by capping barbed ends, and thereby induce depolymerization (for review see ref. 20).

Thus, modulation of actin polymerization by extracellular guidance cues via actin-associated proteins provides a mechanism to regulate the progress of growth cones.

Microtubules and Associated Proteins

Microtubule protofilaments are formed by spontaneous association of α/β tubulin heterodimers. 13 such protofilaments finally associate to form a hollow microtubule of a diameter of 25 nm (for review see ref. 21). Regulation of expression of different α and β tubulin isoforms during axonal development and regeneration regulates MT stability.

The orientation of tubulin monomers makes MTs intrinsically polarized structures with a "plus" and a "minus" end. In the axon, plus ends are oriented distally toward the growth cone. Depolymerization mainly occurs at the minus end, while a constant cycle of polymerization/depolymerization takes place at the plus end.[22,23]

The frequency of polymerization/depolymerization or "rescue/catastrophe" events, as well as the duration of pauses in between, characterizes the "dynamic instability" of microtubules. The term dynamic instability describes an intrinsic property of MTs that allows them to switch abruptly between phases of elongation and rapid shortening.[22] MTs are organized in parallel bundles throughout the axon shaft, and splay out when they enter the growth cone C-domain.[13,24] During pauses, MTs extend loops into the C-domain, and breakage of these loops upon regrowth results in highly dynamic, small polymers capable to enter the P-domain and associate with actin bundles. Indeed, while MTs were previously thought to be restricted to the C-domain, recent progress in imaging techniques has provided evidence for MT-actin interactions within the P-domain. This interaction is fundamental for outgrowth, guidance and branching of the axon.[11,25-28] Local stabilization of MTs is also tightly regulated during these events[29,30] by specific post-translational modifications on MTs, and by their interaction with specific associated proteins:

Post-Translational Modifications

Post-translational modifications of MTs include detyrosination/tyrosination, acetylation, phosphorylation, polyglutamylation and polyglycilation (for review see ref. 21). Unmodified tyrosinated tubulin polymers are enriched in the distal part of the axon, while modified detyrosinated or acetylated isoforms are found in the proximal part of the axon, on "older" and stable MTs.[31] Although these modifications do not have a direct effect on MT stabilization,[32] tubulin modifications are frequently used as markers of MT stability. They may, however, facilitate localization and interaction of microtubule binding proteins, such as plus end-tracking proteins (+TIPS[33]) and Microtubule Associated Proteins (MAPs[34,35]).

Proteins Associated to MTs

Proteins associated to MTs include two groups of proteins that interact with MTs and regulate their dynamic instability.

Structural MAPs such as MAP1A, MAP1B, MAP2, and Tau, bind, bundle, and stabilize MTs. Their association with MTs is regulated by post-translational modifications of tubulin, as well as their own post-translational modification, such as phosphorylation (for review see ref. 36). MAPs are particularly abundant in the nervous system, and their subcellular localization is strictly regulated. Some MAPs are preferentially associated with neuronal processes: MAP2 is concentrated in dendrites, while tau and MAP1B are mainly found in axons (for review see refs. 37,38). As a consequence, the spatial distribution of MAPs defines subcellular zones, in which MTs are more or less stabilized. MAPs are generally important both during development and in the adult nervous system. Maturation of the nervous system is accompanied by a transition from MAPs typically expressed during the phase of axon growth, to other MAPs characteristic of mature neurons.[39,40] Certain MAPs, such as MAP1B and tau, are present in the growth cone, and were shown to play an important role in neurite outgrowth from embryonic neurons in vivo and in vitro.[41-44]

Other MAPs, identified more recently,[46,47] act as potent MT destabilizers. Among these are stathmin and SCG10, members of the same gene family, which are expressed in neurons and promote MT depolymerization by increasing the rate of catastrophes (for review see ref. 48). SCG10 and stathmin are considered as growth-associated proteins, and their expression correlates with neurite outgrowth.

Although the dynamic state of MTs has been shown to be important for neurite elongation and growth cone turning, it is still not clear how MT dynamics are regulated. In fact, MTs are known to be particularly labile within the growth cone,[31,49] despite the rather high concentration of MT-stabilizing MAPs, such as MAP1B and tau. Therefore, it has been proposed that the potent MT destabilizer, such as SCG10, might counteract the activity of stabilizing MAPs, contributing to the regulation of MT dynamics.[46,48]

Recently, a novel type of MT binding proteins called +TIPs has been identified as being specifically associated with the distal ends of growing MTs. These proteins have gained considerable interest with respect to the regulation of MT dynamics and the intracellular transport via MTs (for ref and review, see refs. 50,51), and also due to their anchorage to actin filaments and adhesion sites. A few of them have been detected in neurons, namely, cytoplasmic linker protein-170 (CLIP-170), CLIP-115, end-binding protein 1 (EB1) and EB3. Functions of these proteins in growth cone MTs remains to be determined.

Intermediate Filaments

Neurofilaments (NFs) are the major intermediate filaments in neurons. The NF network is composed of a NF-L (low molecular weight NF, 70 kD) core, associated to NF-M (medium molecular weight NF, 150 kD) and NF-H (high molecular weight NF, 200 kD) chains.[53] The function of NFs in transport of vesicles, membrane material and organelles has been extensively studied (for review see ref. 54). In contrast, even if NFs are found in the C-domain of the growth cone, they do not seem to interact with axonal growth and pathfinding. Thus, transgenic mice lacking axonal NFs are perfectly viable, and do not present any major defect in their neural connections.[55]

Molecular Motors

These molecules present a molecular motor domain capable of generating forces on cytoskeletal polymers by means of ATP hydrolysis. They serve in transporting vesicles back and forth along the axon shaft, and in addition, can generate forces on the cytoskeleton by moving polymers relatively to each other.[56] Kinesin and dynein proteins are microtubule-dependent motors and the polarized structure of MTs induces specificity in the direction of motor molecules. Most kinesins move towards the plus end of MTs, whereas dynein complex moves towards the minus end (for review see ref. 57). Myosin proteins are actin-dependant motors and rather move to the plus end of actin filaments.[58]

The diverse cytoskeletal proteins expressed in the growth cone act in concert to mediate axonal growth and pathfinding. They are regulated by extracellular cues, but also by the axon-intrinsic program. Thus, throughout development and regeneration, the expression of particular components, and their transport and final localization in the axon are tightly regulated.

The subsequent assembly of components and their interactions in the growth cone eventually modulates axonal outgrowth and pathfinding.

Mechanisms of Axonal Elongation

Synthesis of Cytoskeletal Proteins

The bulk of new cytoskeletal proteins is produced in the cell body. Recently, however, there is increasing evidence for a local synthesis in growing and regenerating axons, at least in small amounts (estimated at 5% [59]), which can play a crucial functional role. Thus, it has been demonstrated that ribosomal proteins, translational initiation factors, and ribosome-bound mRNA are present in axons. Moreover, protein synthesis occurs even when processes are separated from their cell bodies.[60,61] The rapidly growing list of identified intra-axonally synthesized proteins includes cytoskeletal proteins (intermediate filaments as well as actin and tubulin), heat shock proteins, endoplasmic reticulum proteins, metabolic proteins, anti-oxidant proteins, and proteins associated with neurodegenerative diseases (see ref. 62). When communication between processes and the cell body is interrupted by axotomy or colchicine treatment, blocking local protein synthesis in regenerating axons results in rapid retraction of growth cones, indicating a physiological importance for local synthesis during axonal regeneration[60] as well as during development[63] to respond to guidance factors.

Cytoskeletal Protein Transport

After their synthesis, cytoskeletal proteins have then to be transported to the site of axonal growth. The first studies on axonal transport were performed in the adult during axonal regeneration, and used radio labeled-methionine for tracing of newly synthesized proteins. They showed a correlation between the rate of axonal regeneration, and the rate of the slow axonal component (SC[64]). Tubulin and actin are transported in two peaks, differing in their velocity and content. In mammals, the slower peak, SCb, is mostly composed of tubulin, while actin moves faster in association with the SCa peak (reviewed in ref. 65).

The polymerization status of actin and tubulin during their transport, as well as the exact mechanism of their transport are still unclear and have been much debated in recent years. Novel methods using fluorescent proteins and time lapse imaging may now yield new insight into this problem.[66] Two models have been proposed: The classical "cargo" model assumes that tubulin and neurofilament polymer transport uses the classical motor molecules. The "sliding filament" model[67,68] suggests that short tubulin polymers can be moved anterogradely on longer MTs by dynein. NF transport was suspected to be linked to this MT transport with the NF "piggy backed" on MTs, but recent evidences suggest that it may rather rely on the classical cargo model.[67,68]

Less is known about anterograde transport of actin. Myosin seems to be the motor for at least a subpopulation of para-axially aligned actin filaments.[68]

Axonal Elongation

During axonal elongation, 3 phases can be distinguished (reviewed in refs. 16,69): In the initial protrusion phase, lamellipodia and filopodia extend from the tip of the axon, forming the growth cone. This phase is mainly governed by actin dynamics, which in turn, are regulated by Rho family GTPases, but is also modulated by MTs dynamics.[70] The engorgement phase that follows protrusion, consists in the invasion of MTs and organelles into the growth cone. It depends on the dynamic instability of MTs, since inhibition of these dynamics leads to a reduction in axonal growth.[27,71,72] During the final consolidation phase, the formation of actin protrusion stops, and MTs become bundled. This phase probably relies on the activity and interaction of microtubule- and actin-associated proteins, although it is still not well elucidated.

Axonal elongation is modulated by extracellular factors that the growth cone senses in the environment. Extracellular guidance cues elicit diverse intracellular signaling cascades. Here we will particularly focus on the signaling mediated by Rho-GTPases, as the consequences of their activation on reorganization of the cytoskeleton has been well characterized.

Regulation of the Cytoskeleton by Extracellular Cues, Role of Rho-GTPases

Regulation of Rho-GTPases

Rho-GTPases act as "molecular switches" by oscillating between an active, GTP-bound and an inactive, GDP-bound state. The three best-characterized members of this family are Rho, Rac and Cdc42. Their regulatory function on the actin cytoskeleton during axon outgrowth and guidance has been extensively demonstrated (for review see refs. 73-75). Rho, Rac and Cdc42 are generally considered to regulate formation of stress fibers (actin- and myosin-rich structures), lamellipodia, and filopodia, respectively.

The activity of Rho-GTPases is itself modulated by three families of factors. GAPs (GTPases activating proteins) facilitate hydrolysis of GTP by GTPases, and hence favor the inactive, GDP-binding state of Rho-GTPases.[76] GEFs (guanine nucleotide exchange factors) activate GTPases by facilitating GDP/GTP exchange.[77] Finally, GDIs (guanine nucleotide dissociation inhibitors) inhibit GDP dissociation and maintain GTPases in an inactive state. GEFs/GDIs/GAPs can be either specific for a given GTPase, or act simultaneously on several molecules.

In addition, Rho-GTPase activity can also be modulated by second messenger cyclic nucleotides. Indeed, cAMP-dependent protein kinase A (PKA) reduces Rho-GEF activity,[78] while activating Rho-GDIs.[79] RhoA is also directly inhibited upon phosphorylation by PKA.[80,81] In addition to its action on Rho-GTPases, PKA can directly act on their downstream targets, including cytoskeletal components (see chapter by Piper et al.).

Binding of permissive or inhibitory factors to their neuronal receptors induces different signaling cascades, which in turn leads to activation/inactivation of Rho-GTPases. Since Rho, Rac and Cdc42 may also interact and thereby modulate themselves, it seems that the balance between the activities of different GTPases, rather than activation of a single group, will control the axonal response.

After binding to their membrane receptors, repulsive guidance cues activate Rho, while inhibiting Cdc42 and Rac activity by acting on GTPase modulators (see Fig. 2). For example ephrines, via Src kinase and RasGAP, inhibit p190RhoGAP.[82,83] In parallel, the RhoGEF ephexin is activated,[84] which leads to RhoA activation.

Semaphorins activate Rho and inhibit Rac via a slightly different mechanism. The semaphorin receptors Plexins are able to directly bind Rac and Rho. This binding then activates Rho, while it sequesters Rac and inhibits its interaction with its downstream effector Pak.[85]

In contrast, outgrowth- and regeneration-permissive factors activate Cdc42 and Rac, while inhibiting Rho. Neurotrophins for example, besides their trophic effect mediating gene transcription in the cell body, activate GEFs via a PI3-K signaling pathway,[86,87] and thereby Cdc42. Binding of the neurotrophin receptor p75 to RhoA inactivates this Rho-GTPase, and further contributes to the attractive effect of neurotrophins.[88] Moreover, it has recently been demonstrated that RhoA-kinase and myosin-II are required for the maintenance of growth cone polarity and guidance mediated by nerve growth factor,[89] suggesting that localized activation of different RhoGTPases is necessary for axonal pathfinding.

Effect of Rho-GTPases on the Cytoskeleton

Activation of Rho-GTPases leads to an important cytoskeletal remodeling. Their effects converge on three main systems: actin polymerization/depolymerization, actin/myosin contractility, and microtubule reorganization, as represented in Figure 2.

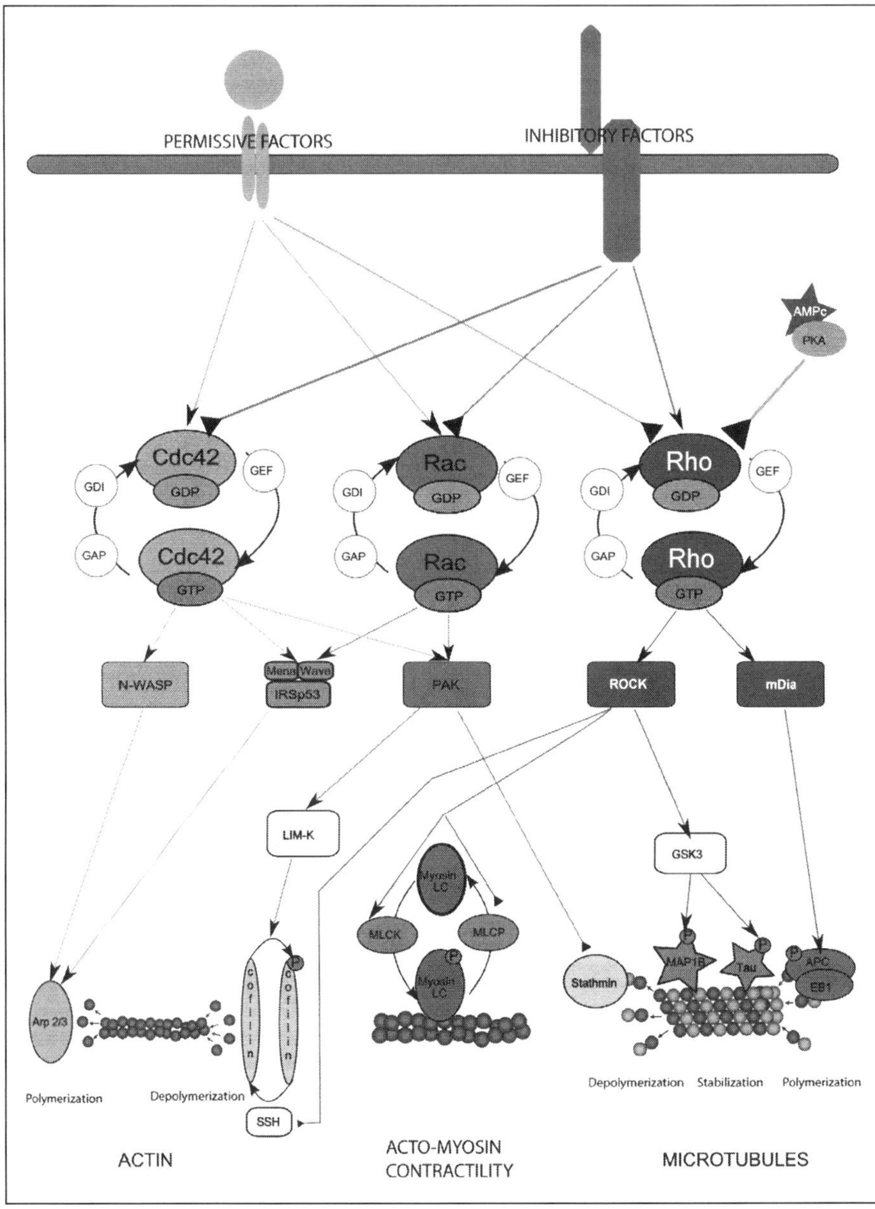

Figure 2. Rho GTPases signalling to the cytoskeleton. Rho-GTPases are controlled via their associated partners by extracellular cues. Specific effectors mediate actions of Rho-GTPases on the cytoskeleton. They focus on three main effects: actin polymerization/depolymerization, regulation of acto/myosin contractility, and microtubules polymerization/depolymerization/ stabilization.

Regulation of Actin Polymerization

Upon Cdc42 activation, N-WASP (N-Wiskott-Aldrich related protein), activates the Arp2/3 factor.[90] Arp2/3 stimulates de novo actin polymerization (for review see ref. 91). This factor is also activated by Rac effectors IRSp53 and WASP related protein SCAR/WAVE.[92]

Conversely, actin depolymerization is induced by cofilin/ADF (Actin Depolymerizing Factor). Cofilin is activated by phosphatase SSH (slingshot), and inhibited by LIM kinase.[93] Rac and Cdc42 activate LIM-K,[94] while Rho and its downstream effector ROCK (Rho-associated kinase) activate SSH, and as a consequence, promote actin depolymerization.[95]

Regulation of Acto-Myosin Contractility

Rho activation of ROCK increases myosin contractility by two converging pathways. ROCK activation induces phosphorylation of myosin light chain 2 (MLC2) by activating myosin light chain kinase, while at the same time inhibiting myosin light chain phosphatase.[96,97] An increase in MLC phosphorylation and myosin activity then leads to contraction of the actin network.

Regulation of Microtubules

Rho-GTPases have been extensively described as actin modulators, but a growing number of studies also suggest a function in regulation of MT dynamics. In particular, it has been described that after Rho activation, the formin mDia not only favors actin nucleation,[98] but also stabilizes MTs via TIPS proteins EB1 and APC.[99] Rac1 activity has a MT stabilizing effect, since the Rac effector PAK inhibits the MT-severing protein Stathmin/Op18.[100,101] These pathways have mainly been characterized in nonneuronal cells, but we assume that the same or similar mechanisms should also regulate neuronal cell motility. In neuronal cell lines, the Rho/ROCK pathway has been shown to induce hyperphosphorylation of two major MAPs, tau and MAP1B, by GSK3β,[102] thereby destabilizing MTs. Interestingly, Rho-GTPase activity is inversely regulated by MT polymerization/depolymerization in nonneuronal cells: MT depolymerization leads to release of a RhoA activating GEF,[103] whereas their polymerization activates Rac1 by a still unknown mechanism.[104]

This "retrograde signaling" from MTs to Rho-GTPases could be one way of coregulating MTs and actin dynamics during motile events. Indeed, a strict coordination of actin and MT systems is required for proper growth cone advance, turning and branching.[13,25,29,105,106] Besides those based on regulatory interactions involving Rho-GTPases, several hypotheses have been put forward for a model of coupling between actin and MTs (for review see refs. 2,8,107). Structural interactions mediated by MT- and actin-binding proteins or protein complexes might physically couple actin and MT movements (for review and see refs. 8,45). For example, in addition to their interaction with MTs, some MAPs directly or indirectly interact with actin filaments. Another hypothesis proposes that actin and MT movements are controlled by the equilibrium between forces generated by molecular motors on the 2 types of filaments. According to this hypothesis, a balance between backward forces generated by myosin and forward forces generated by dynein or kinesins should control advance or retraction of the axon.[56]

In summary, during axonal outgrowth, extracellular signals converge on the reorganization of cytoskeletal proteins, particularly actin filaments and microtubules, and thereby control the advance of the growth cone. On the other hand, specific expression and intrinsic modification of cytoskeletal proteins also modulates the neuronal response to extrinsic factors, allowing for diversity in the response to a specific guidance cue, and underlying the role of the cytoskeleton as a convergence point during axonal outgrowth.

During the past two decades, a huge amount of data has been acquired detailing the molecular mechanisms of axon outgrowth and guidance. Today, it seems possible to exploit these data also in view of a better understanding of phenomena related to axonal plasticity in adult nervous system. Thus, especially our growing knowledge of how exactly extracellular cues and intracellular pathways ultimately converge on the axonal cytoskeleton, is of particular interest for studies of axonal regeneration in the adult following a traumatic lesion.

Growth Cone of Regenerating Axons

In the adult nervous system, particularly of mammals, understanding why certain types of neurons regenerate their axons while others do not, may provide clues to establish a therapy for each type of lesion, be it traumatic, degenerative, or linked to developmental (genetic) anomalies.

In order to regenerate, adult neurons should have the intrinsic capacity to survive a traumatic or degenerative lesion, and to activate a cell-autonomous program that will end in plastic changes in their network. At the same time, this program is influenced by interactions between neurons and neighboring cells (glia in particular), mediated by cell- and substratum- adhesion molecules and their receptors, and by a variety of secreted factors into the extra-cellular space. A number of key players in these regenerative processes have already been identified. However, the relationship between individual molecular events, especially the triggering of gene expression and the corresponding cascade of signaling pathways, are still poorly understood. Here we summarize some relevant findings from studies that were undertaken to understand how the axonal cytoskeleton is reorganized in response to a lesion, particularly in response to axotomy.

Initiation of Axonal Regeneration after Axotomy

The regeneration of an amputated axon involves the transformation of a stable axonal segment, i.e., a stable structure specialized in propagating action potentials, into a highly motile and complex tip, a new growth cone, that will sense the surrounding environment and guide regenerating neurites to their targets. This is a critical step in the process of recovery from neural injury. Most of the studies on the initiation of the regeneration of damaged axon were done on Aplysia neurons[108,109] where it has been shown that cytoskeleton reorganization promotes a growth cone formation, allowing elongation of a new axon.

Gene Expression Recapitulates Developmental Program during Axonal Regeneration

In response to injury, such as axotomy, adult neurons shutdown their specific differentiated functions and activate growth program through local intracellular signaling cascade. Coordinated sequence of gene expression is induced for synthesis and transport of proteins that maintain axonal plasticity, growth cones are formed and ultimately functional synaptic contacts are restored. In vertebrates, these events occur only in the peripheral nervous system (PNS). In contrast, most lesions in the central nervous system (CNS) result in abortive regeneration associated with decrease in protein synthesis and may ultimately induce atrophy or death. The coordination of gene expression pattern after axonal injury is complex and is determined by both intrinsic factors to neurons as well as environment factors.

Cytoskeleton Synthesis in Injury-Induced Axonal Plasticity

The contribution of cytoskeleton proteins to the axonal regeneration process is crucial. Although several studies on neuronal cytoskeleton were undertaken during the development, its regulation during axonal regeneration remains poorly understood. Nevertheless, cytoskeleton proteins in regenerating axon undergo quantitative and qualitative changes in synthesis, organization and protein transport, similar to that of growing axon during development.[110]

In vertebrates, the recapitulation of the developmental cytoskeleton-protein expression has been mainly demonstrated in the peripheral sensory neurons of the dorsal root ganglia (DRG) and motor neurons (MN, in the CNS). These neurons are able to regenerate after peripheral injury.

MN or DRG axotomy is followed by an increase in levels of specific tubulin isoforms, as well as beta actin and peripherin, while levels of neurofilaments (NF), known to regulate the axon caliber[111] decrease. The down regulation of NF gene expression was suggested to facilitate supplying structural elements toward the distal end of the regenerating axon, resulting in a selective acceleration in the transport rate of tubulin and actin (for review see ref. 112).

It is important to note that although the capacity of axonal regeneration is attributed to PNS neurons, it is now well accepted that, in response to CNS injury, there are some neuronal populations able to initiate an axonal growth program to regenerate. This has been observed during the first days after axotomy of rubrospinal neurons, where the amounts of GAP43 and cytoskeleton proteins such actin and tubulin increase. Sustained only in few neurons, a decrease in these proteins occurs thereafter, associated with neuronal atrophy.[113] This study demonstrated that for some CNS neurons, the failure to regenerate after axotomy is not due to the failure to initiate gene-expression changes, but mainly to due to the environment. Depending on extracellular cues, the signals converge in growth cone on cytoskeleton protein reorganization to promote axonal regeneration (PNS) or to impede regeneration (CNS).

Following a traumatic lesion, several inhibitory guidance cues are expressed in the CNS and are partly responsible for the poor regenerative response of axotomized neurons. Besides the inhibitory effect of these molecules, loss of regenerative capacities in the adult nervous system is thought to coincide with myelination. Indeed, several inhibitors of regeneration have been described on the myelinating cells surface, and, in the adult, contribute to the failure of regeneration in the CNS (for review see ref. 114). Furthermore, the effect of these molecules, as well as other extracellular factors, on axonal regeneration is modulated by the intrinsic neuronal state.

How Intrinsic Neuronal Properties Control the Success of Regeneration?

cAMP

The best characterized example of intrinsic neuronal state controlling axonal regeneration comes from demonstration that elevating intracellular cAMP concentration of adult neurons to reach that of young neurons allow them to regenerate on a central myelin substrate.[115-117]

Binding of myelin inhibitors to their receptors induces, repulsive guidance cues during development, an elevation of Rho activity via a RhoGDI.[118,119] The exact mechanisms by which myelin inhibitors inhibit axonal regeneration are still unclear, but probably involve, as during the development, an actin depolymerization/contraction and a MTs destabilization. The precise effect of cAMP in overcoming myelin inhibition is not known. However, for a short phase, PKA action on Rho-GTPases may explain a part of the mechanism (see above). A second, transcription-dependant phase is induced by CREB activation. The multiple targets of this transcription factor are unknown, but one can reasonably consider that it may include cytoskeleton proteins. Moreover, it has been demonstrated that CREB activation leads to polyamines synthesis, which are known to modulate the cytoskeleton.[121,122]

Cytoskeleton Associated Protein, GAP43/CAP32

One of the first and best studied example of a cytoskeletal regeneration-associated protein is GAP43, a phosphoprotein associated with growth cones, whose expression is also induced in adult regenerating axons (for review see ref. 123). GAP43 is one of the final targets of calcium signals. GAP43 is an actin capping protein that blocks microfilament elongation and appears to be an important regulator of growth cone motility during development. Phosphorylation of GAP43 by the protein kinase C (PKC) affects its interaction with actin filament and might therefore trigger actin polymerization and hence regulating axonal outgrowth (for review see ref. 135). Furthermore, it plays a significant role in regeneration, together with CAP23, a functionally related protein that is also upregulated by injury (ref. 124 and references therein).

Cytoskeleton Associated Protein, MAP1B

Several studies strongly suggest an axon growth-related function of MAP1B that is regulated by phosphorylation (for review see ref. 38). Although generally down-regulated in the adult, MAP1B is constitutively highly expressed in adult DRG and MN. After sciatic nerve lesion, the phosphorylated forms of MAP1B (MAP1B-P) is enriched in the more distal portion of the axon and is associated with peripheral regeneration of these neurons.[125,126] In adult CNS, axonal MAP1B-P remains detectable in areas that retain axonal plasticity,[127,128] and can also be reinduced in injury-induced axonal reorganization.[129,130] Adult DRG from MAP1B null allele mutant mice[131] are able to regenerate their axons but exhibit two main abnormalities: (1) the number of terminal and collateral branching is significantly increased and (2) the turning capacity of growth cones, i.e., "choice" of a proper orientation, is impaired.[45] In developing neurons, both growth cone turning[106] and axonal branch formation[25] are known to involve local cross-talk between actin and MTs. MAP1B capacity to bind both actin filaments and microtubules[132-134] suggests that MAP1B is involved in the locally coordinated assembly of cytoskeleton components required for branching and straight directional axon growth.[45] The developmental role of cytoskeleton-associated proteins in the organization of the cross-talk between MTs and actin-filaments[8] appears thus to be maintained during axonal regeneration in the adult.

In conclusion, it seems evident that most, if not all signaling cascades triggered by extracellular stimuli converge onto the cytoskeleton. The subsequent reorganization of actin-filaments and microtubules is a complex phenomenon, and involves numerous cytoskeleton-associated proteins, whose function is fine-tuned via activation or inhibition of particular signaling pathways. Specific expression of some of these cytoskeleton-associated proteins in the neuronal growth cone may, at least in part, explain some unique features of this motile structure. Further studies, such as the one examining the coordinated cross-talk between actin filaments and microtubules during axonal branching and growth cone guidance to the appropriate target, will help determine the precise molecular mechanisms of axonal growth.

In contrast to neural development, the pathways inolved in triggering cytoskeletal reorganization during regeneration are less well known, and this field of research is attracting great interest. Indeed, the ability of regenerating axons to respond to extracellular signals present in their environment depends on both the intrinsic neuronal state, and the presence (or absence) of specific cytoskeleton-associated proteins. It may be particularly interesting to determine a potential central convergence point of inhibitory extrinsic signaling. Modulating the intrinsic state of the neuron, and the response of the cytoskeleton to environmental factors, may provide clues for search of therapeutic targets to promote axonal regeneration after injury.

References

1. Zhou FQ, Cohan CS. How actin filaments and microtubules steer growth cones to their targets. J Neurobiol 2004; 58:84-91.
2. Rodriguez OC, Schaefer AW, Mandato CA et al. Conserved microtubule-actin interactions in cell movement and morphogenesis. Nat Cell Biol 2003; 5:599-609.
3. Kodama A, Lechler T, Fuchs E. Coordinating cytoskeletal tracks to polarize cellular movements. J Cell Biol 2004; 167:203-7.
4. Baas PW, Luo L. Signaling at the growth cone: The scientific progeny of Cajal meet in Madrid. Neuron 2001; 32:981-4.
5. Suter DM, Forscher P. Substrate-cytoskeletal coupling as a mechanism for the regulation of growth cone motility and guidance. J Neurobiol 2000; 44:97-113.
6. Kornack DR, Giger RJ. Probing microtubule +TIPs: Regulation of axon branching. Curr Opin Neurobiol 2005; 15:58-66.
7. Gungabissoon RA, Bamburg JR. Regulation of growth cone actin dynamics by ADF/cofilin. J Histochem Cytochem 2003; 51:411-20.
8. Dehmelt L, Halpain S. Actin and microtubules in neurite initiation: Are MAPs the missing link? J Neurobiol 2004; 58:18-33.
9. Letourneau PC. Differences in the organization of actin in the growth cones compared with the neurites of cultured neurons from chick embryos. J Cell Biol 1983; 97:963-73.
10. Bridgman PC, Dailey ME. The organization of myosin and actin in rapid frozen nerve growth cones. J Cell Biol 1989; 108:95-109.
11. Schaefer AW, Kabir N, Forscher P. Filopodia and actin arcs guide the assembly and transport of two populations of microtubules with unique dynamic parameters in neuronal growth cones. J Cell Biol 2002; 158:139-52.
12. Pollard TD, Borisy GG. Cellular motility driven by assembly and disassembly of actin filaments. Cell 2003; 112:453-65.
13. Forscher P, Smith SJ. Actions of cytochalasins on the organization of actin filaments and microtubules in a neuronal growth cone. J Cell Biol 1988; 107:1505-16.

14. Lewis AK, Bridgman PC. Nerve growth cone lamellipodia contain two populations of actin filaments that differ in organization and polarity. J Cell Biol 1992; 119:1219-43.
15. Brown J, Bridgman PC. Role of myosin II in axon outgrowth. J Histochem Cytochem 2003; 51:421-8.
16. Dent EW, Gertler FB. Cytoskeletal dynamics and transport in growth cone motility and axon guidance. Neuron 2003; 40:209-27.
17. Millard TH, Sharp SJ, Machesky LM. Signalling to actin assembly via the WASP (Wiskott-Aldrich syndrome protein)-family proteins and the Arp2/3 complex. Biochem J 2004; 380:1-17.
18. Kovar DR. Molecular details of formin-mediated actin assembly. Curr Opin Cell Biol 2006; 18:11-7.
19. Huang TY, DerMardirossian C, Bokoch GM. Cofilin phosphatases and regulation of actin dynamics. Curr Opin Cell Biol 2006; 18:26-31.
20. McGough AM, Staiger CJ, Min JK et al. The gelsolin family of actin regulatory proteins: Modular structures, versatile functions. FEBS Lett 2003; 552:75-81.
21. Luduena RF. Multiple forms of tubulin: Different gene products and covalent modifications. Int Rev Cytol 1998; 178:207-75.
22. Mitchison T, Kirschner M. Dynamic instability of microtubule growth. Nature 1984; 312:237-42.
23. Baas PW. Microtubules and axonal growth. Curr Opin Cell Biol 1997; 9:29-36.
24. Tanaka EM, Kirschner MW. Microtubule behavior in the growth cones of living neurons during axon elongation. J Cell Biol 1991; 115:345-63.
25. Dent EW, Kalil K. Axon branching requires interactions between dynamic microtubules and actin filaments. J Neurosci 2001; 21:9757-69.
26. Gordon-Weeks PR. Growth cones: The mechanism of neurite advance. Bioessays 1991; 13:235-9.
27. Tanaka E, Ho T, Kirschner MW. The role of microtubule dynamics in growth cone motility and axonal growth. J Cell Biol 1995; 128:139-55.
28. Tanaka E, Kirschner MW. The role of microtubules in growth cone turning at substrate boundaries. J Cell Biol 1995; 128:127-37.
29. Buck KB, Zheng JQ. Growth cone turning induced by direct local modification of microtubule dynamics. J Neurosci 2002; 22:9358-67.
30. Mack TG, Koester MP, Pollerberg GE. The microtubule-associated protein MAP1B is involved in local stabilization of turning growth cones. Mol Cell Neurosci 2000; 15:51-65.
31. Baas PW, Ahmad FJ, Pienkowski TP et al. Sites of microtubule stabilization for the axon. J Neurosci 1993; 13:2177-85.
32. Khawaja S, Gundersen GG, Bulinski JC. Enhanced stability of microtubules enriched in detyrosinated tubulin is not a direct function of detyrosination level. J Cell Biol 1988; 106:141-9.
33. Erck C, Peris L, Andrieux A et al. A vital role of tubulin-tyrosine-ligase for neuronal organization. Proc Natl Acad Sci USA 2005; 102:7853-8.
34. Bonnet C, Boucher D, Lazereg S et al. Differential binding regulation of microtubule-associated proteins MAP1A, MAP1B, and MAP2 by tubulin polyglutamylation. J Biol Chem 2001; 276:12839-48.
35. Goold RG, Owen R, Gordon-Weeks PR. Glycogen synthase kinase 3beta phosphorylation of microtubule-associated protein 1B regulates the stability of microtubules in growth cones. J Cell Sci 1999; 112(Pt 19):3373-84.
36. Gelfand VI, Bershadsky AD. Microtubule dynamics: Mechanism, regulation, and function. Annu Rev Cell Biol 1991; 7:93-116.
37. Matus A. Microtubule-associated proteins: Their potential role in determining neuronal morphology. Annu Rev Neurosci 1988; 11:29-44.
38. Gordon-Weeks PR, Fischer I. MAP1B expression and microtubule stability in growing and regenerating axons. Microsc Res Tech 2000; 48:63-74.
39. Matus A. Microtubule-associated proteins and neuronal morphogenesis. J Cell Sci Suppl 1991; 15:61-7.
40. Hirokawa N. Microtubule organization and dynamics dependent on microtubule-associated proteins. Curr Opin Cell Biol 1994; 6:74-81.
41. Takei Y, Teng J, Harada A et al. Defects in axonal elongation and neuronal migration in mice with disrupted tau and map1b genes. J Cell Biol 2000; 150:989-1000.
42. Takei Y, Kondo S, Harada A et al. Delayed development of nervous system in mice homozygous for disrupted microtubule-associated protein 1B (MAP1B) gene. J Cell Biol 1997; 137:1615-26.
43. Gonzalez-Billault C, Owen R, Gordon-Weeks PR et al. Microtubule-associated protein 1B is involved in the initial stages of axonogenesis in peripheral nervous system cultured neurons. Brain Res 2002; 943:56-67.
44. DiTella MC, Feiguin F, Carri N et al. MAP-1B/TAU functional redundancy during laminin-enhanced axonal growth. J Cell Sci 1996; 109:467-77.
45. Bouquet C, Soares S, von Boxberg Y et al. Microtubule-associated protein 1B controls directionality of growth cone migration and axonal branching in regeneration of adult dorsal root ganglia neurons. J Neurosci 2004; 24:7204-13.

46. Riederer BM, Pellier V, Antonsson B et al. Regulation of microtubule dynamics by the neuronal growth-associated protein SCG10. Proc Natl Acad Sci USA 1997; 94:741-5.
47. Belmont LD, Mitchison TJ. Identification of a protein that interacts with tubulin dimers and increases the catastrophe rate of microtubules. Cell 1996; 84:623-31.
48. Grenningloh G, Soehrman S, Bondallaz P et al. Role of the microtubule destabilizing proteins SCG10 and stathmin in neuronal growth. J Neurobiol 2004; 58:60-9.
49. Tanaka E, Sabry J. Making the connection: Cytoskeletal rearrangements during growth cone guidance. Cell 1995; 83:171-6.
50. Stepanova T, Slemmer J, Hoogenraad CC et al. Visualization of microtubule growth in cultured neurons via the use of EB3-GFP (end-binding protein 3-green fluorescent protein). J Neurosci 2003; 23:2655-64.
51. Galjart N. CLIPs and CLASPs and cellular dynamics. Nat Rev Mol Cell Biol 2005; 6:487-98.
52. Jimenez-Mateos EM, Paglini G, Gonzalez-Billault C et al. End binding protein-1 (EB1) complements microtubule-associated protein-1B during axonogenesis. J Neurosci Res 2005; 80:350-9.
53. Lee MK, Cleveland DW. Neuronal intermediate filaments. Annu Rev Neurosci 1996; 19:187-217.
54. Gallant PE. Axonal protein synthesis and transport. J Neurocytol 2000; 29:779-82.
55. Eyer J, Peterson A. Neurofilament-deficient axons and perikaryal aggregates in viable transgenic mice expressing a neurofilament-beta-galactosidase fusion protein. Neuron 1994; 12:389-405.
56. Baas PW, Ahmad FJ. Force generation by cytoskeletal motor proteins as a regulator of axonal elongation and retraction. Trends Cell Biol 2001; 11:244-9.
57. Hirokawa N, Takemura R. Molecular motors in neuronal development, intracellular transport and diseases. Curr Opin Neurobiol 2004; 14:564-73.
58. Medeiros NA, Burnette DT, Forscher P. Myosin II functions in actin-bundle turnover in neuronal growth cones. Nat Cell Biol 2006; 8:215-26.
59. Lee SK, Hollenbeck PJ. Organization and translation of mRNA in sympathetic axons. J Cell Sci 2003; 116:4467-78.
60. Zheng JQ, Kelly TK, Chang B et al. A functional role for intra-axonal protein synthesis during axonal regeneration from adult sensory neurons. J Neurosci 2001; 21:9291-303.
61. Hanz S, Perlson E, Willis D et al. Axoplasmic importins enable retrograde injury signaling in lesioned nerve. Neuron 2003; 40:1095-104.
62. Willis D, Li KW, Zheng JQ et al. Differential transport and local translation of cytoskeletal, injury-response, and neurodegeneration protein mRNAs in axons. J Neurosci 2005; 25:778-91.
63. Ming GL, Wong ST, Henley J et al. Adaptation in the chemotactic guidance of nerve growth cones. Nature 2002; 417:411-8.
64. Wujek JR, Lasek RJ. Correlation of axonal regeneration and slow component B in two branches of a single axon. J Neurosci 1983; 3:243-51.
65. Galbraith JA, Gallant PE. Axonal transport of tubulin and actin. J Neurocytol 2000; 29:889-911.
66. Wang L, Brown A. Rapid movement of microtubules in axons. Curr Biol 2002; 12:1496-1501.
67. He Y, Francis F, Myers KA et al. Role of cytoplasmic dynein in the axonal transport of microtubules and neurofilaments. J Cell Biol 2005; 168:697-703.
68. Baas PW, Buster DW. Slow axonal transport and the genesis of neuronal morphology. J Neurobiol 2004; 58:3-17.
69. Gallo G, Letourneau PC. Neurotrophins and the dynamic regulation of the neuronal cytoskeleton. J Neurobiol 2000; 44:159-73.
70. Gallo G, Letourneau PC. Axon guidance: GTPases help axons reach their targets. Curr Biol 8:R80-2.
71. Challacombe JF, Snow DM, Letourneau PC. Dynamic microtubule ends are required for growth cone turning to avoid an inhibitory guidance cue. J Neurosci 1997; 17:3085-95.
72. Gallo G, Letourneau PC. Different contributions of microtubule dynamics and transport to the growth of axons and collateral sprouts. J Neurosci 1999; 19:3860-73.
73. Hall A. Rho GTPases and the actin cytoskeleton. Science 1998; 279:509-14.
74. Luo L. Rho GTPases in neuronal morphogenesis. Nat Rev Neurosci 2000; 1:173-80.
75. Luo L. Actin cytoskeleton regulation in neuronal morphogenesis and structural plasticity. Annu Rev Cell Dev Biol 2002; 18:601-35.
76. Moon SY, Zheng Y. Rho GTPase-activating proteins in cell regulation. Trends Cell Biol 2003; 13:13-22.
77. Schmidt A, Hall A. Guanine nucleotide exchange factors for Rho GTPases: Turning on the switch. Genes Dev 2002; 16:1587-609.
78. Manganello JM, Huang JS, Kozasa T et al. Protein kinase A-mediated phosphorylation of the Galpha13 switch I region alters the Galphabetagamma13-G protein-coupled receptor complex and inhibits Rho activation. J Biol Chem 2003; 278:124-30.
79. Ellerbroek SM, Wennerberg K, Burridge K. Serine phosphorylation negatively regulates RhoA in vivo. J Biol Chem 2003; 278:19023-31.

80. Feoktistov I, Goldstein AE, Biaggioni I. Cyclic AMP and protein kinase A stimulate Cdc42: Role of A(2) adenosine receptors in human mast cells. Mol Pharmacol 2000; 58:903-10.
81. O'Connor KL, Mercurio AM. Protein kinase A regulates Rac and is required for the growth factor-stimulated migration of carcinoma cells. J Biol Chem 2001; 276:47895-900.
82. Brambilla R, Schnapp A, Casagranda F et al. Membrane-bound LERK2 ligand can signal through three different Eph-related receptor tyrosine kinases. EMBO J 1995; 14:3116-26.
83. Bruckner K, Klein R. Signaling by Eph receptors and their ephrin ligands. Curr Opin Neurobiol 1998; 8:375-82.
84. Wahl S, Barth H, Ciossek T et al. Ephrin-A5 induces collapse of growth cones by activating Rho and Rho kinase. J Cell Biol 2000; 149:263-70.
85. Hu H, Marton TF, Goodman CS. Plexin B mediates axon guidance in Drosophila by simultaneously inhibiting active Rac and enhancing RhoA signaling. Neuron 2001; 32:39-51.
86. Ozdinler PH, Erzurumlu RS. Regulation of neurotrophin-induced axonal responses via Rho GTPases. J Comp Neurol 2001; 438:377-87.
87. Yamauchi J, Chan JR, Miyamoto Y et al. The neurotrophin-3 receptor TrkC directly phosphorylates and activates the nucleotide exchange factor Dbs to enhance Schwann cell migration. Proc Natl Acad Sci USA 2005; 102:5198-203.
88. Yamashita T, Tucker KL, Barde YA. Neurotrophin binding to the p75 receptor modulates Rho activity and axonal outgrowth. Neuron 24:585-93
89. Loudon R, Silver LD, Yee Jr HF et al. RhoA-kinase and myosin II are required for the maintenance of growth cone polarity and guidance by nerve growth factor. J Neurobiol 2006; 66:847-67.
90. Rohatgi R, Ma L, Miki H et al. The interaction between N-WASP and the Arp2/3 complex links Cdc42-dependent signals to actin assembly. Cell 1999; 97:221-31.
91. Welch MD, Mullins RD. Cellular control of actin nucleation. Annu Rev Cell Dev Biol 2002; 18:247-88.
92. Miki H, Yamaguchi H, Suetsugu S et al. IRSp53 is an essential intermediate between Rac and WAVE in the regulation of membrane ruffling. Nature 2000; 408:732-5.
93. Arber S, Barbayannis FA, Hanser H et al. Regulation of actin dynamics through phosphorylation of cofilin by LIM-kinase. Nature 1998; 393:805-9.
94. Edwards DC, Sanders LC, Bokoch GM et al. Activation of LIM-kinase by Pak1 couples Rac/Cdc42 GTPase signalling to actin cytoskeletal dynamics. Nat Cell Biol 1999; 1:253-9.
95. Hsieh SH, Ferraro GB, Fournier AE. Myelin-associated inhibitors regulate cofilin phosphorylation and neuronal inhibition through LIM kinase and Slingshot phosphatase. J Neurosci 2006; 26:1006-15.
96. Hirose M, Ishizaki T, Watanabe N et al. Molecular dissection of the Rho-associated protein kinase (p160ROCK)-regulated neurite remodeling in neuroblastoma N1E-115 cells. J Cell Biol 1998; 141:1625-36.
97. Amano M, Chihara K, Nakamura N et al. Myosin II activation promotes neurite retraction during the action of Rho and Rho-kinase. Genes Cells 1998; 3:177-88.
98. Pruyne D, Evangelista M, Yang C et al. Role of formins in actin assembly: Nucleation and barbed-end association. Science 2002; 297:612-5.
99. Palazzo AF, Cook TA, Alberts AS et al. mDia mediates Rho-regulated formation and orientation of stable microtubules. Nat Cell Biol 2001; 3:723-9.
100. Daub H, Gevaert K, Vandekerckhove J et al. Rac/Cdc42 and p65PAK regulate the microtubule-destabilizing protein stathmin through phosphorylation at serine 16. J Biol Chem 2001; 276:1677-80.
101. Wittmann T, Waterman-Storer CM. Cell motility: Can Rho GTPases and microtubules point the way? J Cell Sci 2001; 114:3795-803.
102. Sayas CL, Avila J, Wandosell F. Glycogen synthase kinase-3 is activated in neuronal cells by Galpha12 and Galpha13 by Rho-independent and Rho-dependent mechanisms. J Neurosci 2002; 22:6863-75.
103. Krendel M, Zenke FT, Bokoch GM. Nucleotide exchange factor GEF-H1 mediates cross-talk between microtubules and the actin cytoskeleton. Nat Cell Biol 2002; 4:294-301.
104. Waterman-Storer CM, Worthylake RA, Liu BP et al. Microtubule growth activates Rac1 to promote lamellipodial protrusion in fibroblasts. Nat Cell Biol 1999; 1:45-50.
105. Zhou FQ, Waterman-Storer CM, Cohan CS. Focal loss of actin bundles causes microtubule redistribution and growth cone turning. J Cell Biol 2002; 157:839-49.
106. Challacombe JF, Snow DM, Letourneau PC. Actin filament bundles are required for microtubule reorientation during growth cone turning to avoid an inhibitory guidance cue. J Cell Sci 1996; 109:2031-40.
107. Owen R, Gordon-Weeks PR. Inhibition of glycogen synthase kinase 3beta in sensory neurons in culture alters filopodia dynamics and microtubule distribution in growth cones. Mol Cell Neurosci 2003; 23:626-37.

108. Ziv NE, Spira ME. Induction of growth cone formation by transient and localized increases of intracellular proteolytic activity. J Cell Biol 1998; 140:223-32.
109. Spira ME, Oren R, Dormann A et al. Critical calpain-dependent ultrastructural alterations underlie the transformation of an axonal segment into a growth cone after axotomy of cultured Aplysia neurons. J Comp Neurol 2003; 457:293-312.
110. Hoffman PN, Cleveland DW. Neurofilament and tubulin expression recapitulates the developmental program during axonal regeneration: Induction of a specific beta-tubulin isotype. Proc Natl Acad Sci USA 1988; 85:4530-3.
111. Hoffman PN, Thompson GW, Griffin JW. Changes in neurofilament transport coincide temporally with alterations in the caliber of axons in regenerating motor fibers. J Cell Biol 1985; 101:1332-40.
112. Tetzlaff W, Leonard C, Krekoski CA et al. Reductions in motoneuronal neurofilament synthesis by successive axotomies: A possible explanation for the conditioning lesion effect on axon regeneration. Exp Neurol 1996; 139:95-106.
113. Tetzlaff W, Alexander SW, Miller FD et al. Response of facial and rubrospinal neurons to axotomy: Changes in mRNA expression for cytoskeletal proteins and GAP-43. J Neurosci 1991; 11:2528-44.
114. Domeniconi M, Filbin MT. Overcoming inhibitors in myelin to promote axonal regeneration. J Neurol Sci 2005; 233:43-7.
115. Qiu J, Cai D, Dai H et al. Spinal axon regeneration induced by elevation of cyclic AMP. Neuron 2002; 34:895-903.
116. Neumann S, Bradke F, Tessier-Lavigne M et al. Regeneration of sensory axons within the injured spinal cord induced by intraganglionic cAMP elevation. Neuron 2002; 34:885-93.
117. Cai D, Qiu J, Cao Z et al. Neuronal cyclic AMP controls the developmental loss in ability of axons to regenerate. J Neurosci 2001; 21:4731-9.
118. Fournier AE, Takizawa BT, Strittmatter SM. Rho kinase inhibition enhances axonal regeneration in the injured CNS. J Neurosci 2003; 23:1416-23.
119. Niederost B, Oertle T, Fritsche J et al. Nogo-A and myelin-associated glycoprotein mediate neurite growth inhibition by antagonistic regulation of RhoA and Rac1. J Neurosci 2002; 22:10368-76.
120. Mimura F, Yamagishi S, Arimura N et al. Myelin-associated glycoprotein inhibits microtubule assembly by a Rho-kinase-dependent mechanism. J Biol Chem 2006; 281:15970-9.
121. Gao Y, Deng K, Hou J et al. Activated CREB is sufficient to overcome inhibitors in myelin and promote spinal axon regeneration in vivo. Neuron 2004; 44:609-21.
122. Cai D, Deng K, Mellado W et al. Arginase I and polyamines act downstream from cyclic AMP in overcoming inhibition of axonal growth MAG and myelin in vitro. Neuron 2002; 35:711-9.
123. Skene JH. Axonal growth-associated proteins. Annu Rev Neurosci 1989; 12:127-56.
124. Bomze HM, Bulsara KR, Iskandar BJ et al. Spinal axon regeneration evoked by replacing two growth cone proteins in adult neurons. Nat Neurosci 2001; 4:38-43.
125. Ma D, Nothias, F, Boyne LJ et al. Differential regulation of microtubule-associated protein 1B (MAP1B) in rat CNS and PNS during development. J Neurosci Res 1997; 49:319-32.
126. Ma D, Connors T, Nothias F. et al. Regulation of the expression and phosphorylation of microtubule-associated protein 1B during regeneration of adult dorsal root ganglion neurons. Neuroscience 2000; 99:157-70.
127. Nothias F, Vernier P, von Boxberg Y et al. Modulation of NCAM polysialylation is associated with morphofunctional modifications in the hypothalamo-neurohypophysial system during lactation. Eur J Neurosci 1997; 9:1553-65.
128. Nothias F, Fischer I, Murray M et al. Expression of a phosphorylated isoform of MAP1B is maintained in adult central nervous system areas that retain capacity for structural plasticity. J Comp Neurol 1996; 368:317-34.
129. Soares S, von Boxberg Y, Lombard MC et al. Phosphorylated MAP1B is induced in central sprouting of primary afferents in response to peripheral injury but not in response to rhizotomy. Eur J Neurosci 2002; 16:593-606.
130. Soares S, Fischer I, Ravaille-Veron M. Induction of MAP1B phosphorylation in target-deprived afferent fibers after kainic acid lesion in the adult rat. J Comp Neurol 1998; 396:193-210.
131. Meixner A, Haverkamp S, Wassle H et al. MAP1B is required for axon guidance and Is involved in the development of the central and peripheral nervous system. J Cell Biol 2000; 151:1169-78.
132. Togel M, Wiche G, Propst F. Novel features of the light chain of microtubule-associated protein MAP1B: Microtubule stabilization, self interaction, actin filament binding, and regulation by the heavy chain. J Cell Biol 1998; 143:695-707.
133. Pedrotti B, Islam K. Dephosphorylated but not phosphorylated microtubule associated protein MAP1B binds to microfilaments. FEBS Lett 1996; 388:131-3.
134. Noiges R, Eichinger R, Kutschera W et al. Microtubule-associated protein 1A (MAP1A) and MAP1B: Light chains determine distinct functional properties. J Neurosci 2002; 22:2106-14.
135. Bolsover, SR. Calcium signalling in growth cone migration. Cell Calcium 2005; 37:395-402.

Chapter 2

Netrins and Their Receptors

Simon W. Moore, Marc Tessier-Lavigne and Timothy E. Kennedy*

Abstract

Netrins are a family of proteins that direct cell and axon migration during development. Three secreted netrins (netrin-1, -3 and -4) have been identified in mammals, in addition to two GPI-anchored membrane proteins, netrin-G1 and G2. Orthologues of netrin-1 play a highly conserved role as guidance cues at the midline of the developing CNS of vertebrates and some bilaterally symmetric invertebrates. In vertebrates, floor plate cells at the ventral midline of the embryonic neural tube secrete netrin-1, generating a circumferential gradient of netrin protein in the neuroepithelium. This protein gradient is bifunctional, attracting some axons to the midline and repelling others. Receptors for the secreted netrins include DCC (deleted in colorectal cancer) and the UNC5 homologues: UNC5A, B, C and D in mammals. DCC mediates chemoattraction, while repulsion requires an UNC5 homologue and, in some cases, DCC. The netrin-G proteins bind NGLs (netrin G ligands), single pass transmembrane proteins unrelated to either DCC or the UNC5 homologues. Netrin function is not limited to the developing CNS midline. Various netrins direct cell and axon migration throughout the embryonic CNS, and in some cases continue to be expressed in the mature nervous system. Furthermore, although initially identified for their ability to guide axons, functional roles for netrins have now been identified outside the nervous system where they influence tissue morphogenesis by directing cell migration and regulating cell-cell and cell-matrix adhesion.

Introduction

The discovery of netrins can be traced back to insights provided by Santiago Ramón y Cajal at the end of the 19th century, when he proposed that axons may be guided by diffusible cues.[1] Upon observing, in fixed sections, the projections of spinal commissural neuron axons towards the ventral midline of the embryonic spinal cord, he hypothesized that floor plate cells at the midline secreted a diffusible cue that established a chemotropic gradient in the neuroepithelium (Fig. 1A). Direct evidence of chemotropic axon guidance began to accumulate in the 1980s through single cell turning assays and coculture of explanted embryonic neural tissue.[2] Notably, explants of embryonic rat spinal floor plate, when cultured at a distance from explants of dorsal spinal cord, evoked commissural axon outgrowth (Fig. 1D),[3] and an ectopic floor plate cocultured alongside an embryonic spinal cord attracted commissural axons, deflecting them away from their normal dorsal-ventral trajectory (Fig. 1E).[4] These findings provided strong evidence for the existence of a chemotropic axon guidance factor(s) secreted by the floor plate.

In parallel, studies in the nematode *Caenorhabditis elegans* identified genes required for circumferential axon guidance.[5,6] One of the genes identified, *unc-6*, encoded a secreted

*Corresponding Author: Timothy E. Kennedy—Centre for Neuronal Survival, Montreal Neurological Institute, McGill University, 3801 University Avenue, Montreal, Quebec, H3A 2B4, Canada. Email: timothy.kennedy@mcgill.ca

Axon Growth and Guidance, edited by Dominique Bagnard. ©2007 Landes Bioscience and Springer Science+Business Media.

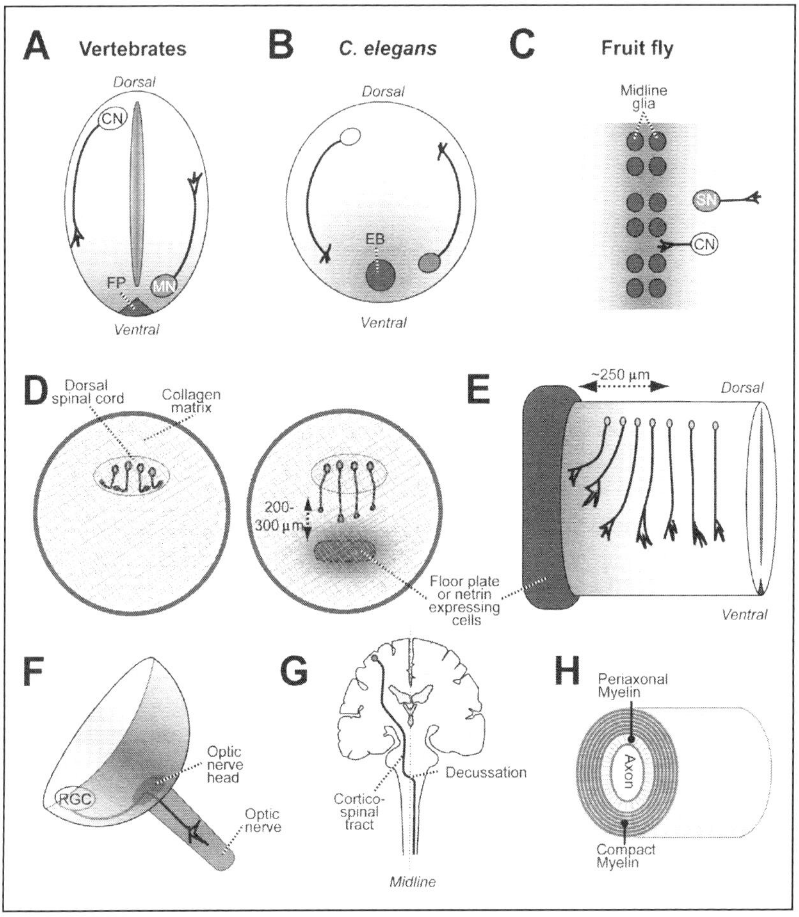

Figure 1. Netrins are important midline axon guidance cues: A) Netrin-1 secreted by the floor plate (FP) attracts commissural neuron (CN) axons and repels motoneuron (MN) axons from the ventral midline. B) During early neural development in *C. elegans*, axons are guided towards and away from a row of epidermoblasts (EB) expressing the netrin homologue UNC-6 at the ventral midline. C) Netrin-A and -B emanating from midline glia guides commissural (CN) axons to and segmental nerve (SN) axons away from the *D. melanogaster* midline. D) Embryonic spinal commissural axon outgrowth assay: An explant of dorsal embryonic rat spinal cord containing the commissural neuron cell bodies is embedded in a collagen matrix. In the absence of a source of netrin-1, such as the floor plate, the extending axons remain within the explant. In the presence of netrin-1, the axons emerge from the explant and grow into the collagen. E) Embryonic spinal commissural axon turning assay: A segment of embryonic rat spinal cord is embedded into a collagen matrix and an explant of the floor plate is grafted onto one end. Neurons within ~250 µm of the ectopic floor plate turn away from their normal dorsal to ventral trajectory and grow toward the grafted floor plate. F) Netrin-1, expressed at the optic nerve head, is required for retinal ganglion cell (RGC) axons to exit from the retina into optic nerve. G) Netrin and its receptors DCC and UNC5C are required for the decussation of the corticospinal tract at the spinal medulla boundary. H) In the mature mammalian CNS, netrin-1 is localized to periaxonal myelin suggesting a role regulating interactions between axonal and oligodendroglial membranes. Panels A, D, E and H have been reprinted from Current Opinions in Neurobiology 16:529-534 with permission from Elsevier, ©2006.[124]

protein with sequence homology to laminins.[7] In 1994, using commissural axon outgrowth from explants of embryonic rat dorsal spinal cord as a functional assay, two proteins were purified from homogenates of embryonic chick brain and discovered to be homologous to UNC-6.[8] They were named netrin-1 and netrin-2 based on the Sanskrit word 'netr' meaning 'one who guides'. Netrin-1 is expressed by floor plate cells[9] and forms a gradient in the spinal neuroepithelium as commissural axons extend to the floor plate.[10] Engineering an aggregate of cells to express either netrin-1 or netrin-2, mimicked the commissural axon guidance activity of the floor plate (Fig. 1D-E).[9] Identification of the mouse ortholog of netrin-1, and generation of netrin-1 mutant mice, demonstrated that netrin-1 is essential for appropriate spinal commissural axon extension in the embryonic spinal cord.[11] In parallel, *C. elegans unc-6* was shown to be expressed at the ventral midline.[12] and to function as a long-range midline attractant guidance cue.[13] Furthermore, two netrins, Netrin-A and Netrin-B, were implicated in midline attraction in *Drosophila,*[14,15] although in this case netrin mediated attraction is apparently only essential at short-range close to the midline.[16] Thus, a century after chemotropic mechanisms were proposed to direct axon guidance, netrins were identified as diffusible chemotropic cues that guide spinal commissural axon extension, with homologues implicated in long- and short-range guidance in worms and flies. Netrins are now known to function not only as attractants, but also as repellents, and to be essential for the development of numerous axonal tracts.

Netrin Structure

Netrins are highly conserved in the course of animal evolution. Illustrating this, a netrin homologue has recently been identified in the sea anemone *Nematostella vectensis*, an organism thought to exhibit some of the earliest hallmarks of bilateral symmetry (Fig. 2A).[17] Vertebrate species express the secreted netrins, netrins 1-4, and two related GPI-anchored membrane proteins, netrin-G1 and -G2 (Fig. 2A). All netrins are composed of approximately 600 amino acids, and have a molecular mass of approximately 70 kilodaltons. They share two characteristic amino terminal domains, V and VI, that are homologous to domains V and VI found at the amino terminal ends of laminins (Fig. 3A). Laminins are large secreted heterotrimers made up of α, β, and γ subunits.[18] Domains V and VI of netrin-4 and netrin-Gs are most similar to β subunits of laminins, while those of netrins 1-3 are more similar to the γ subunits (Fig. 3C).[19]

Netrins 1, 3, 4, G1 and G2 are expressed in mammals, including rats, mice and humans, whereas orthologues of netrin-2 have thus far only been identified in chicken[8] and zebrafish.[20] The amino acid sequences of netrins 1-3 are highly similar (Fig. 3C) and, consistent with this, cellular sources of any of these proteins mimic the chemoattractant function of the floor plate.[8,9,21] The sequences of netrin-4 and netrin-Gs are substantially divergent, notably exhibiting a higher degree of homology to laminins than to netrins 1-3 (Fig. 3C).[22-25] Orthologues of netrin-4 or the netrin-Gs have thus far only been found in vertebrates, while orthologues of netrins 1-3 have been identified in distantly related animals, including the nematode worm *C. elegans,*[7] the flatworm *Schmidea mediterranea,*[26] the fruit fly *Drosophila melanogaster,*[14,15] the leech *Hirudo medicinalis*[27] and the sea anemone *Nematostella vectensis* (Fig. 2A).[17]

In laminins, domain VI, approximately 300 amino acids in length, is capable of binding heparin, cell surface receptors and ECM proteins[28,29] and is required for calcium-dependent multimerization between laminin molecules.[30] Mutational studies carried out in *C. elegans* indicate that domain VI of netrin is critical for both axon attraction and repulsion.[31] The motif SXDXGXS/TW is present in domain VI of all netrins and mutation of these residues in the *C. elegans* netrin UNC-6 disrupts guidance functions.[19,31] Interestingly, only the β subunits of laminin contain this motif. This is noteworthy because, as described above, netrins 1 through 3 are most homologous to the γ chain. Domain VI of netrins 1-3 also contains two cysteine residues not present in other netrins or laminins. One of these cysteines replaces a tryptophan that is strictly conserved among laminin subunits.[19] Domain V of netrins contains three tandem arrays of cysteine-rich epidermal growth factor (EGF) repeats named V-1, V-2 and V-3, and is approximately 150 amino acids in size.[7] Mutation of domain V-3 in the *C. elegans* netrin

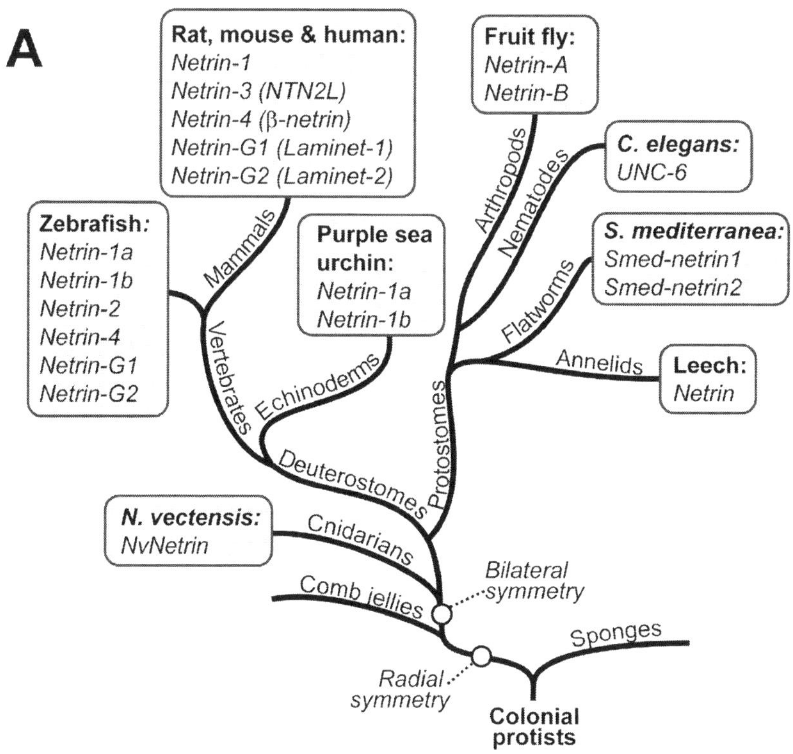

B

	DCC	UNC5	NGL
Mammals	Neogenin DCC	UNC5A *(UNC5h1)* UNC5B *(UNC5h2)* UNC5C *(UNC5h3, RCM)* UNC5D *(UNC5h4)*	NGL-1 NGL-2
Chicken	Neogenin cDCC	UNC5h2 cUNC5h3	
Zebrafish	zDCC zNeo	UNC5a UNC5b	
X. laevis	xDCC	xUNC5	
Fruit fly	Frazzled	UNC5	
C. elegans	UNC-40	UNC-5	
S. mediterranea	Smed-netR		

Figure 2. Netrins and their receptors in various organisms: A) Evolutionary tree diagram highlighting the presence of netrin homologues in a wide variety of bilaterally symmetrical organisms. B) Netrin 1-3 receptors (DCC and UNC5) and the netrin-G receptors (NGL) in various organisms.

UNC-6 disrupts attractant mechanisms, whereas repulsion is lost following mutation of either V-2 or V-3 domains.[12,31]

Netrins 1-4 contain a conserved carboxyl terminal domain, domain C (Fig. 3A), that has a predicted α-helical secondary structure and is homologous to domains found in the complement C3, 4 and 5 protein family (CC3, 4 and 5), secreted frizzled-related proteins (sFRP), type

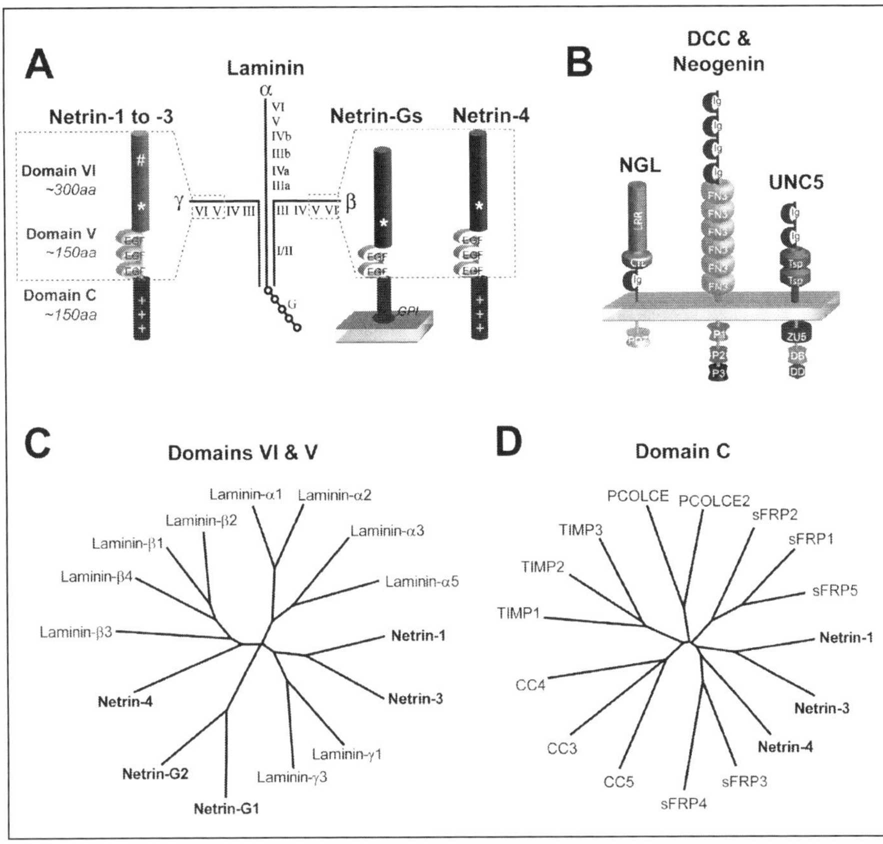

Figure 3. Netrin and netrin receptor structure: A) All netrins contain amino terminal domains V and VI related to corresponding amino terminal domains of laminins. Domain V is composed of cysteine-rich epidermal growth factor (EGF) repeats. Domain C in secreted netrins contains many positively charged, basic residues. B) DCC and UNC5 are receptors for netrin-1 to -3. NGL1 and NGL2 are receptors for netrin-G_1 and -G_2, respectively. C) Tree illustrating a phylogenetic relationship based on sequence of the VI and V domains in human netrins and laminins. D) Phylogenetic tree based on human protein sequences related to the C domain of netrin-1 (see text for details). Panels A and B have been reprinted with permission from 'A830: Netrins' in the Encyclopedia of Life Sciences by John Wiley & Sons, Ltd.

I C-proteinase enhancer proteins (PCOLCEs) and tissue inhibitors of metalloproteinases (TIMPs) (Fig. 3D). Deletion of domain C from UNC-6 netrin in *C. elegans* does not appear to disrupt axon guidance, although increased axon branching has been detected.[32] Most netrin-1 protein in the vertebrate CNS is not freely soluble, but bound to cell surfaces or extracellular matrix.[33,34] A notable feature of the netrin C domain is that it contains many basic amino acids. It has been hypothesized that these may bind to negatively charged sugars associated with proteoglycans on cell surfaces, such as heparin sulfate proteoglycans and chondroitin sulfate proteoglycans.[8,35,36] Presentation of netrins closely associated with cell surfaces may be a common mode of action in the netrin family. Although the C domain is not conserved in the netrin-Gs, a C terminal GPI-link anchors them to cell surfaces.

Functional Roles for Netrins during Nervous System Development

During embryogenesis in *C. elegans* and *D. melanogaster*, secretion of the netrin UNC-6 and netrins A/B respectively, are essential for orienting cell and axon migration with respect to the ventral midline of the developing nervous system (Fig. 1B,C).[6,7,15,37,38] Similarly, netrin-1 expressed by the floor plate in mouse plays an essential role directing axon extension relative to the ventral midline of the embryonic spinal cord. Netrin-1 deficiency in mouse also disrupts the formation of major axon projections to the midline in brain, including the corpus callosum and hippocampal commissure,[11] indicating that numerous axon tracts require netrin-1 to cross from one side of the CNS to the other. Acting as a repellent, netrin-1 directs axon extension by subsets of motoneurons, including: trochlear motoneurons,[39] cranial motoneurons[40] and spinal accessory motoneurons.[41]

Away from the midline, netrin-1 expression at the optic nerve head is required for the axons of retinal ganglion cells to exit the retina and enter the optic nerve (Fig. 1F).[42] Netrin-1 is also implicated in the guidance of dopaminergic axons within the ventral midbrain,[43] in the thalamocortical projection,[44] as well as in the formation of axon projections within the hippocampus.[45]

In contrast to netrin-1, the function of other netrin family members in vertebrates is relatively poorly understood. Netrin-3 can mimic the ability of netrin-1 to attract spinal commissural axons and repel trochlear motor neuron axons in vitro,[21] however, netrin-3 expression in the spinal cord begins after the initial commissural axons have pioneered the path to the floor plate. Netrin-3 is, however, expressed in dorsal root ganglia in the developing PNS, and by mesodermal cells that may influence axon guidance to peripheral targets.[46] Netrin-4 is widely expressed in the developing nervous system, including in the olfactory bulb, retina, dorsal root ganglia, as well as by cerebellar granule, hippocampal, and cortical neurons.[22] In the developing spinal cord, a relatively low level of netrin-4 is expressed adjacent to floor plate cells; however, like netrin-3, this begins after the first commissural axons have crossed the midline. Both netrin-G1 and -G2 are expressed primarily by neurons, with very limited expression outside the nervous system.[25,47] Netrin-G1 is expressed in the dorsal thalamus, olfactory bulb and inferior colliculus, while netrin-G2 is expressed in the cerebral cortex. *Netrin-G1* gene mutations in humans produce symptoms similar to Rett syndrome,[48] characterized by normal early development followed by loss of purposeful use of the hands, distinctive hand movements, slowed brain and head growth, gait abnormalities, seizures, and mental retardation. Netrin G1-deficient mice have no obvious abnormalities in gross anatomy and neural circuitry, but exhibit altered synaptic responses and defects in sensorimotor gating behavior.[49] These findings led to the suggestion that the major role for netrin-G proteins may be in the maturation, refinement, and maintenance of synapses, rather than axonal outgrowth and guidance. Consistent with this, the netrin-G receptor NGL-2 influences the formation of glutamatergic synapses through an interaction with the post-synaptic scaffold protein PSD-95.[50]

Netrin Signal Transduction

The signal transduction mechanisms regulated by netrins are currently the subject of intense scrutiny. The majority of the studies carried out have focused on the role of netrin-1 as a chemoattractant axon guidance cue and comparatively little is known regarding signal transduction by other netrins. The following provides an overview of signal transduction events implicated in the response to netrin-1, for a detailed (for a detailed review see ref. 51,52).

Netrin receptors in vertebrates include DCC (deleted in colorectal cancer), the DCC paralogue neogenin, and four UNC5 proteins, UNC5A-D (Fig. 2B). Although DCC, neogenin, and the UNC5 proteins all bind netrin-1, the majority of studies of netrin signaling have focused on DCC. Attractant responses to netrin-1 require DCC. In contrast, repellent responses require expression of an UNC5 protein, with coexpression of DCC in some cases. Interestingly, neogenin also interacts with a GPI-linked protein called Repulsive Guidance Molecule.[53]

Netrin-1 Mediated Chemoattraction

Unc-40 encodes the *C. elegans* orthologue of DCC.[6,54] *C. elegans unc-40* mutants predominantly exhibit defects in ventrally-directed migration of cells and axons, in contrast to *unc-6 (netrin)* mutants in which migrations both toward and away from the ventral midline are disrupted. Consistent with the *unc-40* mutant phenotype in the nematode, application of DCC function blocking antibodies to explants of embryonic mouse spinal cord blocked netrin-1 induced commissural axon outgrowth.[55] Furthermore, *dcc* gene knockout produced a phenotype very similar to that generated by loss of netrin-1 function, including loss of the spinal ventral commissure, corpus callosum and hippocampal commissure.[56]

The extracellular domain of DCC is composed of six fibronectin type 3 (FN3) repeats and four immunoglobulin (Ig) repeats (Fig. 3B). The DCC FN3 domains are implicated as netrin-1 binding sites, but exactly which FN3 domain binds netrin-1 remains controversial.[57-59] The DCC intracellular domain has no known intrinsic catalytic activity, but contains several putative protein binding and phosphorylation sites. Based on particularly strong identity between DCC family members, three regions of the intracellular domain of DCC, termed domains, P1, P2 and P3, have been identified (Figs. 3B and 4A).[60] The P1 domain is a highly conserved 17 amino acid motif, the P2 domain is rich in proline residues, containing four PXXP putative SH3 domain-binding motifs (Fig. 4A), and the P3 domain contains several highly conserved possible phosphorylation sites.

The ability of a cue to attract axon growth is thought to reflect its capacity to regulate membrane protrusions made by the growth cone. Rho GTPases are a family of intracellular proteins that coordinate cytoskeletal organization and adhesive interactions.[61] In particular, the activation of the Rho GTPases Rac and Cdc42 has been shown to be essential for attractant responses to a number of guidance cues,[62,63] including netrin-1.[64,65] The exact sequence of events linking DCC to Rho GTPase activation, and their downstream effectors, remains unclear. Multimerization of the DCC P3 domain following binding to netrin-1 is implicated as an initial event in mediating chemoattraction.[66,67] The DCC intracellular domain associates with the adaptor protein Nck1,[68] the tyrosine kinases Fak[69] and Fyn,[70] the serine/threonine kinase Pak,[64] as well as the actin binding proteins Ena/Vasp[71] and N-WASP.[64] In addition to Rac and Cdc42 activation, application of netrin-1 leads to production of phosphoinositides by recruitment of phosphatidylinositol transfer protein-α,[72] activation of phosphatidylinositol-3 kinase,[73] and the breakdown of phosphoinositides by phospholipase C into IP3 and diacylglycerol (DAG).[73] IP3 promotes intracellular calcium release from intracellular stores and DAG activates protein kinase C.[74] Supporting a role for IP3 production in netrin-1 mediated chemoattraction, elevating intracellular calcium is required for turning to netrin-1.[75] Notably, such calcium increases can contribute to Rac and Cdc42 activation.[76] Figure 4C presents a speculative model of how these events may contribute to netrin-1 mediated axonal chemoattraction.

Netrin-1 Mediated Chemorepulsion

UNC5 netrin receptors were first implicated as mediators of repellent responses to the netrin UNC-6 from studies in *C. elegans.*[6,77] *Unc-5* mutants exhibit defects in dorsally-directed migrations, away from the ventral midline source of UNC-6 netrin, and misexpression of *unc-5* by neurons caused their axons to be redirected along a dorsal trajectory.[37] As in *C. elegans,* a single UNC5 family member has been identified in *D. melanogaster.*[38] Four have been found in mammals: UNC5A, B, C and D (Fig. 2B).[78-81] UNC5s are composed of two extracellular Ig domains, that bind netrin, and two extracellular Tsp (thrombospondin) domains (Fig. 3B).[58] The UNC5 intracellular domain is made up of three conserved domains: a ZU5 domain, a DCC-binding (DB) domain and a death domain (DD, Fig. 3B). The function of the ZU5 domain is unknown, however it is homologous to a sequence in the scaffolding protein Zona Occludens-1 found at tight junctions.[82]

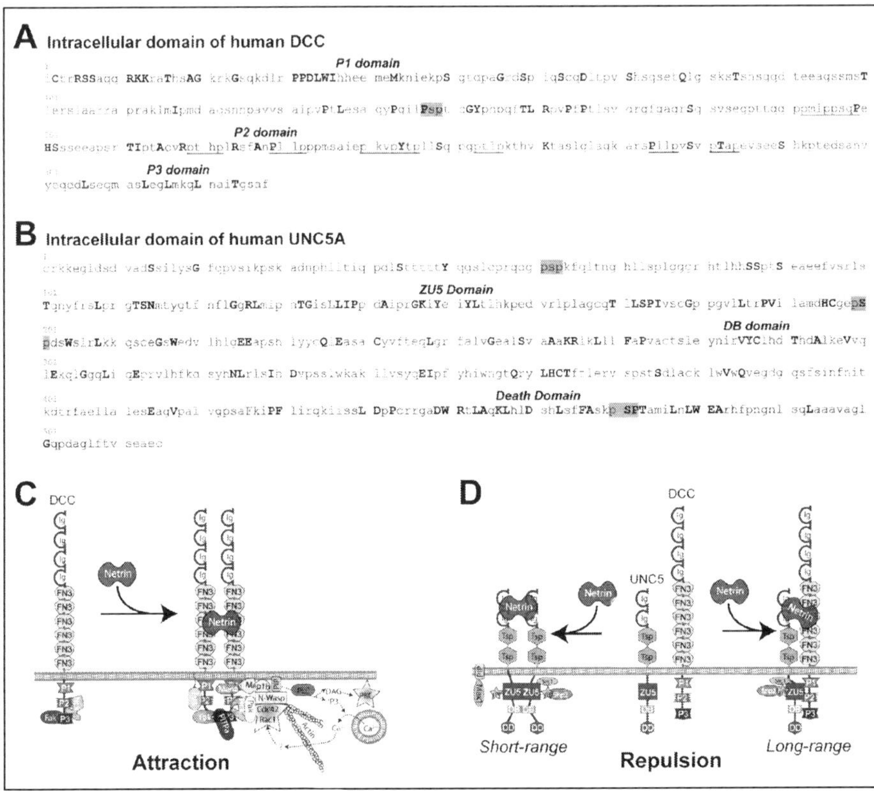

Figure 4. Model of netrin signal transduction: Amino acid sequences of the intracellular domains of human DCC (A) and UNC5A (B). Amino acids conserved between C. elegans, X. laevis and humans are in bold capital letters. Assigned domains are lightly shaded, while WW class IV motifs (PSP) are more darkly shaded. Core SH3 PXXP motifs are underlined. Panel C and D summarize signaling events involved in attractive and repellent responses, respectively (see text for details). Panels A and B have been reprinted with permission from 'A830: Netrins' in the Encyclopedia of Life Sciences by John Wiley & Sons, Ltd.

Studies in worms, flies and vertebrates suggest that long-range repulsion to netrin requires the cooperation of UNC5 and DCC, but that UNC5 without DCC is sufficient for short-range repulsion.[38,66] Although the reason for this difference is not clear, it may be the case that DCC and UNC5 together form a more sensitive netrin receptor complex that is able to respond to lower concentrations of protein found at a greater distance from a source of netrin secretion. At long-range, direct association between the cytoplasmic domains of UNC5 and DCC appears to be essential.[66,83] While mediating short-range responses to netrin independently of DCC, genetic studies in *C. elegans* have stressed the importance of the region between UNC5 cytoplasmic ZU5 and DD domains.[84] Several proteins have been proposed to interact with UNC5 family members in mediating a repellent response, including: the tyrosine kinase Src1, the tyrosine phosphatase Shp2,[85] the F-actin anti-capping protein Mena,[86] the structural protein ankryn, and the adaptor protein Max1.[87] Repellent responses to netrin-1 are thought to involve tyrosine phosphorylation of UNC5's intracellular domains at multiple sites.[85] Figure 4D outlines a speculative model of the intracellular events occurring during short and long-range repulsion.

Regulating the Response to Netrin-1

Growth cones respond rapidly to local guidance cues and exhibit substantial autonomy from the neuronal cell body. Growth cones react to netrin along a continuum that ranges from repulsion to unresponsiveness to attraction. The mechanisms that control this shift in netrin responsiveness are just beginning to be understood.

Many of the factors shown to regulate the response of growth cones to netrin can be correlated with changes in the expression of either UNC5 or DCC. At the transcriptional level, mis-expressing the homeobox transcription factor even-skipped in *D. melanogaster* resulted in disruption of *unc5* expression and motoneuron axon guidance defects.[88] Local protein synthesis within the growth cone is required for chemoattraction of cultured *X. laevis* neurons to netrin.[89] The newly synthesized proteins have been suggested to influence either the recovery of growth cones from desensitization, or netrin signal transduction directly.[90] Conversely, DCC function is negatively regulated by proteolysis, including both extracellular metalloproteinase implicated in shedding of the DCC ectodomain,[91] and ubiquitination of the DCC intracellular domain through an interaction with Siah-1, a RING domain containing protein that promotes DCC degradation via the ubiquitin-proteasome pathway.[92,93] In mammals, the intracellular domains of UNC5 proteins are substrates for caspases.[94]

Intracellular concentrations of cyclic nucleotides are key regulators of growth cone responsiveness to several guidance cues (see Chapter 10 for further discussion). Manipulating the intracellular concentration of cAMP, thereby activating protein kinase A (PKA), regulates the response of growth cones to netrin-1. Initial experiments demonstrated that axons of cultured *X. laevis* spinal neurons attracted to a pipette puffing netrin-1, were instead repelled when PKA was inhibited.[95] These studies led to the proposal that PKA can control the direction of growth cone turning by regulating intracellular signal transduction pathways downstream of netrin-1. PKA activation has been shown to selectively recruit DCC from an intracellular vesicular pool to the plasma membrane of commissural neuron growth cones, and the increased levels of DCC potentiate the outgrowth and turning response of these neurons to netrin-1.[96,97] Interestingly, activation of protein kinase C (PKC) induces endocytosis of UNC5 homologues resulting in cultured cerebellar granule cell neurons switching from chemorepellent to chemoattractant responses to netrin-1.[98] These findings suggest that extracellular factors that regulate PKA and PKC will influence axon outgrowth by determining which receptors are presented by the growth cone.

Other Potential Netrin Receptors

Other receptors, in addition to DCC and UNC5 proteins, have been suggested for netrins 1-3. The G-protein coupled adenosine receptor, A2B, was reported to bind netrin-1 and cooperate with DCC in spinal commissural axon guidance.[99] However, subsequent studies provided evidence that A2B is neither expressed by these neurons nor required for commissural axon guidance in response to netrin-1.[67] The $\alpha6\beta4$ and $\alpha3\beta1$ integrins bind netrin-1 and these interactions have been implicated in the development of the pancreas.[100] Given the homology of the N-terminus of netrin-1 to laminins, it might be predicted that netrins would bind integrins through N-terminal domains; but surprisingly $\alpha6\beta3$ and $\alpha3\beta1$ integrins interact with a highly charged sequence of basic amino acids at the C-terminus of netrin-1 that is not homologous to laminins. While these findings raise the exciting possibility that integrins may function as netrin receptors in other contexts, the significance of netrin-integrin interactions in vivo remains to be demonstrated. In contrast to the secreted netrins, netrin-Gs bind transmembrane proteins called the netrin-G ligands (NGL) (Fig. 3B) and netrin-Gs do not appear to interact with DCC, neogenin, or the UNC5 proteins.[24,101]

Netrin in the Adult Nervous System

Netrins and netrin receptors are expressed in the adult vertebrate nervous system.[9,21,22,24,25,33,102-108] Netrin-1 is expressed by many types of neurons and by myelinating glia: oligodendrocytes in the CNS[33] and Schwann cells in the PNS.[105,107] Subcellular fractionation of CNS white matter indicated that netrin-1 is enriched in periaxonal myelin membranes (Fig. 1H),[33] suggesting that it may normally mediate interactions between axonal and oligodendrocyte membranes. Expression by mature myelinating oligodendrocytes raises the possibility that netrin-1 may influence axon regeneration. Notably, netrin-1, DCC and UNC5s influence the development of the corticospinal tract, which transmits information controlling voluntary limb movements, suggesting that netrin-1 might play an important role following spinal cord injury (Fig. 1G).[109,110] During maturation of the mammalian spinal cord, DCC expression is downregulated, while UNC5 homologue expression increases,[108] indicating that UNC5 repellent signaling may be the dominant response to netrin in the adult spinal cord.

An examination of the consequences of spinal cord injury in the adult rat found that levels of netrin-1 mRNA and protein were substantially reduced at the site of injury itself, and this decreased expression persisted for at least 7 months.[111] Netrin-1 was not associated with the glial scar, but netrin-1 was expressed in an apparently normal distribution by neurons and oligodendrocytes adjacent to the lesion. The expression of DCC and UNC5 proteins was also reduced after injury. Although DCC expression remained low, UNC5 expression recovered and subsets of neurites adjacent to the lesion exhibited elevated UNC5 immunoreactivity. These findings are consistent with earlier studies carried out in the optic nerve, indicating that both DCC and UNC5B continue to be expressed by retinal ganglion cells following axotomy, albeit at reduced levels, as their axons attempt to extend along either the injured optic nerve itself or into a growth permissive peripheral nerve graft.[106,107] While a role for netrin-1 in axon regeneration remains to be demonstrated directly, these findings suggest a role for netrin-1 as a component of CNS myelin that inhibits axon regeneration by neurons expressing UNC5 following injury.

Although the functional significance of netrin-1 expression in the adult CNS remains unknown, an intriguing hypothesis is that netrins may contribute to maintaining appropriate connections in the intact CNS by restraining inappropriate axonal sprouting. A consequence of this may be that netrins subsequently inhibit the reestablishment of connections following injury. In line with this hypothesis, studies carried out in lamprey, a primitive vertebrate with the ability to recover significant function following spinal cord transaction,[112] indicate a correlation between UNC5 expression and poor axonal regeneration following lesion.[113] Importantly, it may be possible to reverse such an inhibitory role for netrin in the adult mammalian CNS by manipulating cAMP levels within regenerating axons. As described above, increasing cAMP converts netrin-mediated repulsion to attraction, and encouraging findings indicate that increasing the concentration of cAMP in neurons promotes axon regeneration in the mature CNS following injury.[114,115]

Conclusion and Perspectives

Since their discovery a little over a decade ago, significant insight has been gained into netrin function. Extending axons have been found to be directed by netrins in multiple contexts. Netrins also direct the migration of numerous cell types during development, including: inferior olivary,[116] basilar pontine[117] and LHRH neurons,[118] as well as, striatal neuronal precursors,[119] cerebellar granule cells,[120] spinal accessory neurons[41] and oligodendrocyte precursor cells.[121,122] An exciting new avenue of research has identified roles for netrins in the morphogenesis of a variety of tissues.[123,124] Netrins are now implicated in the development of the lung,[125,126] mammary gland[127] and vascular networks.[128-131] Although aspects of this work is in its initial stages, the studies described here identify roles for netrins in axon guidance, cell migration, tissue morphogenesis, and the maintanance of appropriate cell-cell interactions, supporting the conclusion that netrins influence development in a broad range of biological contexts.

References

1. Ramón y Cajal S. Texture of the Nervous System of Man and the Vertebrates. Vienna/New York: Springer, 1999.
2. Tessier-Lavigne M, Goodman CS. The molecular biology of axon guidance. Science 1996; 274:1123-1133.
3. Tessier-Lavigne M, Placzek M, Lumsden AG et al. Chemotropic guidance of developing axons in the mammalian central nervous system. Nature 1988; 336:775-778.
4. Placzek M, Tessier-Lavigne M, Jessell T et al. Orientation of commissural axons in vitro in response to a floor plate-derived chemoattractant. Development 1990; 110:19-30.
5. Brenner S. The genetics of Caenorhabditis elegans. Genetics 1974; 77:71-94.
6. Hedgecock EM, Culotti JG, Hall DH. The unc-5, unc-6, and unc-40 genes guide circumferential migrations of pioneer axons and mesodermal cells on the epidermis in C. elegans. Neuron 1990; 4:61-85.
7. Ishii N, Wadsworth WG, Stern BD et al. UNC-6, a laminin-related protein, guides cell and pioneer axon migrations in C. elegans. Neuron 1992; 9:873-881.
8. Serafini T, Kennedy TE, Galko MJ et al. The netrins define a family of axon outgrowth-promoting proteins homologous to C. elegans UNC-6. Cell 1994; 78:409-424.
9. Kennedy TE, Serafini T, de Jr IT et al. Netrins are diffusible chemotropic factors for commissural axons in the embryonic spinal cord. Cell 1994; 78:425-435.
10. Kennedy TE, Wang H, Marshall W et al. Axon guidance by diffusible chemoattractants: A gradient of netrin protein in the developing spinal cord. J Neurosci 2006; 26:8866-8874.
11. Serafini T, Colamarino SA, Leonardo ED et al. Netrin-1 is required for commissural axon guidance in the developing vertebrate nervous system. Cell 1996; 87:1001-1014.
12. Wadsworth WG, Bhatt H, Hedgecock EM. Neuroglia and pioneer neurons express UNC-6 to provide global and local netrin cues for guiding migrations in C. elegans. Neuron 1996; 16:35-46.
13. Adler CE, Fetter RD, Bargmann CI. UNC-6/Netrin induces neuronal asymmetry and defines the site of axon formation. Nat Neurosci 2006; 9:511-518.
14. Mitchell KJ, Doyle JL, Serafini T et al. Genetic analysis of Netrin genes in Drosophila: Netrins guide CNS commissural axons and peripheral motor axons. Neuron 1996; 17:203-215.
15. Harris R, Sabatelli LM, Seeger MA. Guidance cues at the Drosophila CNS midline: Identification and characterization of two Drosophila Netrin/UNC-6 homologs. Neuron 1996; 17:217-228.
16. Brankatschk M, Dickson BJ. Netrins guide Drosophila commissural axons at short range. Nat Neurosci 2006; 9:188-194.
17. Matus DQ, Pang K, Marlow H et al. Molecular evidence for deep evolutionary roots of bilaterality in animal development. Proc Natl Acad Sci USA 2006; 103:11195-11200.
18. Miner JH, Yurchenco PD. Laminin functions in tissue morphogenesis. Annu Rev Cell Dev Biol 2004; 20:255-284.
19. Yurchenco PD, Wadsworth WG. Assembly and tissue functions of early embryonic laminins and netrins. Curr Opin Cell Biol 2004; 16:572-579.
20. Park KW, Urness LD, Senchuk MM et al. Identification of new netrin family members in zebrafish: Developmental expression of netrin2 and netrin4. Dev Dyn 2005; 234(3):726-731.
21. Wang H, Copeland NG, Gilbert DJ et al. Netrin-3, a mouse homolog of human NTN2L, is highly expressed in sensory ganglia and shows differential binding to netrin receptors. J Neurosci 1999; 19:4938-4947.
22. Koch M, Murrell JR, Hunter DD et al. A novel member of the netrin family, beta-netrin, shares homology with the beta chain of laminin: Identification, expression, and functional characterization. J Cell Biol 2000; 151:221-234.
23. Yin Y, Sanes JR, Miner JH. Identification and expression of mouse netrin-4. Mech Dev 2000; 96:115-119.
24. Nakashiba T, Ikeda T, Nishimura S et al. Netrin-G1: A novel glycosyl phosphatidylinositol-linked mammalian netrin that is functionally divergent from classical netrins. J Neurosci 2000; 20:6540-6550.
25. Nakashiba T, Nishimura S, Ikeda T et al. Complementary expression and neurite outgrowth activity of netrin-G subfamily members. Mech Dev 2002; 111:47-60.
26. Cebria F, Newmark PA. Planarian homologs of netrin and netrin receptor are required for proper regeneration of the central nervous system and the maintenance of nervous system architecture. Development 2005; 132:3691-3703.
27. Gan WB, Wong VY, Phillips A et al. Cellular expression of a leech netrin suggests roles in the formation of longitudinal nerve tracts and in regional innervation of peripheral targets. J Neurobiol 1999; 40:103-115.

28. Colognato H, MacCarrick M, O'Rear JJ et al. The laminin alpha2-chain short arm mediates cell adhesion through both the alpha1beta1 and alpha2beta1 integrins. J Biol Chem 1997; 272:29330-29336.
29. Ettner N, Gohring W, Sasaki T et al. The N-terminal globular domain of the laminin alpha1 chain binds to alpha1beta1 and alpha2beta1 integrins and to the heparan sulfate-containing domains of perlecan. FEBS Lett 1998; 430:217-221.
30. Paulsson M, Saladin K, Landwehr R. Binding of Ca2+ influences susceptibility of laminin to proteolytic digestion and interactions between domain-specific laminin fragments. Eur J Biochem 1988; 177:477-481.
31. Lim YS, Wadsworth WG. Identification of domains of netrin UNC-6 that mediate attractive and repulsive guidance and responses from cells and growth cones. J Neurosci 2002; 22:7080-7087.
32. Wang Q, Wadsworth WG. The C domain of netrin UNC-6 silences calcium/calmodulin-dependent protein kinase- and diacylglycerol-dependent axon branching in Caenorhabditis elegans. J Neurosci 2002; 22:2274-2282.
33. Manitt C, Colicos MA, Thompson KM et al. Widespread expression of netrin-1 by neurons and oligodendrocytes in the adult mammalian spinal cord. J Neurosci 2001; 21:3911-3922.
34. Manitt C, Kennedy TE. Where the rubber meets the road: Netrin expression and function in developing and adult nervous systems. Prog Brain Res 2002; 137:425-442.
35. Kappler J, Franken S, Junghans U et al. Glycosaminoglycan-binding properties and secondary structure of the C-terminus of netrin-1. Biochem Biophys Res Commun 2000; 271(2):287-291.
36. Suzuki N, Toyoda H, Sano M et al. Chondroitin acts in the guidance of gonadal distal tip cells in C. elegans. Dev Biol 2006.
37. Hamelin M, Zhou Y, Su MW et al. Expression of the UNC-5 guidance receptor in the touch neurons of C. elegans steers their axons dorsally. Nature 1993; 364:327-330.
38. Keleman K, Dickson BJ. Short- and long-range repulsion by the Drosophila Unc5 netrin receptor. Neuron 2001; 32:605-617.
39. Colamarino SA, Tessier-Lavigne M. The axonal chemoattractant netrin-1 is also a chemorepellent for trochlear motor axons. Cell 1995; 81:621-629.
40. Varela-Echavarria A, Tucker A, Puschel AW et al. Motor axon subpopulations respond differentially to the chemorepellents netrin-1 and semaphorin D. Neuron 1997; 18:193-207.
41. Dillon AK, Fujita SC, Matise MP et al. Molecular control of spinal accessory motor neuron/axon development in the mouse spinal cord. J Neurosci 2005; 25:10119-10130.
42. Deiner MS, Kennedy TE, Fazeli A et al. Netrin-1 and DCC mediate axon guidance locally at the optic disc: Loss of function leads to optic nerve hypoplasia. Neuron 1997; 19:575-589.
43. Lin L, Rao Y, Isacson O. Netrin-1 and slit-2 regulate and direct neurite growth of ventral midbrain dopaminergic neurons. Mol Cell Neurosci 2005; 28:547-555.
44. Braisted JE, Catalano SM, Stimac R et al. Netrin-1 promotes thalamic axon growth and is required for proper development of the thalamocortical projection. J Neurosci 2000; 20:5792-5801.
45. Barallobre MJ, Del Rio JA, Alcantara S et al. Aberrant development of hippocampal circuits and altered neural activity in netrin 1-deficient mice. Development 2000; 127:4797-4810.
46. Puschel AW. Divergent properties of mouse netrins. Mech Dev 1999; 83:65-75.
47. Yin Y, Miner JH, Sanes JR. Laminets: Laminin- and netrin-related genes expressed in distinct neuronal subsets. Mol Cell Neurosci 2002; 19:344-358.
48. Borg I, Freude K, Kubart S et al. Disruption of Netrin G1 by a balanced chromosome translocation in a girl with Rett syndrome. Eur J Hum Genet 2005; 13:921-927.
49. Inaki K, Nishimura S, Nakashiba T et al. Laminar organization of the developing lateral olfactory tract revealed by differential expression of cell recognition molecules. J Comp Neurol 2004; 479:243-256.
50. Kim S, Burette A, Chung HS et al. NGL family PSD-95-interacting adhesion molecules regulate excitatory synapse formation. Nat Neurosci 2006; 9:1294-1301.
51. Barallobre MJ, Pascual M, Del Rio JA et al. The Netrin family of guidance factors: Emphasis on Netrin-1 signalling. Brain Res Brain Res Rev 2005; 49:22-47.
52. Huber AB, Kolodkin AL, Ginty DD et al. Signaling at the growth cone: Ligand-receptor complexes and the control of axon growth and guidance. Annu Rev Neurosci 2003; 26:509-563.
53. Rajagopalan S, Deitinghoff L, Davis D et al. Neogenin mediates the action of repulsive guidance molecule. Nat Cell Biol 2004; 6:756-762.
54. Chan SS, Zheng H, Su MW et al. UNC-40, a C. elegans homolog of DCC (Deleted in Colorectal Cancer), is required in motile cells responding to UNC-6 netrin cues. Cell 1996; 87:187-195.
55. Keino-Masu K, Masu M, Hinck L et al. Deleted in Colorectal Cancer (DCC) encodes a netrin receptor. Cell 1996; 87:175-185.

56. Fazeli A, Dickinson SL, Hermiston ML et al. Phenotype of mice lacking functional Deleted in colorectal cancer (Dcc) gene. Nature 1997; 386:796-804.
57. Bennett KL, Bradshaw J, Youngman T et al. Deleted in colorectal carcinoma (DCC) binds heparin via its fifth fibronectin type III domain. J Biol Chem 1997; 272:26940-26946.
58. Geisbrecht BV, Dowd KA, Barfield RW et al. Netrin binds discrete subdomains of DCC and UNC5 and mediates interactions between DCC and heparin. J Biol Chem 2003; 278:32561-32568.
59. Kruger RP, Lee J, Li W et al. Mapping netrin receptor binding reveals domains of Unc5 regulating its tyrosine phosphorylation. J Neurosci 2004; 24:10826-10834.
60. Kolodziej PA, Timpe LC, Mitchell KJ et al. Frazzled encodes a Drosophila member of the DCC immunoglobulin subfamily and is required for CNS and motor axon guidance. Cell 1996; 87:197-204.
61. Hall A. Rho GTPases and the actin cytoskeleton. Science 1998; 279:509-514.
62. Luo L. Rho GTPases in neuronal morphogenesis. Nat Rev Neurosci 2000; 1:173-180.
63. Dickson BJ. Rho GTPases in growth cone guidance. Curr Opin Neurobiol 2001; 11:103-110.
64. Shekarabi M, Moore SW, Tritsch NX et al. Deleted in colorectal cancer binding netrin-1 mediates cell substrate adhesion and recruits Cdc42, Rac1, Pak1, and N-WASP into an intracellular signaling complex that promotes growth cone expansion. J Neurosci 2005; 25:3132-3141.
65. Causeret F, Hidalgo-Sanchez M, Fort P et al. Distinct roles of Rac1/Cdc42 and Rho/Rock for axon outgrowth and nucleokinesis of precerebellar neurons toward netrin 1. Development 2004; 131:2841-2852.
66. Hong K, Hinck L, Nishiyama M et al. A ligand-gated association between cytoplasmic domains of UNC5 and DCC family receptors converts netrin-induced growth cone attraction to repulsion. Cell 1999; 97:927-941.
67. Stein E, Zou Y, Poo M et al. Binding of DCC by netrin-1 to mediate axon guidance independent of adenosine A2B receptor activation. Science 2001; 291:1976-1982.
68. Li X, Meriane M, Triki I et al. The adaptor protein Nck-1 couples the netrin-1 receptor DCC (deleted in colorectal cancer) to the activation of the small GTPase Rac1 through an atypical mechanism. J Biol Chem 2002; 277:37788-37797.
69. Li W, Lee J, Vikis HG et al. Activation of FAK and Src are receptor-proximal events required for netrin signaling. Nat Neurosci 2004; 7:1213-1221.
70. Meriane M, Tcherkezian J, Webber CA et al. Phosphorylation of DCC by Fyn mediates Netrin-1 signaling in growth cone guidance. J Cell Biol 2004; 167:687-698.
71. Lebrand C, Dent EW, Strasser GA et al. Critical role of Ena/VASP proteins for filopodia formation in neurons and in function downstream of netrin-1. Neuron 2004; 42:37-49.
72. Xie Y, Ding YQ, Hong Y et al. Phosphatidylinositol transfer protein-alpha in netrin-1-induced PLC signalling and neurite outgrowth. Nat Cell Biol 2005; 7:1124-1132.
73. Ming G, Song H, Berninger B et al. Phospholipase C-gamma and phosphoinositide 3-kinase mediate cytoplasmic signaling in nerve growth cone guidance. Neuron 1999; 23:139-148.
74. Rhee SG. Regulation of phosphoinositide-specific phospholipase C. Annu Rev Biochem 2001; 70:281-312.
75. Hong K, Nishiyama M, Henley J et al. Calcium signalling in the guidance of nerve growth by netrin-1. Nature 2000; 403:93-98.
76. Jin M, Guan CB, Jiang YA et al. Ca2+-dependent regulation of rho GTPases triggers turning of nerve growth cones. J Neurosci 2005; 25:2338-2347.
77. Leung-Hagesteijn C, Spence AM, Stern BD et al. UNC-5, a transmembrane protein with immunoglobulin and thrombospondin type 1 domains, guides cell and pioneer axon migrations in C. elegans. Cell 1992; 71:289-299.
78. Ackerman SL, Kozak LP, Przyborski SA et al. The mouse rostral cerebellar malformation gene encodes an UNC-5-like protein. Nature 1997; 386:838-842.
79. Leonardo ED, Hinck L, Masu M et al. Vertebrate homologues of C. elegans UNC-5 are candidate netrin receptors. Nature 1997; 386:833-838.
80. Przyborski SA, Knowles BB, Ackerman SL. Embryonic phenotype of Unc5h3 mutant mice suggests chemorepulsion during the formation of the rostral cerebellar boundary. Development 1998; 125:41-50.
81. Engelkamp D. Cloning of three mouse Unc5 genes and their expression patterns at mid-gestation. Mech Dev 2002; 118:191-197.
82. Itoh M, Nagafuchi A, Moroi S et al. Involvement of ZO-1 in cadherin-based cell adhesion through its direct binding to alpha catenin and actin filaments. J Cell Biol 1997; 138:181-192.
83. Merz DC, Zheng H, Killeen MT et al. Multiple signaling mechanisms of the UNC-6/netrin receptors UNC-5 and UNC-40/DCC in vivo. Genetics 2001; 158:1071-1080.

84. Killeen M, Tong J, Krizus A et al. UNC-5 function requires phosphorylation of cytoplasmic tyrosine 482, but its UNC-40-independent functions also require a region between the ZU-5 and death domains. Dev Biol 2002; 251:348-366.
85. Tong J, Killeen M, Steven R et al. Netrin stimulates tyrosine phosphorylation of the UNC-5 family of netrin receptors and induces Shp2 binding to the RCM cytodomain. J Biol Chem 2001; 276:40917-40925.
86. Colavita A, Culotti JG. Suppressors of ectopic UNC-5 growth cone steering identify eight genes involved in axon guidance in Caenorhabditis elegans. Dev Biol 1998; 194:72-85.
87. Huang X, Cheng HJ, Tessier-Lavigne M et al. MAX-1, a novel PH/MyTH4/FERM domain cytoplasmic protein implicated in netrin-mediated axon repulsion. Neuron 2002; 34:563-576.
88. Labrador JP, O'keefe D, Yoshikawa S et al. The homeobox transcription factor even-skipped regulates netrin-receptor expression to control dorsal motor-axon projections in Drosophila. Curr Biol 2005; 15:1413-1419.
89. Campbell DS, Holt CE. Chemotropic responses of retinal growth cones mediated by rapid local protein synthesis and degradation. Neuron 2001; 32:1013-1026.
90. Ming GL, Wong ST, Henley J et al. Adaptation in the chemotactic guidance of nerve growth cones. Nature 2002; 417:411-418.
91. Galko MJ, Tessier-Lavigne M. Function of an axonal chemoattractant modulated by metalloprotease activity. Science 2000; 289:1365-1367.
92. Hu G, Zhang S, Vidal M et al. Mammalian homologs of seven in absentia regulate DCC via the ubiquitin-proteasome pathway. Genes Dev 1997; 11:2701-2714.
93. Kim TH, Lee HK, Seo IA et al. Netrin induces down-regulation of its receptor, Deleted in Colorectal Cancer, through the ubiquitin-proteasome pathway in the embryonic cortical neuron. J Neurochem 2005; 95:1-8.
94. Tanikawa C, Matsuda K, Fukuda S et al. p53RDL1 regulates p53-dependent apoptosis. Nat Cell Biol 2003; 5:216-223.
95. Ming GL, Song HJ, Berninger B et al. cAMP-dependent growth cone guidance by netrin-1. Neuron 1997; 19:1225-1235.
96. Bouchard JF, Moore SW, Tritsch NX et al. Protein kinase A activation promotes plasma membrane insertion of DCC from an intracellular pool: A novel mechanism regulating commissural axon extension. J Neurosci 2004; 24:3040-3050.
97. Moore SW, Kennedy TE. Protein kinase A regulates the sensitivity of spinal commissural axon turning to netrin-1 but does not switch between chemoattraction and chemorepulsion. J Neurosci 2006; 26:2419-2423.
98. Bartoe JL, McKenna WL, Quan TK et al. Protein interacting with C-kinase 1/protein kinase Calpha-mediated endocytosis converts netrin-1-mediated repulsion to attraction. J Neurosci 2006; 26:3192-3205.
99. Corset V, Nguyen-Ba-Charvet KT, Forcet C et al. Netrin-1-mediated axon outgrowth and cAMP production requires interaction with adenosine A2b receptor. Nature 2000; 407:747-750.
100. Yebra M, Montgomery AM, Diaferia GR et al. Recognition of the neural chemoattractant Netrin-1 by integrins alpha6beta4 and alpha3beta1 regulates epithelial cell adhesion and migration. Dev Cell 2003; 5:695-707.
101. Lin JC, Ho WH, Gurney A et al. The netrin-G1 ligand NGL-1 promotes the outgrowth of thalamocortical axons. Nat Neurosci 2003; 6:1270-1276.
102. Livesey FJ, Hunt SP. Netrin and netrin receptor expression in the embryonic mammalian nervous system suggests roles in retinal, striatal, nigral, and cerebellar development. Mol Cell Neurosci 1997; 8:417-429.
103. Volenec A, Bhogal RK, Moorman JM et al. Differential expression of DCC mRNA in adult rat forebrain. Neuroreport 1997; 8:2913-2917.
104. Volenec A, Zetterstrom TS, Flanigan TP. 6-OHDA denervation substantially decreases DCC mRNA levels in rat substantia nigra compacta. Neuroreport 1998; 9:3553-3556.
105. Madison RD, Zomorodi A, Robinson GA. Netrin-1 and peripheral nerve regeneration in the adult rat. Exp Neurol 2000; 161:563-570.
106. Petrausch B, Jung M, Leppert CA et al. Lesion-induced regulation of netrin receptors and modification of netrin-1 expression in the retina of fish and grafted rats. Mol Cell Neurosci 2000; 16:350-364.
107. Ellezam B, Selles-Navarro I, Manitt C et al. Expression of netrin-1 and its receptors DCC and UNC-5H2 after axotomy and during regeneration of adult rat retinal ganglion cells. Exp Neurol 2001; 168:105-115.

108. Manitt C, Thompson KM, Kennedy TE. Developmental shift in expression of netrin receptors in the rat spinal cord: Predominance of UNC-5 homologues in adulthood. J Neurosci Res 2004; 77:690-700.
109. Finger JH, Bronson RT, Harris B et al. The netrin 1 receptors Unc5h3 and Dcc are necessary at multiple choice points for the guidance of corticospinal tract axons. J Neurosci 2002; 22:10346-10356.
110. Harel NY, Strittmatter SM. Can regenerating axons recapitulate developmental guidance during recovery from spinal cord injury? Nat Rev Neurosci 2006; 7:603-616.
111. Manitt C, Wang D, Kennedy TE et al. Positioned to inhibit: Netrin-1 and netrin receptor expression after spinal cord injury. J Neurosci Res 2006; 84, (in press).
112. Cohen AH, Mackler SA, Selzer ME. Behavioral recovery following spinal transection: Functional regeneration in the lamprey CNS. Trends Neurosci 1988; 11:227-231.
113. Shifman MI, Selzer ME. Expression of the netrin receptor UNC-5 in lamprey brain: Modulation by spinal cord transection. Neurorehabil Neural Repair 2000; 14:49-58.
114. Neumann S, Bradke F, Tessier-Lavigne M et al. Regeneration of sensory axons within the injured spinal cord induced by intraganglionic cAMP elevation. Neuron 2002; 34:885-893.
115. Qiu J, Cai D, Filbin MT. A role for cAMP in regeneration during development and after injury. Prog Brain Res 2002; 137:381-387.
116. Bloch-Gallego E, Ezan F, Tessier-Lavigne M et al. Floor plate and netrin-1 are involved in the migration and survival of inferior olivary neurons. J Neurosci 1999; 19:4407-4420.
117. Yee KT, Simon HH, Tessier-Lavigne M et al. Extension of long leading processes and neuronal migration in the mammalian brain directed by the chemoattractant netrin-1. Neuron 1999; 24:607-622.
118. Schwarting GA, Raitcheva D, Bless EP et al. Netrin 1-mediated chemoattraction regulates the migratory pathway of LHRH neurons. Eur J Neurosci 2004; 19:11-20.
119. Hamasaki T, Goto S, Nishikawa S et al. A role of netrin-1 in the formation of the subcortical structure striatum: Repulsive action on the migration of late-born striatal neurons. J Neurosci 2001; 21:4272-4280.
120. Alcantara S, Ruiz M, De Castro F et al. Netrin 1 acts as an attractive or as a repulsive cue for distinct migrating neurons during the development of the cerebellar system. Development 2000; 127:1359-1372.
121. Jarjour AA, Manitt C, Moore SW et al. Netrin-1 is a chemorepellent for oligodendrocyte precursor cells in the embryonic spinal cord. J Neurosci 2003; 23:3735-3744.
122. Tsai HH, Tessier-Lavigne M, Miller RH. Netrin 1 mediates spinal cord oligodendrocyte precursor dispersal. Development 2003; 130:2095-2105.
123. Hinck L. The versatile roles of "axon guidance" cues in tissue morphogenesis. Dev Cell 2004; 7:783-793.
124. Baker KA, Moore SW, Jarjour AA et al. When a diffusible axon guidance cue stops diffusing: Roles for netrins in adhesion and morphogenesis. Curr Opin Neurobiol 2006; 16:529-534.
125. Dalvin S, Anselmo MA, Prodhan P et al. Expression of Netrin-1 and its two receptors DCC and UNC5H2 in the developing mouse lung. Gene Expr Patterns 2003; 3:279-283.
126. Liu Y, Stein E, Oliver T et al. Novel role for Netrins in regulating epithelial behavior during lung branching morphogenesis. Curr Biol 2004; 14:897-905.
127. Srinivasan K, Strickland P, Valdes A et al. Netrin-1/neogenin interaction stabilizes multipotent progenitor cap cells during mammary gland morphogenesis. Dev Cell 2003; 4:371-382.
128. Park KW, Crouse D, Lee M et al. The axonal attractant Netrin-1 is an angiogenic factor. Proc Natl Acad Sci USA 2004; 101:16210-16215.
129. Lu X, Le Noble F, Yuan L et al. The netrin receptor UNC5B mediates guidance events controlling morphogenesis of the vascular system. Nature 2004; 432:179-186.
130. Klagsbrun M, Eichmann A. A role for axon guidance receptors and ligands in blood vessel development and tumor angiogenesis. 2005; 16:535-548.
131. Wilson BD, Ii M, Park KW et al. Netrins promote developmental and therapeutic angiogenesis. Science 2006; 313:640-644.

CHAPTER 3

Eph Receptors and Ephrin Ligands in Axon Guidance

Michael Reber,* Robert Hindges and Greg Lemke

Introduction

The Eph tyrosine kinase receptors (a receptor family named for the expression of Eph in an erythropoietin-producing human hepatocellular carcinoma cell line) make up the largest family of receptor tyrosine kinases. In vertebrates, 14 Eph receptor members have been identified, divided in two sub-groups, the EphAs (EphA1 to A8) and EphBs (EphB1 to B6). Their nine membrane-bound ligands, the ephrins, are also subdivided into the ephrin-As (ephrin-A1 to A6) and ephrin-Bs (ephrin-B1 to B3). The first Eph receptor (EphA1) was identified in 1987, whereas the ephrin ligands were cloned in the mid-90s.[1] Eph receptors and ephrins have been found in all animal species analyzed so far, from *C. elegans* to humans, and are highly conserved through evolution.[2] Ephs and ephrins are involved in numerous developmental processes, such as boundary formation, angiogenesis and cell migration. Within the nervous system, Eph signaling regulates the migration pattern of neural crest cells, the boundary formation between hindbrain segments (rhombomeres), the proper formation of the corticospinal tract, the establishment of neural topographic maps and the formation and functional properties of neuronal synapses.[1,3-6] It is therefore not surprising that nature built a complicated and detailed network of proteins interacting with each other to fine tune each of these important processes. The identity of the receptor or ligand molecule is as important as the structure of the receptor-ligand complex to activate a specific signaling pathway and ultimately elicit the right cell decision.

Molecular Structure

Eph Receptors (Table 1)

In general, all Eph receptors are composed of conserved structural domains, including an extracellular domain (between 500 and 600 amino acid - aa), a transmembrane domain of ~20 aa and a cytoplasmic domain of 430 aa. Their extracellular domains include a unique N-terminal ephrin-binding domain forming a globular β-barrel, a cysteine-rich region with 19 conserved cysteines that includes an EGF-like region and two fibronectin type III repeats. The cytoplasmic part contains an uninterrupted tyrosine kinase domain and several protein-protein interaction modules including SH2-docking sites, a sterile-α-motif (SAM) and a C-terminal PDZ binding motif (Fig. 1A).

*Corresponding Author: Michael Reber—INSERM U.575, Centre de Neurochimie, 5, rue Blaise Pascal, 67084 Strasbourg, France. Email: Michael.reber@inserm.u-strasbg.fr

Axon Growth and Guidance, edited by Dominique Bagnard. ©2007 Landes Bioscience and Springer Science+Business Media.

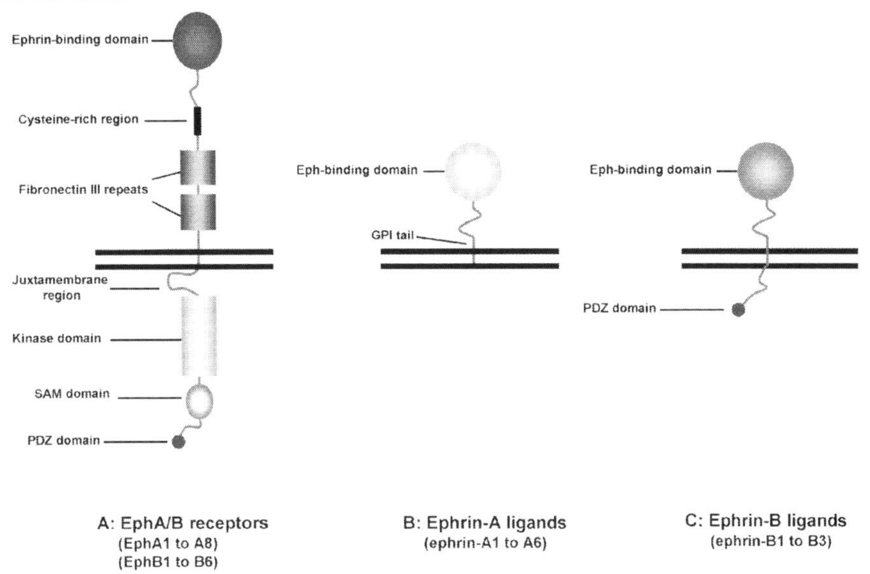

Figure 1. Structure of the membrane-bound Eph receptors and ephrin ligands.

Table 1. Mouse Eph receptors

Gene Name	mRNA (kb)	Amino Acid	Protein (kD)
EphA1	3.2	977	108
EphA2	3.9	977	108
EphA3	3.9	983	109
EphA4	5.5	986	109
EphA5	4.3	877	97
EphA6	3.9	1035	116
EphA7	4.3	998	111
EphA8	4.7	1004	110
EphB1	4.6	984	109
EphB2	4.8	994	110
EphB3	4.1	993	109
EphB4	4.2	987	108
EphB5	nd	nd	nd
EphB6	3.6	1014	110

nd, not determined.

Ephrin Ligands (Table 2)

Ephrin ligands share a conserved extracellular core sequence of approximatively 125 amino acids, including 4 invariant cysteine residues corresponding to the receptor binding domain. The difference between type A and type B ephrin ligands is their attachment to the cell membrane. Ephrin-As are linked through a glycosyl phophatidylinositol (GPI) group at their C-terminus (Fig. 1B), whereas ephrin-Bs contain a single transmembrane and a short cytoplasmic

Table 2. Mouse ephrins ligands

Gene Name	mRNA (kb)	Amino Acid	Protein (kD)
Ephrin-A1	1.5	205	23
Ephrin-A2	1.7	209	23
Ephrin-A3	1	206	21
Ephrin-A4	0.8	187	22
Ephrin-A5	3.4	228	26
Ephrin-A6	1.7	209	23
Ephrin-B1	3.2	345	38
Ephrin-B2	4.3	336	37
Ephrin-B3	3.1	340	36

domain. This domain is highly conserved and contains tyrosine residues as well as a C-terminal PDZ interaction domain (Fig. 1C).

Binding studies revealed that EphA molecules bind preferentially to type A ephrin ligand counterparts, whereas EphB receptors bind to ephrin-Bs, with two exceptions: the EphA4 receptor displays moderate binding affinity to ephrin-B2 and ephrin-B3 ligands while EphB2 can also interact with ephrin-A5.[7,8]

Mechanisms of Signal Transduction

Both the Eph receptors and the ephrin ligands are membrane attached. As a result, complexes thereof are assembled only at sites of cell-cell contact and transduce forward signals ("forward signaling") through the Eph receptors and reverse signals ("reverse signaling") through the ephrin ligands, influencing the behavior of both interacting cells.[9] Despite their large number, Eph receptors have a remarkably conserved domain structure and are activated through interaction with clustered ephrin ligands. Structural characterization of a ligand-receptor complex showed a high affinity interaction, mediated by an ephrin loop inserted into an Eph receptor cleft, and a lower affinity interface mediating assembly of tetrameric complexes comprising two receptor molecules and two ligand molecules.[10-12] However recent studies show that activation of the EphA signaling (forward signaling) by the ephrin-As requires the assembly of high order oligomer through a third contact region.[13,14] This engagement induces a conformational change in the cytoplasmic portion of the Eph receptor. In an inactive form, the juxtamembrane segment inhibits the kinase domain and upon ephrin binding, tyrosine phosphorylation of two residues in the juxtamembrane segment relieves inhibition of the kinase domain.

EphAs "Forward" Signaling

Rho GTPases

Activation of the EphA receptors by clustered ephrin-As (forward signaling) (Fig. 2) triggers a repulsive response leading to a turning or a collapse of the growth cone. The analysis of the underlying signal transduction network allowed identification of small GTPases of the Rho family (RhoA, Rac1 and Cdc42) that link the EphA receptors to the actin/microtubule dynamics. Rho GTPases cycle between an active, GTP-bound conformation, and an inactive GDP-bound conformation. In neurons, RhoA activation and its downstream effectors Rho-associated kinases (ROCK) promote growth cone collapse and axon retraction.[15,16] Rac1 and Cdc42, through their major downstream effector Pak, play an antagonistic role to Rho by promoting growth cone lamellipodia (Rac1) and filopodia extension (Cdc42).[17,16]

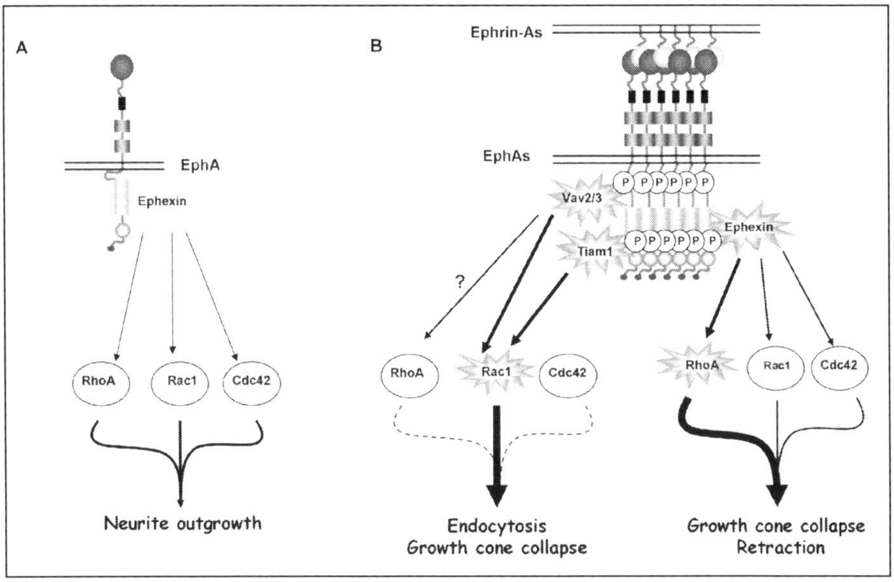

Figure 2. EphAs forward signalling. A) inactive EphAs. B) ephrinAs-activated EphAs.

EphA receptors directly activate Rho GTPases through exchange factors Ephexin1, Vav2/3 and Tiam1, which catalyze the replacement of GDP with GTP.[18-21] In the absence of ephrin stimulation, EphA kinase domain-bound ephexin1 is tethered at the plasma membrane of the growth cone and activates RhoA, Rac1 and Cdc42 leading to a balance of GTPase activation that promotes axonal outgrowth (Fig. 2A).[19] Upon ephrin stimulation, activated EphAs induce Src-dependent phosphorylation of ephexin1 on an evolutionarily conserved tyrosine (the amino-terminal Tyr87).[22] Tyrosine phosphorylation of ephexin1 selectively potentiates its exchange activity toward RhoA leading to a growth cone collapse/repulsion, while leaving the basal activity of Rac1 and Cdc42 unchanged (Fig. 2B).

In parallel to the activation of RhoA, EphAs activation by ephrins-As in retinal and cortical neurons transiently inhibits Rac1 and its downstream effector Pak.[18,23,24] The initial downregulation of Rac1 activity is followed by a rapid recovery, consistent with the requirement of Rac1 activity for Ephs-ephrins endocytosis.[23] The Rho family GEF proteins of the Vav family represent the molecular link between activated Ephs and Rac-dependent endocytosis. Upon ephrin activation, Vav2 and Vav3 are recruited to the intracellular domain of Ephs and transiently phosphorylated. They activate Rac1 leading to growth cone collapse and Eph-ephrin endocystosis (Fig. 2B).[20] Moreover, Tiam 1 is a specific Rac1-guanine-nucleotide exchange factor homogenously expressed throughout the cytoplasm in the absence of ephrin stimulation. After stimulation with ephrin-As, Tiam is translocated and phosphorylated to patches-containing EphAs resulting in Rac1 activation (Fig. 2B).[21] However, interaction of Tiam 1 with EphAs does not seem to be modified by the phoshorylation status of the EphAs.

Ras Pathway

In addition to regulate the Rho family proteins, the EphAs-ephrinAs also regulate the activity of Ras family proteins. The Ras/MAPK pathway is a key regulator of cell proliferation and transformation but also influences axon outgrowth and cell migration. Interestingly, under some circumstances, activation of several Ephs by ephrins can stimulate or attenuate the Ras/MAP kinase pathway. Some studies report that EphA2 stimulation by ephrin-A1 activates the

Ras/MAPK pathway in a variety of cell lines. In contrast, the GTPase-activating protein p120RasGAP, which inactivates H-Ras (the prototype of the Ras family proteins) is recruited to activated EphA2 and is believed to turn off the signal within minutes after ephrin treatment.[25,26]

Ephrin-As "Reverse" Signaling

Class A ephrins are tethered to the plasma membrane by a GPI anchor and are localized within discrete plasma membrane microdomains or rafts.[27] Ephrin-A2 and A5 have been shown to signal through these microdomains when bound to their cognate Eph receptors.[28,29] The physiological response induced by this signaling event leads to increased adhesion of the ephrin-A-expressing cell to fibronectin and laminin, as well as to changes in cell morphology. Both effects are dependant on the activation of β1-integrin and require the activity of the Src-tyrosine kinase Fyn.[28-30]

Ephrin-As Mediated EphAs Inactivation

EphAs and ephrin-As are present as overlapping countergradients in several areas of the central nervous system (CNS) (retina, optic tectum-OT/superior colliculus-SC, thalamus and cortex) and peripheral nervous system (PNS) (motorneurons and limb mesenchyme). The presence of both EphA receptors and ephrin-A ligands on the same axon seem to be important for regulating EphA receptors activity. In the RGCs, ephrin-As present at high level in the nasal pole of the retina desensitize the EphA receptors.[31,32] Two different mechanisms seem to be involved, a cis-masking of the ligand-binding domain of the EphAs[33] and a recently identified cis-interaction through the proximal fibronectin III domain of the EphA receptors.[34] This cis-interaction leads to an abolishment of the tyrosine phosphorylation of EphAs and renders them insensitive to the ephrin-As in trans. In contrast, another study performed in chick motor neurons suggests that coexpressed EphAs and ephrin-As are localized to different membrane domains, therefore avoiding any cis-interaction. They signal independently with EphAs mediating growth cone collapse and repulsion and ephrin-As directing motor axon growth and attraction.[27] These divergent results might be due to the differences between RGCs and motor neurons.

EphAs/Ephrin-As Cleavage and/or Endocytosis

During cell-cell communication, Eph receptors activation by ephrins causes cells to retract or to migrate in opposite direction despite the fact that ephrins and Ephs form membrane-bound high affinity complexes. Two mechanisms have been identified that might explain how EphAs/ephrin-As high-affinity binding could be terminated to allow cellular repulsion and axon withdrawal.[20,35,36] At the surface of neuroblastoma cells, in the absence of stimulation, ephrin-A2 forms a stable complex with ADAM10/Kuzbanian protease. In the presence of clustered EphA3 receptors, ADAM10/Kuzbanian cleaves and releases ephrin-A2, involving sequences in its juxtamembrane region.[35] Contrasting results demonstrate that ADAM10/Kuzbanian constitutively associates with the ephrin-binding domain of the EphA3 receptor and this association is enhanced by the presence of the ephrin-A5 ligand. Interestingly, fragments of ADAM10 bind to ephrin-A1 and -A2 but not to ephrin-A5. ADAM10 mediates cleavage of ephrin-A5 from its membrane tether only in trans. This trans proteolysis ensures that the cleavage of the ephrins will occur only upon binding to its receptor on another cell.[36] The second mechanism identified is a Vav2-dependent EphA receptor endocytosis. Retinal ganglion cells treated with clustered ephrin-A1 showed an increased internalization of ephrin-A1/EphA complexes allowing growth cones to collapse.[20] Which one of the mechanisms predominates in vivo is still unknown but it might depend on the neuronal cell type and/or on the class of EphA/ephrin-A complexes engaged.

EphBs Forward Signaling (Fig. 3)

Rho GTPases

As for the EphA receptors, EphB receptors activate members of the Rho family of GTPases, Rac1 and Cdc42, but through different exchange factors. Besides Vav2/3,[20] two other proteins, Intersectin and Kalirin, that bind to EphB2 have been identified recently.[37,38] This signaling pathway is involved in regulating dendritic spine morphogenesis through EphB2. Intersectin can also be activated in cooperation with N-WASP and provides thus a link for EphB receptors activation to actin filament assembly and branching through Arp2/3 and Cdc42.

Ras Pathway

Like EphAs, the EphB receptors also regulate the Ras-MAP-kinase pathway through downregulation of H-Ras and MAPK phosphorylation, and therefore have a negative effect on neurite outgrowth or cell migration.[39] Alternatively, activated EphBs also bind the Ras-GTPase-activating protein RasGAP and have the potential to regulate the RAS-MAPK pathway like this.[40]

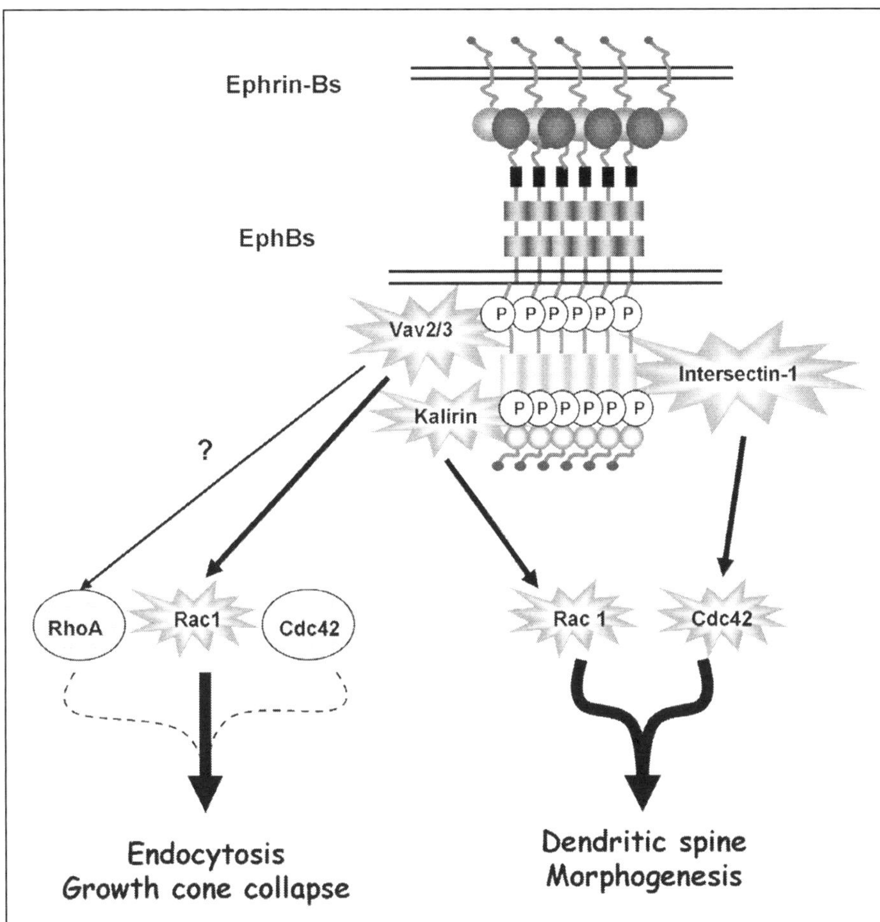

Figure 3. EphrinBs-actived EphBs forward signaling.

EphB signaling can also have an effect on integrin-mediated cell adhesion. This can be negative, through phosphorylation of R-Ras by EphB2[41] or through regulation of Rap1,[42] or positive, through binding of Nck/NIK, which activates JNK.[43,44] Another way to regulate cell adhesion is through the binding of the low molecular weight phophostyrosine phosphatase LMW-PTP.[45] The interaction with adaptor proteins like Nck, Shc or Grb family members, which themselves then interact with more downstream signaling proteins link EphB signaling to cell migration or changes in the cytoskeleton.

Ephrin-Bs Reverse Signaling (Fig. 4)

The existence of reverse signaling through ephrin-B ligands has initially been identified biochemically.[46,47] However subsequent in vivo work linked this pathway to important developmental processes like axon targeting, vascular development, cell migration and cellular compartmentalization.[1,4,48] Reverse signaling through the ephrin-B intracellular domain works either through phosphorylation-dependent mechanisms that recruit signaling effectors to transduce the signal into the ephrins expressing cell. On the other hand, several alternatives exist that provide phosphorylation-independent ways to signal. Here, we will only summarize the main proteins involved in the different pathways and where appropriate, point to more extensive reviews for a detailed description of the signaling processes.

Phosphorylation-Dependent Signaling (Fig. 4A)

In contrast to Eph receptors that become autophosphorylated upon activation, the intracellular tyrosine residues of ephrin-Bs are phosphorylated by Src-family kinases.[46,47] Recently, it was shown that the SH2/SH3 domain adaptor protein Grb4 binds the intracellular domain of ephrin-Bs upon tyrosine phosphorylation and subsequently initiates signaling cascades that regulate cytoskeleton dynamics.[49] Grb4 is related to Nck, an adaptor protein that interacts with the EphB2 receptor and to the *Drosophila* protein dreadlocks (dock), which was shown to be important for axon guidance in the visual system.[50] Upon ephrin-B binding through its

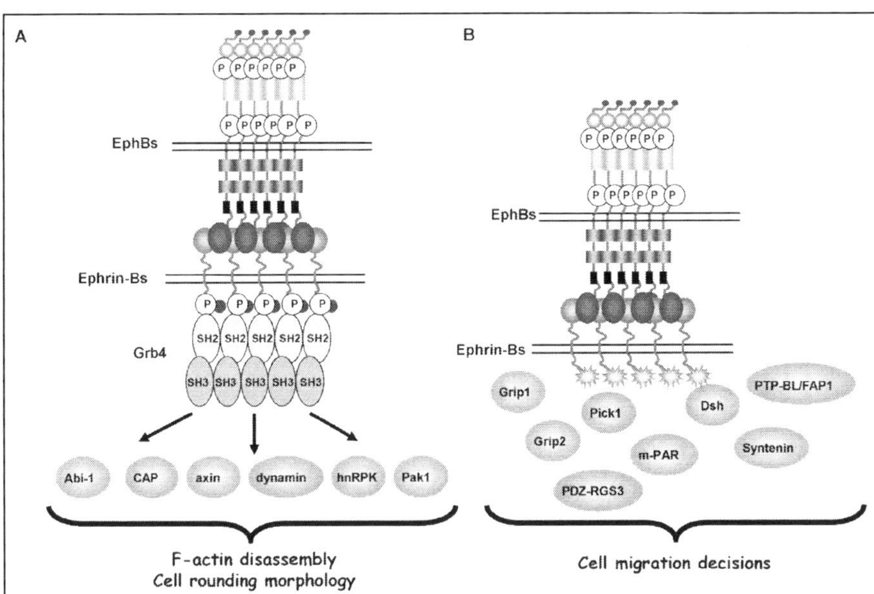

Figure 4. Ephrin-Bs reverse signaling. A) phosphorylation dependent signaling. B) phosphorylation-dependent signaling (interaction through the PDZ domain of ephrinBs).

SH2 domain, Grb4 recruits several other proteins through its SH3 domains: Abl interacting protein-1 (Abi-1), the Cbl-associated protein (CAP), axin, a regulator in the Wnt/β-catenin pathway, dynamin, hnRNPK and Pak1. These proteins then are involved in further pathways leading to the disassembly of F-actin and rounding cell morphology, which are typical features of cell collapse.

Phosphorylation-Independent Signaling (Fig. 4B)

Ephrin-Bs signaling can also be transduced in phosphorylation-independent manner. One possibility is through the interaction of the C-termini of ephrin-Bs with PDZ domain proteins like GRIP1, GRIP2, Pick1, mPAR-3, PDZ-RGS3, Syntenin and PTP-BL/FAP-1.[4,51] Whereas for example the mode of action is known for PDZ-RGS3, which selectively inhibits G-protein coupled chemoattraction,[52] for some of the other proteins mentioned above the exact mechanisms of the reverse signal propagation remain speculative. Secondly, ephrin-B reverse signaling was linked to activation of JNK through TAK1 and MMK4/MMK7.[53] A very recent addition to this list of pathways is the discovery of disheveled (dsh) as a mediator of reverse signaling. Through its different protein domains and either binding directly to ephrin-Bs or through interaction with Grb4 it involves both the canonical Wnt pathway[54] and also the PCP pathway, which controls cell migration decisions.[55]

Despite the major advancements in recent years, many details about the transduction mechanism of ephrin-B reverse signaling remain unclear. More has to be done to understand the highly complex network of signaling components with all its players.

EphBs/Ephrin-Bs Endocytosis

Like EphA/ephrin-A, endocytosis of EphB/ephrin-B complexes is a mechanism for switching from cell-cell adhesion to repulsion following plasma membrane contact. Rapid and bidirectional internalization (forward in EphBs- and reverse in ephrin-Bs-expressing cells) of EphB/ephrin-B complexes into vesicles from sites of cell-cell contact is essential for detachment and unilateral cell retraction.[56,57] The role of this endocytotic process of active cell surface EphB/ephrin-B signaling is destabilization of the adhesive interface between two cells, permitting rapid growth cone retraction in primary neurons[56] and repulsion of EphB4-expressing venous endothelial cells from those expressing ephrin-B2.[57] The fate of the internalized complexes is unknown but may involve recycling to the cell surface or proteolytic degradation. Forward endocytosis of EphB4/ephrin-B2 complexes requires a functional EphB tyrosine kinase domain and rearrangement of the actin cytoskeleton via Rac-GTPase.[57] Reverse endocytosis of EphB2/ephrin-B1 seems independent of the conserved ephrin-B tyrosine residues and C-terminal PDZ domain. For both forward and reverse endocytosis, the determinants of internalization are located in the cytoplasmic domains of both EphB receptors and ephrin-B ligands. Interestingly, blocking reverse endocytosis of ephrin-B1 leads to strong retractive response of EphB2-expressing cells, whereas blocking bidirectional endocytosis reverts the retractive response to an adhesive one.[56]

Functional Implications during Nervous System Development

Within the nervous system, Ephs and ephrins have been implicated in many biological processes during development and in the adult, like hindbrain segmentation, neural crest cell migration, axon guidance and topographic mapping as well as dendritic spine formation, synaptic plasticity and neuroregeneration.[1,5,6,48,58,59]

Hindbrain Segmentation

The vertebrate hindbrain develops a segmental pattern of seven segments called rhombomeres, which are molecularly and neuroanatomically distinct. The rhombomere boundaries are formed by cell intermingling restriction. Eph receptors and ephrin ligands are expressed in a complemenatery pattern in the hindbrain, where EphA4 is found in the even-numbered rhombomeres in mouse, *Xenopus*, zebrafish and chick, and ephrin-B2 is found in the odd-numbered segments in mouse and *Xenopus*.[60] Interestingly, only bidirectional activation of receptor and ligand together restricts the intermingling of cells resulting in clear segment borders, since experiments in zebrafish showed that unidirectional activation of either side does not have this effect.[61] This sorting out process involves cooperation between an EphA4-dependent repulsive interaction between rhombomeres and an EphA4-dependent adhesion within rhombomeres.[58,62] Recent data suggest also an involvement of a B-type receptor, EphB4, in the segmentation control of the caudal hindbrain through interaction with ephrin-B2.[63]

Axon Guidance at the Midline

The midline is one of the major choice points for axons. Axon growth cones must decide whether to cross the midline or not and furthermore, once this decision is made, what route to take for reaching their target. Here we provide examples of Ephs and ephrins and their involvement in these decisions at different midline structures: the optic chiasm, the forebrain commissures and the ventral midline of the CNS. The formation of the corticospinal tract, a long fiber tract connecting the brain and the spinal cord and providing central control of body movements, requires EphA4 forward signaling through its ligand ephrin-B3. This long fiber tract takes a complex route starting from one lobe of the neocortex, crossing the midline at the brain-spinal cord junction and then continuing down in the opposite site of the spinal cord to connect with motor neurons. Ephrin-B3 expressed at the midline in the spinal cord repels the EphA4-expressing neurons avoiding any aberrant recrossing.[4,58,64-66]

At the optic chiasm, some retinal ganglion cell axons cross the midline to the contralateral side of the brain, whereas—in species with binocular vision—another population of axons stays on the ipsilateral side. A number of different molecules (e.g., slits) have been found at the chiasm that control in part the guidance of axons at this midline structure. However, it is ephrin-B2 that plays a crucial role in whether to let RGC axons cross or not. This ligand is expressed by midline glia cells and repels EphB1 expressing RGCs to send them ipsilaterally and therefore ensuring binocular vision.[67,68] The spatiotemporal expression pattern of the receptor EphB1 in the retina is highly dynamic and tightly regulated through a genetic hierarchy of transcriptional regulators.[69,70]

Several members of the Eph family are expressed in the corpus callosum or the anterior commissure during mouse forebrain development. Mutant analysis showed initially that the loss of EphB2 and EphB3 result in a malformation or even loss (in the case of double mutant mice) of these midline structures.[71,72] However, recent work suggest that multiple receptors (EphB1, B2, B3, A4 and A5) and ligands (ephrin-B1, B2 and B3) are involved together in the correct formation of the corpus callosum.[73,74]

Topographic Mapping

A topographic map is an ordered set of connections in which the spatial relationship of the projecting neurons is preserved within the receiving set of neurons. In the nervous system, topographic maps are dynamically formed at successive levels of sensory information processing over an extended time frame during late embryonic and early postnatal development, and are a fundamental organizational feature of our brains. As proposed by Sperry with his Chemoaffinity Hypothesis almost 50 years ago, gradients of axon guidance molecules control the formation of neural connections into topographic maps.[75] Eph and ephrins have been shown to be one of the most important families of molecules to control guidance decisions during map development.

Retinotectal Mapping (Fig. 5)

Projections of retinal ganglion cells (RGCs) to their main target in the midbrain, the OT in chick, frogs and fish (or SC in mammals) is the best-studied model system for map development. We therefore describe this system and its developmental control through Eph and ephrins in more detail. The spatial ordering of axonal arborizations of RGCs within the OT/SC maps the retina, and therefore visual space, along two sets of orthogonally oriented axes. These are

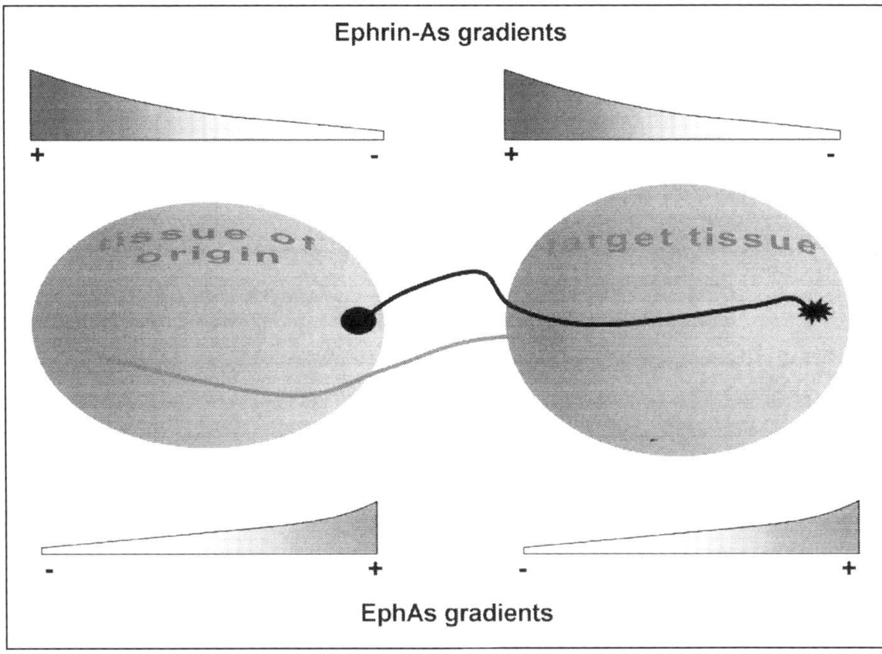

Figure 5. Map formation controlled by EphAs/ephrin-As gradients in retinocollicular, retinogeniculate and thalamocortical projections.

the temporal-nasal (TN) axis of the retina along the anterior-posterior (AP) axis of the OT/SC and the dorsal-ventral (DV) retinal axis along the lateral-medial (LM) OT/SC axis. The formation of the mature visual map in chick and rodents is based on axon overshoot and interstitial branching.[6] The different Eph subfamilies are involved in various aspects of this developmental process and along the two different axes of the map.

EphAs/Ephrin-As Control AP Mapping

Several studies demonstrate that the low-to-high NT gradients of the receptors EphA4, A5 and A6 in the retina and the low to high AP gradients of the ligands ephrin-A2 and A5 in the OT/SC control retinotectal map formation.[1,5,6,75-78] The mechanism involved is based on axon repulsion and neuronal competition for termination sites in the OT/SC determined by ratiometric rather than absolute difference comparison in EphA signaling along NT axis of the eye.[5,79,80] The repulsive effect of ephrin-As through EphA signaling does restrict the initital axon overshoot and subsequently controls the AP location of the interstitial branch formation. Countergradients of ephrin-As in the retina and EphAs in the OT/SC have recently been shown to be required for proper retinotectal map development.[81,82]

EphBs/Ephrin-Bs Control LM Mapping

Countergradients also exist along the DV axis of the retina with EphBs expressed in an overall low-to-high DV gradient, and ephrin-Bs in a high-to-low DV gradient, as well as the ML axis of the midbrain with high-to-low ML gradients for ephrin-B1 and low-to-high ML gradient for EphBs.[83,84] Mice lacking EphB2 and EphB3 show an aberrant mapping along the ML axis of the SC, consistent with an attractive guidance role of ephrin-B1 through EphB forward signaling.[83] On the other hand ephrin-B1 can also act as a repulsive cue through EphB receptors as shown by overexpression experiments in chick tecta.[84] This bifunctionality of ephrin-B1 as an attractant and as a repellent can therefore guide interstitial axon branches to the correct ML location of the termination zone, depending where the main axon shaft is located.[83,84] In contrast to the chick and rodents, experiments in *Xenopus* suggest that ephrin-B reverse signaling is the critical mechanism to control ML mapping.[85] It remains unclear however if reverse signaling also plays a role for map development in other species. The fact that the primary axon growth cones in mouse and chick (in contrast to fish and frog) do not respond to the EphBs/ephrin-B1 gradient present in the OT/SC, but only the interstitial branches do, remains unexplained.[83,84] One intriguing possibility is the selective localization and translation of mRNAs for EphB receptors along the axon or in branches, a mechanism that has been demonstrated recently for other members of the Eph family.[86]

Retinogeniculate Mapping (Fig. 5)

The lateral geniculate nucleus (LGN), including its dorsal (dLGN) and ventral (vLGN) divisions is the primary thalamic recipient of visual projections from RGCs. Retino-LGN projections are retinotopic and provide local-eye specific segregation. Gradients of ephrin-As (ephrin-A2, A3 and A5) are present in the LGN and participate to the formation of the retinotopic map.[87-89] Moreover, ephrin-As are also required for proper placement of eye-specific inputs in the dLGN but are not essential for the segregation of these inputs.[89] Recent studies in ferret demonstrate that the mechanism of retino-LGN projections formation, similarly to the one described for retinotectal projections, takes place during early postnatal period and is based on repulsion and competition between RGCs.[90]

Auditory Mapping

Auditory circuitry conveys detailed information about the timing, intensity and frequency of sounds. Complex computation of these features requires precise arrangement of the auditory circuitry—the tonotopy- in primary and high-order projection levels. EphAs and ephrin-As are required for the formation and organization of auditory neural maps at both primary and higher order level. Ephrin-A2 and one of its receptor, EphA4 showed complementary expression in the embryonic and mature ear.[91] In the chick, the receptors EphA3, A5 and A6 are present in the brainstem (inferior olive) and their ligands ephrin-A2 and A5 are present in the cerebellum.[92] Thalamic targets of the auditory pathway also show complementary EphAs/ephrin-As expression and projection defects are observed in ephrin-A2/A5 double mutant mice.[93,94] Recent findings show also expression of EphBs and ephrin-Bs in the auditory system of chick and mouse.[95-98] In vitro studies suggest that EphA4 provides repulsive signals through ephrin-B2 and -B3 for cochlear ganglion neuritis,[99] whereas EphB receptors mediate repulsive cues through ephrin-B1 for statoacoustic ganglion fibers.[95] Mutations in EphB2 cause vestibular defects in mouse.[96]

Olfactory Sensory Mapping

Olfactory sensory neurons expressing a given odorant receptor (OR) project with precision to specific glomeruli in the olfactory bulb, generating a discontinuous topographic map of receptor activation. Unlike the visual system that has simple and stable gradients of EphAs/ephrin-As, the olfactory system demonstrates complex dynamic spatio-temporal expression patterns without gradients of expression. Sensory neurons express EphA4 and EphA5 receptors and both ephrin-A3 and ephrin-A5 ligands whereas target cells in the olfactory bulb express EphA5, EphA7 as well as ephrin-A2 and A4.[100] Moreover, neurons expressing different olfactory receptors also express different levels of ephrin-As. In mutant mice alterations in the level of ephrin-As lead to a significant perturbation of the glomerular map.[101] Therefore, ephrin-A molecules play an instructive role and act together with ORs as guidance molecules that direct the formation of the precise sensory map in the olfactory bulb.

The vomeronasal axons projection within the accessory olfactory bulb (AOB) also requires EphAs/ephrin-As signaling system. Ephrin-A5 follows a basal to apical expression gradient in the vomeronasal organ while EphA6 follows a posterior to anterior expression gradient in the accessory olfactory bulb. The projections is such that axons with high ligand concentration project onto regions of the AOB with high receptor concentration and conversely. Therefore, the mechanism of development of this projection is the opposite of the repellent interaction between EphAs and ephrin-As observed in other systems.[102]

Mapping between Thalamus and Cortex (Fig. 5)

Most thalamic nuclei project to specific cortical areas, providing the first level of inter-areal specificity of thalamocortical connections. A second level is achieved within each area, where projections from each individual thalamic nucleus display a precise intra-areal topographic organization. Gradients of EphA3, EphA4 and EphA7 receptors, expressed in the dorsal thalamus and ephrin-A5 gradient, present in the ventral telencephalon, control the inter-areal specificity of thalamocortical projections through early topographic sorting of thalamocortical axons in an intermediate target, the ventral telencephalon, around E15-16 in mouse.[103,104] Later during development, EphA/ephrin-A gradients are also involved in the formation of the intra-areal topographic connections.

The topography of the somatosensory projections is disrupted in ephrin-A5/EphA4 double mutant mice with the formation of ectopic projections. Ephrin-A5 in the cortex acts as a graded repulsive cue for thalamocortical axons expressing graded levels of EphA receptors to generate a precise somatosensory map.[103,104] Misexpression of EphA7 in the sensory cortex disrupts the topography of corticothalamic projections to sensory nuclei in the thalamus. Surprisingly, corresponding thalamocortical projections are normal suggesting that EphA/ephrin-A signaling independently controls the precision of the thalamocortical and the corticothalamic projections.[105]

Within the visual cortex, ephrin-A2, -A3 and -A5 control map formation in the V1 area. Without those ligand, dorsal lateral geniculate axons misproject in the V1 area leading to a shifted, rotated and compressed map.[106] Cortical ephrin-A5 has a dual action on cortical neurons, acting as a repulsive signal for layers 2/3 neurons without axonal branching effect, and as a branch-promoting signal for layer 6 pyramidal cell axons without an effect on axon guidance.[107,108] Ephrin-A5 also induces branching for thalamic axons and acts as a repulsive axonal guidance cue for limbic thalamic axons.[85,108]

In addition, ephrin-B3 was shown to contribute in the sorting of limbic and neocortical projections through repulsive interaction with EphA4-expressing axons from lateral thalamus.[109]

Motor Axon Projections

The anteroposterior and mediolateral positions of motor neurons in the spinal cord correlate with the pathway followed by their axons and with the location of their target muscle in the limb.[110] EphA4 is required and sufficient for lateral motor column neurons to project dorsally in the hindlimb, to innervate dorsal muscle.[64,111,112] Distinct subsets of motor neurons that express EphA4 respond differently to ephrin-A5 present in the hindlimb mesoderm. EphA4-positive lateral motor column axons are repelled by ephrin-A5 whereas medial motor column axons expressing EphA4 extend through ephrin-A5 positive domains.[113] Topographic innervation of axial muscle targets are mediated by ephrin-A5 and EphA3 in addition to other EphA receptors.[114]

Dendritic Spine Remodeling

Dendritic spines are small protrusions found on the surface of dendrites and are postsynaptic targets of most excitatory synapses in the brain. Abnormal development of dendritic spines has been associated with various defects in neurodevelopmental disorders. EphA receptors are localized in synapses and dendritic spines.[115,116] They participate in cell-cell communication in the extrasynaptyic regions of the dendritic spine, where spines are in close proximity with astrocytes. The EphA4 receptor, present on dendritic spines of mouse hippocampal pyramidal neurons is activated by the astrocytic ephrin-A3 ligand leading to changes in spine morphology. Dendritic spines become shorter and even collapse upon ephrin-A3 stimulation. This suggests that neuron-glia interactions involving EphA4/ephrin-A3 forward signaling stabilize spine morphology and possibly synapses in adult hippocampus.[59,117]

As mentioned above, signaling through EphB receptors plays an important role for dendritic spine formation and synapse modulation. EphBs and ephrin-Bs are found in lipid rafts, which are important structural components of synaptic junctions. The direct interaction of EphBs and ephrin-Bs has been shown to cluster EphB and NMDA receptors at the synapse.[118] Furthermore, this interaction was also shown to modulate NMDA-receptor-dependent calcium influx.[119] EphB1/B2 and B3 all have an effect on spine development, although to different extent.[59,120] Most work has been done on EphB2, which regulates spine development through intersectin, cdc42 and N-WASP[38] and leads to interaction and clustering with syndecan-2.[121] It is unclear however which ephrin-B ligand activates the EphB receptors.

Neurogenesis

Regulation of neural progenitor cell proliferation, differentiation, survival and migration controls the number of new neurons that integrate the brain and therefore the brain size. EphA/ephrin-A signaling has been implicated in neural progenitor proliferation acting as negative regulators.[122,123] Some authors describe a reverse signaling through ephrin-A2 in the neural stem cell niche in the adult brain inhibiting neural progenitor proliferation.[122] In contrast, upon activation by ephrin-A5, EphA7 forward signaling in the developing neocortex leads to a transient wave of neural progenitor cell apoptosis and subsequent dramatic decrease in cortical size.[123] These results suggest that EphA/ephrin-A signaling negatively regulate neural progenitors cell proliferation in the brain but the mechanism might be depending on the cellular context.

Neuroregeneration

The adult CNS in mammals possesses little ability for self-repair after injury due to multiple inhibitory processes. It is believed that guidance molecules involved in axon targeting during development are reused after lesion to reestablish proper neuronal circuits.[124] Up-regulation of ephrin-A ligand expression has been shown in the OT/SC following nerve crush/section or brachial lesions.[44,125-127] This graded expression pattern in the OT/SC after lesion is very similar to the expression patterns found during development. Moreover EphA expression in the SC is severely reduced after lesion, allowing guidance function by collicular ephrin-As.[102] In fish, EphA/ephrin-A interactions restore precise topography and regulate ephrin-A2 expression during regeneration.[128] These observations suggest that ephrin-As topographic guidance information for a successful reestablishment of the retinocollicular projection is available after injury. Moreover, EphA4 seems to regulate axonal inhibition and astrocytic gliosis after spinal cord injury. EphA4 mutant mice recovered the ability to walk and grasp after spinal cord hemisection whereas astrocytic gliosis and the glial scar were greatly reduced in these mutants.[129] Recent data in rats suggest that upregulation of EphA3 in reactive astrocytes may contribute to the repulsive environment for neurite outgrowth after spinal cord injury.[130]

Conclusion and Perspectives

Eph receptors and their ephrin ligands are essential for a variety of biological processes, and signaling though these molecules is critical to the regulation of attractive or repulsive behaviors in cell migration, cell adhesion, axon guidance and branching, as well as for certain events outside the nervous system. Indeed Ephs/ephrins have been implicated in blood vessel formation, where they provide repulsive signals for the establishment and maintenance of boundaries at the arterial-venous interface, reminiscent of the repulsive guidance of axonal growth cones.[131] The remodeling of the lymphatic vasculature into a hierarchically organized vessel network consisting of lymphatic capillaries and collecting lymphatic vessels depends on ephrin-B function.[132] Ephs/ephrins also have a role in T-cell function and development by facilitating T-cell co-stimulation and proliferation, and reducing the response threshold of T-cell receptor activation.[133] Moreover, up-regulated expression and de-regulated function of Ephs and ephrins in a large variety of human cancers (lung, breast, colon, prostate and neuroblastomas) may promote a more aggressive and metastatic tumour phenotype. In contrast to other RTKs, Ephs do not act as classical proto-oncogenes and do not affect cell proliferation or differentiation but rather influence cell movements and positioning. High levels of Ephs/ephrins are correlated with tumor progression/metastasis, tumor angiogenesis and poor clinical prognosis.[134,135] Unexpectedly, ephrin-B2 and ephrin-B3 have been identified as the entry receptor for an emergent deadly virus, the Nipah virus that causes fatal encephalitis.[136,137] Bidirectional EphBs/ephrin-B signaling controls urorectal development.[138] Considering the involvement of the Eph/ephrin proteins in these many different biological processes, very much attention has been given to their functional role but still much has to be done to get a clear picture of the network of interacting proteins involved in Eph/ephrin signaling, especially how the choice between Eph/ephrin contact repulsion and stable adhesion is made at the molecular level.

References

1. Flanagan JG, Vanderhaeghen P. The ephrins and Eph receptors in neural development. Annu Rev Neurosci 1998; 21:309-345.
2. Drescher U. Eph family functions from an evolutionary perspective. Curr Opin Genet Dev 2002; 12:397-402.
3. Dodelet VC, Pasquale EB. Eph receptors and ephrin ligands: Embryogenesis to tumorigenesis. Oncogene 2000; 19:5614-5619.
4. Kullander K, Klein R. Mechanisms and functions of Eph and ephrin signalling. Nat Rev Mol Cell Biol 2002; 3:475-486.
5. Lemke G, Reber M. Retinotectal mapping: New insights from molecular genetics. Annu Rev Cell Dev Biol 2005; 21:551-580.
6. McLaughlin T, O'Leary DD. Molecular gradients and development of retinotopic maps. Annu Rev Neurosci 2005; 28:327-355.
7. Gale NW, Holland SJ, Valenzuela DM et al. Eph receptors and ligands comprise two major specificity subclasses and are reciprocally compartmentalized during embryogenesis. Neuron 1996; 17:9-19.
8. Himanen JP, Chumley MJ, Lackmann M et al. Repelling class discrimination: Ephrin-A5 binds to and activates EphB2 receptor signaling. Nat Neurosci 2004; 7:501-509.
9. Murai KK, Pasquale EB. 'Eph'ective signaling: Forward, reverse and crosstalk. J Cell Sci 2003; 116:2823-2832.
10. Koolpe M, Dail M, Pasquale EB. An ephrin mimetic peptide that selectively targets the EphA2 receptor. J Biol Chem 2002; 277:46974-46979.
11. Himanen JP, Nikolov DB. Eph signaling: A structural view. Trends Neurosci 2003; 26:46-51.
12. Himanen JP, Nikolov DB. Eph receptors and ephrins. Int J Biochem Cell Biol 2003; 35:130-134.
13. Day B, To C, Himanen JP et al. Three distinct molecular surfaces in ephrin-A5 are essential for a functional interaction with EphA3. J Biol Chem 2005; 280:26526-26532.
14. Egea J, Nissen UV, Dufour A et al. Regulation of EphA 4 kinase activity is required for a subset of axon guidance decisions suggesting a key role for receptor clustering in Eph function. Neuron 2005; 47:515-528.
15. Dickson BJ. Rho GTPases in growth cone guidance. Curr Opin Neurobiol 2001; 11:103-110.
16. Yuan XB, Jin M, Xu X et al. Signalling and crosstalk of Rho GTPases in mediating axon guidance. Nat Cell Biol 2003; 5:38-45.
17. Kozma R, Sarner S, Ahmed S et al. Rho family GTPases and neuronal growth cone remodelling: Relationship between increased complexity induced by Cdc42Hs, Rac1, and acetylcholine and collapse induced by RhoA and lysophosphatidic acid. Mol Cell Biol 1997; 17:1201-1211.
18. Shamah SM, Lin MZ, Goldberg JL et al. EphA receptors regulate growth cone dynamics through the novel guanine nucleotide exchange factor ephexin. Cell 2001; 105:233-244.
19. Sahin M, Greer PL, Lin MZ et al. Eph-dependent tyrosine phosphorylation of ephexin1 modulates growth cone collapse. Neuron 2005; 46:191-204.
20. Cowan CW, Shao YR, Sahin M et al. Vav family GEFs link activated Ephs to endocytosis and axon guidance. Neuron 2005; 46:205-217.
21. Tanaka M, Ohashi R, Nakamura R et al. Tiam1 mediates neurite outgrowth induced by ephrin-B1 and EphA2. EMBO J 2004; 23:1075-1088.
22. Knöll B, Drescher U. Src family kinases are involved in EphA receptor-mediated retinal axon guidance. J Neurosci 2004; 24:6248-6257.
23. Jurney WM, Gallo G, Letourneau PC et al. Rac1-mediated endocytosis during ephrin-A2- and semaphorin 3A-induced growth cone collapse. J Neurosci 2002; 22:6019-6028.

24. Wahl S, Barth H, Ciossek T et al. Ephrin-A5 induces collapse of growth cones by activating Rho and Rho kinase. J Cell Biol 2000; 149:263-270.
25. Noren NK, Pasquale EB. Eph receptor-ephrin bidirectional signals that target Ras and Rho proteins. Cell Signal 2004; 16:655-666.
26. Miao H, Wei BR, Peehl DM et al. Activation of EphA receptor tyrosine kinase inhibits the Ras/MAPK pathway. Nat Cell Biol 2001; 3:527-530.
27. Marquardt T, Shirasaki R, Ghosh S et al. Coexpressed EphA receptors and ephrin-A ligands mediate opposing actions on growth cone navigation from distinct membrane domains. Cell 2005; 121:127-139.
28. Davy A, Gale NW, Murray EW et al. Compartmentalized signaling by GPI-anchored ephrin-A5 requires the Fyn tyrosine kinase to regulate cellular adhesion. Genes Dev 1999; 13:3125-3135.
29. Davy A, Robbins SM. Ephrin-A5 modulates cell adhesion and morphology in an integrin-dependent manner. EMBO J 2000; 19:5396-5405.
30. Huai J, Drescher U. An ephrin-A-dependent signaling pathway controls integrin function and is linked to the tyrosine phosphorylation of a 120-kDa protein. J Biol Chem 2001; 276:6689-6694.
31. Connor RJ, Menzel P, Pasquale EB. Expression and tyrosine phosphorylation of Eph receptors suggest multiple mechanisms in patterning of the visual system. Dev Biol 1998; 193:21-35.
32. Hornberger MR, Dütting D, Ciossek T et al. Modulation of EphA receptor function by coexpressed ephrinA ligands on retinal ganglion cell axons. Neuron 1999; 22:731-742.
33. Yin Y, Yamashita Y, Noda H et al. EphA receptor tyrosine kinases interact with coexpressed ephrin-A ligands in cis. Neurosci Res 2004; 48:285-296.
34. Carvalho RF, Beutler M, Marler KJ et al. Silencing of EphA3 through a cis interaction with ephrinA5. Nat Neurosci 2006; 9:322-330.
35. Hattori M, Osterfield M, Flanagan JG. Regulated cleavage of a contact-mediated axon repellent. Science 2000; 289:1360-1365.
36. Janes PW, Saha N, Barton WA et al. Adam meets Eph: An ADAM substrate recognition module acts as a molecular switch for ephrin cleavage in trans. Cell 2005; 123:291-304.
37. Penzes P, Beeser A, Chernoff J et al. Rapid induction of dendritic spine morphogenesis by trans-synaptic ephrinB-EphB receptor activation of the Rho-GEF kalirin. Neuron 2003; 37:263-274.
38. Irie F, Yamaguchi Y. EphB receptors regulate dendritic spine development via intersectin, Cdc42 and N-WASP. Nat Neurosci 2002; 5:1117-1118.
39. Murai KK, Pasquale EB. Eph receptors, ephrins, and synaptic function. Neuroscientist 2004; 10:304-314.
40. Tong J, Elowe S, Nash P et al. Manipulation of EphB2 regulatory motifs and SH2 binding sites switches MAPK signaling and biological activity. J Biol Chem 2003; 278:6111-6119.
41. Zou JX, Wang B, Kalo MS et al. An Eph receptor regulates integrin activity through R-Ras. Proc Natl Acad Sci USA 1999; 96:13813-13818.
42. Nagashima K, Endo A, Ogita H et al. Adaptor protein Crk is required for ephrin-B1-induced membrane ruffling and focal complex assembly of human aortic endothelial cells. Mol Biol Cell 2002; 13:4231-4242.
43. Stein E, Lane AA, Cerretti DP et al. Eph receptors discriminate specific ligand oligomers to determine alternative signaling complexes, attachment, and assembly responses. Genes Dev 1998; 12:667-678.
44. Becker E, Huynh-Do U, Holland S et al. Nck-interacting Ste20 kinase couples Eph receptors to c-Jun N-terminal kinase and integrin activation. Mol Cell Biol 2000; 20:1537-1545.
45. Stein E, Huynh-Do U, Lane AA et al. Nck recruitment to Eph receptor, EphB1/ELK, couples ligand activation to c-Jun kinase. J Biol Chem 1998; 273:1303-1308.
46. Holland SJ, Gale NW, Mbamalu G et al. Bidirectional signalling through the EPH-family receptor Nuk and its transmembrane ligands. Nature 1996; 383:722-725.
47. Brückner K, Pasquale EB, Klein R. Tyrosine phosphorylation of transmembrane ligands for Eph receptors. Science 1997; 275:1640-1643.
48. Wilkinson DG. Multiple roles of EPH receptors and ephrins in neural development. Nat Rev Neurosci 2001; 2:155-164.
49. Cowan CA, Henkemeyer M. The SH2/SH3 adaptor Grb4 transduces B-ephrin reverse signals. Nature 2001; 413:174-179.
50. Garrity PA, Rao Y, Salecker I et al. Drosophila photoreceptor axon guidance and targeting requires the dreadlocks SH2/SH3 adapter protein. Cell 1996; 85:639-650.
51. Cowan CA, Henkemeyer M. Ephrins in reverse, park and drive. Trends Cell Biol 2002; 12:339-346.
52. Lu Q, Sun EE, Klein RS et al. Ephrin-B reverse signaling is mediated by a novel PDZ-RGS protein and selectively inhibits G protein-coupled chemoattraction. Cell 2001; 105:69-79.

53. Xu Z, Lai KO, Zhou HM et al. Ephrin-B1 reverse signaling activates JNK through a novel mechanism that is independent of tyrosine phosphorylation. J Biol Chem 2003; 278:24767-24775.
54. Tanaka M, Kamo T, Ota S et al. Association of Dishevelled with Eph tyrosine kinase receptor and ephrin mediates cell repulsion. EMBO J 2003; 22:847-858.
55. Lee HS, Bong YS, Moore KB et al. Dishevelled mediates ephrinB1 signalling in the eye field through the planar cell polarity pathway. Nat Cell Biol 2006; 8:55-63.
56. Zimmer M, Palmer A, Köhler J et al. EphB-ephrinB bi-directional endocytosis terminates adhesion allowing contact mediated repulsion. Nat Cell Biol 2003; 5:869-878.
57. Marston DJ, Dickinson S, Nobes CD. Rac-dependent trans-endocytosis of ephrinBs regulates Eph-ephrin contact repulsion. Nat Cell Biol 2003; 5:879-888.
58. Palmer A, Klein R. Multiple roles of ephrins in morphogenesis, neuronal networking, and brain function. Genes Dev 2003; 17:1429-1450.
59. Yamaguchi Y, Pasquale EB. Eph receptors in the adult brain. Curr Opin Neurobiol 2004; 14:288-296.
60. Cooke JE, Moens CB. Boundary formation in the hindbrain: Eph only it were simple. Trends Neurosci 2002; 25:260-267.
61. Mellitzer G, Xu Q, Wilkinson DG. Eph receptors and ephrins restrict cell intermingling and communication. Nature 1999; 400:77-81.
62. Cooke JE, Kemp HA, Moens CB. EphA4 is required for cell adhesion and rhombomere-boundary formation in the zebrafish. Curr Biol 2005; 15:536-542.
63. Cooke J, Moens C, Roth L et al. Eph signalling functions downstream of Val to regulate cell sorting and boundary formation in the caudal hindbrain. Development 2001; 128:571-580.
64. Dottori M, Hartley L, Galea M et al. EphA4 (Sek1) receptor tyrosine kinase is required for the development of the corticospinal tract. Proc Natl Acad Sci USA 1998; 95:13248-13253.
65. Kullander K, Croll SD, Zimmer M et al. Ephrin-B3 is the midline barrier that prevents corticospinal tract axons from recrossing, allowing for unilateral motor control. Genes Dev 2001; 15:877-888.
66. Kullander K, Mather NK, Diella F et al. Kinase-dependent and kinase-independent functions of EphA4 receptors in major axon tract formation in vivo. Neuron 2001; 29:73-84.
67. Nakagawa S, Brennan C, Johnson KG et al. Ephrin-B regulates the Ipsilateral routing of retinal axons at the optic chiasm. Neuron 2000; 25:599-610.
68. Williams SE, Mann F, Erskine L et al. Ephrin-B2 and EphB1 mediate retinal axon divergence at the optic chiasm. Neuron 2003; 39:919-935.
69. Pak W, Hindges R, Lim YS et al. Magnitude of binocular vision controlled by islet-2 repression of a genetic program that specifies laterality of retinal axon pathfinding. Cell 2004; 119:567-578.
70. Williams SE, Mason CA, Herrera E. The optic chiasm as a midline choice point. Curr Opin Neurobiol 2004; 14:51-60.
71. Henkemeyer M, Orioli D, Henderson JT et al. Nuk controls pathfinding of commissural axons in the mammalian central nervous system. Cell 1996; 86:35-46.
72. Orioli D, Henkemeyer M, Lemke G et al. Sek4 and Nuk receptors cooperate in guidance of commissural axons and in palate formation. EMBO J 1996; 15:6035-6049.
73. Hu Z, Yue X, Shi G et al. Corpus callosum deficiency in transgenic mice expressing a truncated ephrin-A receptor. J Neurosci 2003; 23:10963-10970.
74. Mendes SW, Henkemeyer M, Liebl DJ. Multiple Eph receptors and B-class ephrins regulate midline crossing of corpus callosum fibers in the developing mouse forebrain. J Neurosci 2006; 26:882-892.
75. Sperry RW. Chemoaffinity in the orderly growth of nerve fiber patterns and connections. Proc Natl Acad Sci USA 1963; 50:703-710.
76. Cheng HJ, Nakamoto M, Bergemann AD et al. Complementary gradients in expression and binding of ELF-1 and Mek4 in development of the topographic retinotectal projection map. Cell 1995; 82:371-381.
77. Drescher U, Kremoser C, Handwerker C et al. In vitro guidance of retinal ganglion cell axons by RAGS, a 25 kDa tectal protein related to ligands for Eph receptor tyrosine kinases. Cell 1995; 82:359-370.
78. Feldheim DA, Kim YI, Bergemann AD et al. Genetic analysis of ephrin-A2 and ephrin-A5 shows their requirement in multiple aspects of retinocollicular mapping. Neuron 2000; 25:563-574.
79. Brown A, Yates PA, Burrola P et al. Topographic mapping from the retina to the midbrain is controlled by relative but not absolute levels of EphA receptor signaling. Cell 2000; 102:77-88.
80. Reber M, Burrola P, Lemke G. A relative signalling model for the formation of a topographic neural map. Nature 2004; 431:847-853.
81. Hansen MJ, Dallal GE, Flanagan JG. Retinal axon response to ephrin-as shows a graded, concentration-dependent transition from growth promotion to inhibition. Neuron 2004; 42:717-730.

82. Rashid T, Upton AL, Blentic A et al. Opposing gradients of ephrin-As and EphA7 in the superior colliculus are essential for topographic mapping in the mammalian visual system. Neuron 2005; 47:57-69.
83. Hindges R, McLaughlin T, Genoud N et al. EphB forward signaling controls directional branch extension and arborization required for dorsal-ventral retinotopic mapping. Neuron 2002; 35:475-487.
84. McLaughlin T, Hindges R, Yates PA et al. Bifunctional action of ephrin-B1 as a repellent and attractant to control bidirectional branch extension in dorsal-ventral retinotopic mapping. Development 2003; 130:2407-2418.
85. Mann F, Peuckert C, Dehner F et al. Ephrins regulate the formation of terminal axonal arbors during the development of thalamocortical projections. Development 2002; 129:3945-3955.
86. Brittis PA, Lu Q, Flanagan JG. Axonal protein synthesis provides a mechanism for localized regulation at an intermediate target. Cell 2002; 110:223-235.
87. Feldheim DA, Vanderhaeghen P, Hansen MJ et al. Topographic guidance labels in a sensory projection to the forebrain. Neuron 1998; 21:1303-1313.
88. Ellsworth CA, Lyckman AW, Feldheim DA et al. Ephrin-A2 and -A5 influence patterning of normal and novel retinal projections to the thalamus: Conserved mapping mechanisms in visual and auditory thalamic targets. J Comp Neurol 2005; 488:140-151.
89. Pfeiffenberger C, Cutforth T, Woods G et al. Ephrin-As and neural activity are required for eye-specific patterning during retinogeniculate mapping. Nat Neurosci 2005; 8:1022-1027.
90. Huberman AD, Murray KD, Warland DK et al. Ephrin-As mediate targeting of eye-specific projections to the lateral geniculate nucleus. Nat Neurosci 2005; 8:1013-1021.
91. Bianchi LM, Gale NW. Distribution of Eph-related molecules in the developing and mature cochlea. Hear Res 1998; 117:161-172.
92. Nishida K, Flanagan JG, Nakamoto M. Domain-specific olivocerebellar projection regulated by the EphA-ephrin-A interaction. Development 2002; 129:5647-5658.
93. Lyckman AW, Jhaveri S, Feldheim DA et al. Enhanced plasticity of retinothalamic projections in an ephrin-A2/A5 double mutant. J Neurosci 2001; 21:7684-7690.
94. Cramer KS. Eph proteins and the assembly of auditory circuits. Hear Res 2005; 206:42-51.
95. Bianchi LM, Gray NA. EphB receptors influence growth of ephrin-B1-positive statoacoustic nerve fibers. Eur J Neurosci 2002; 16:1499-1506.
96. Cowan CA, Yokoyama N, Bianchi LM et al. EphB2 guides axons at the midline and is necessary for normal vestibular function. Neuron 2000; 26:417-430.
97. Person AL, Cerretti DP, Pasquale EB et al. Tonotopic gradients of Eph family proteins in the chick nucleus laminaris during synaptogenesis. J Neurobiol 2004; 60:28-39.
98. Siddiqui SA, Cramer KS. Differential expression of Eph receptors and ephrins in the cochlear ganglion and eighth cranial nerve of the chick embryo. J Comp Neurol 2005; 482:309-319.
99. Brors D, Bodmer D, Pak K et al. EphA4 provides repulsive signals to developing cochlear ganglion neurites mediated through ephrin-B2 and -B3. J Comp Neurol 2003; 462:90-100.
100. St John JA, Pasquale EB, Key B. EphA receptors and ephrin-A ligands exhibit highly regulated spatial and temporal expression patterns in the developing olfactory system. Brain Res Dev Brain Res 2002; 138:1-14.
101. Cutforth T, Moring L, Mendelsohn M et al. Axonal ephrin-As and odorant receptors: Coordinate determination of the olfactory sensory map. Cell 2003; 114:311-322.
102. Knöll B, Zarbalis K, Wurst W et al. A role for the EphA family in the topographic targeting of vomeronasal axons. Development 2001; 128:895-906.
103. Vanderhaeghen P, Polleux F. Developmental mechanisms patterning thalamocortical projections: Intrinsic, extrinsic and in between. Trends Neurosci 2004; 27:384-391.
104. Dufour A, Seibt J, Passante L et al. Area specificity and topography of thalamocortical projections are controlled by ephrin/Eph genes. Neuron 2003; 39:453-465.
105. Torii M, Levitt P. Dissociation of corticothalamic and thalamocortical axon targeting by an EphA7-mediated mechanism. Neuron 2005; 48:563-575.
106. Cang J, Kaneko M, Yamada J et al. Ephrin-as guide the formation of functional maps in the visual cortex. Neuron 2005; 48:577-589.
107. Castellani V, Yue Y, Gao PP et al. Dual action of a ligand for Eph receptor tyrosine kinases on specific populations of axons during the development of cortical circuits. J Neurosci 1998; 18:4663-4672.
108. Bolz J, Uziel D, Mühlfriedel S et al. Multiple roles of ephrins during the formation of thalamocortical projections: Maps and more. J Neurobiol 2004; 59:82-94.
109. Takemoto M, Fukuda T, Sonoda R et al. Ephrin-B3-EphA4 interactions regulate the growth of specific thalamocortical axon populations in vitro. Eur J Neurosci 2002; 16:1168-1172.

110. Pfaff S, Kintner C. Neuronal diversification: Development of motor neuron subtypes. Curr Opin Neurobiol 1998; 8:27-36.
111. Helmbacher F, Schneider-Maunoury S, Topilko P et al. Targeting of the EphA4 tyrosine kinase receptor affects dorsal/ventral pathfinding of limb motor axons. Development 2000; 127:3313-3324.
112. Kania A, Jessell TM. Topographic motor projections in the limb imposed by LIM homeodomain protein regulation of ephrin-A:EphA interactions. Neuron 2003; 38:581-596.
113. Eberhart J, Barr J, O'Connell S et al. Ephrin-A5 exerts positive or inhibitory effects on distinct subsets of EphA4-positive motor neurons. J Neurosci 2004; 24:1070-1078.
114. Vaidya A, Pniak A, Lemke G et al. EphA3 null mutants do not demonstrate motor axon guidance defects. Mol Cell Biol 2003; 23:8092-8098.
115. Gerlai R, McNamara A. Anesthesia induced retrograde amnesia is ameliorated by ephrinA5-IgG in mice: EphA receptor tyrosine kinases are involved in mammalian memory. Behav Brain Res 2000; 108:133-143.
116. Murai KK, Pasquale EB. Can Eph receptors stimulate the mind? Neuron 2002; 33:159-162.
117. Murai KK, Nguyen LN, Irie F et al. Control of hippocampal dendritic spine morphology through ephrin-A3/EphA4 signaling. Nat Neurosci 2003; 6:153-160.
118. Dalva MB, Takasu MA, Lin MZ et al. EphB receptors interact with NMDA receptors and regulate excitatory synapse formation. Cell 2000; 103:945-956.
119. Takasu MA, Dalva MB, Zigmond RE et al. Modulation of NMDA receptor-dependent calcium influx and gene expression through EphB receptors. Science 2002; 295:491-495.
120. Henkemeyer M, Itkis OS, Ngo M et al. Multiple EphB receptor tyrosine kinases shape dendritic spines in the hippocampus. J Cell Biol 2003; 163:1313-1326.
121. Ethell IM, Irie F, Kalo MS et al. EphB/syndecan-2 signaling in dendritic spine morphogenesis. Neuron 2001; 31:1001-1013.
122. Holmberg J, Armulik A, Senti KA et al. Ephrin-A2 reverse signaling negatively regulates neural progenitor proliferation and neurogenesis. Genes Dev 2005; 19:462-471.
123. Depaepe V, Suarez-Gonzalez N, Dufour A et al. Ephrin signalling controls brain size by regulating apoptosis of neural progenitors. Nature 2005; 435:1244-1250.
124. Koeberle PD, Bähr M. Growth and guidance cues for regenerating axons: Where have they gone? J Neurobiol 2004; 59:162-180.
125. Rodger J, Bartlett CA, Beazley LD et al. Transient upregulation of the rostrocaudal gradient of ephrin A2 in the tectum coincides with reestablishment of orderly projections during optic nerve regeneration in goldfish. Exp Neurol 2000; 166:196-200.
126. Symonds AC, Rodger J, Tan MM et al. Reinnervation of the superior colliculus delays down-regulation of ephrin A2 in neonatal rat. Exp Neurol 2001; 170:364-370.
127. Rodger J, Lindsey KA, Leaver SG et al. Expression of ephrin-A2 in the superior colliculus and EphA5 in the retina following optic nerve section in adult rat. Eur J Neurosci 2001; 14:1929-1936.
128. Rodger J, Vitale PN, Tee LB et al. EphA/ephrin-A interactions during optic nerve regeneration: Restoration of topography and regulation of ephrin-A2 expression. Mol Cell Neurosci 2004; 25:56-68.
129. Goldshmit Y, Galea MP, Wise G et al. Axonal regeneration and lack of astrocytic gliosis in EphA4-deficient mice. J Neurosci 2004; 24:10064-10073.
130. Irizarry-Ramírez M, Willson CA, Cruz-Orengo L et al. Upregulation of EphA3 receptor after spinal cord injury. J Neurotrauma 2005; 22:929-935.
131. Klagsbrun M, Eichmann A. A role for axon guidance receptors and ligands in blood vessel development and tumor angiogenesis. Cytokine Growth Factor Rev 2005; 16:535-548.
132. Tammela T, Petrova TV, Alitalo K. Molecular lymphangiogenesis: New players. Trends Cell Biol 2005; 15:434-441.
133. Wu J, Luo H. Recent advances on T-cell regulation by receptor tyrosine kinases. Curr Opin Hematol 2005; 12:292-297.
134. Wimmer-Kleikamp SH, Lackmann M. Eph-modulated cell morphology, adhesion and motility in carcinogenesis. IUBMB Life 2005; 57:421-431.
135. Surawska H, Ma PC, Salgia R. The role of ephrins and Eph receptors in cancer. Cytokine Growth Factor Rev 2004; 15:419-433.
136. Negrete OA, Levroney EL, Aguilar HC et al. EphrinB2 is the entry receptor for Nipah virus, an emergent deadly paramyxovirus. Nature 2005; 436:401-405.
137. Negrete OA, Wolf MC, Aguilar HC et al. Two key residues in Ephrinb3 are critical for its use as an alternative receptor for Nipah virus. PLoS Pathog 2006; 2:e7.
138. Dravis C, Yokoyama N, Chumley MJ et al. Bidirectional signaling mediated by ephrin-B2 and EphB2 controls urorectal development. Dev Biol 2004; 271:272-290.

Chapter 4

Role of Semaphorins during Axon Growth and Guidance

Eric Koncina, Lise Roth, Bertand Gonthier and Dominique Bagnard*

Introduction

During development, neuronal growth cones navigate over long distances to reach their target and establish appropriate connections. This process is usually described as a step-by-step mechanism of growth recruiting several guidance cues with attractive and repulsive properties. Among these signals, the semaphorins define a large family of more than 20 members divided into eight classes according to their phylogenic relationship and the existence of differential structural domains or sequence motifs[1] (Fig. 1). Classes I and II are found in invertebrates and classes III, IV and VII in vertebrates while class V contains members from both vertebrates and invertebrates. The class VIII corresponds to viral semaphorins. Semaphorins can be soluble proteins (classes II and III), transmembrane proteins (classes I, IV, V and VI) or membrane-bound through a glycosylphosphatidylinositol anchor (class VII). Together with plexins and scatter factor receptors, semaphorins belong to the semaphorin superfamily whose defining feature is the existence of a conserved domain: the semaphorin domain. This domain of more than 500 amino acids located at the mature protein N-terminus was first identified by Kolodkin and collaborators in 1993.[2] All semaphorins also contain N-glycosylation sites and their C-terminal part is differing from one class to another. Indeed, class V contains seven copies of thrombospondin repeats, classes II-V and VII contain an immunoglobulin-like domain while class III has an N-terminus basic motif. Semaphorins are widely expressed in the developing nervous system. Initially described as guidance cues ensuring axon targeting, further studies showed that they are also key regulators of cell migration, cell death or synapse formation during nervous system development. Like other guidance cues, semaphorins are also clearly implicated in various aspects of organogenesis[3] (including lung and kidney formation or angiogenesis) and during tumor progression.[4] Here, we review the major functions of semaphorins in the nervous system together with the signaling mechanisms involved both at the level of receptor complex formation and recruitment of selective intracellular pathways.

Functional Roles of Semaphorins during Axon Guidance

The most studied semaphorin is certainly the soluble Sema3A. The first functional description was the ability of Sema3A to act as a repulsive factor on chicken DRG neurons by inducing the collapse and retraction of their growth cones.[5] Further studies showed a repulsive effect of Sema3A on other neuronal cells, such as sensory, sympathetic and cortical neurons (Table 1 and for review see ref. 6). Most of the semaphorins identified so far mediate axon repulsion.

*Corresponding Author: Dominique Bagnard—INSERM U575, Physiopathologie du Système Nerveux, 5 rue Blaise Pascal, 67084 Strasbourg, France. Email: bagnard@neurochem.u-strasbg.fr

Axon Growth and Guidance, edited by Dominique Bagnard. ©2007 Landes Bioscience and Springer Science+Business Media.

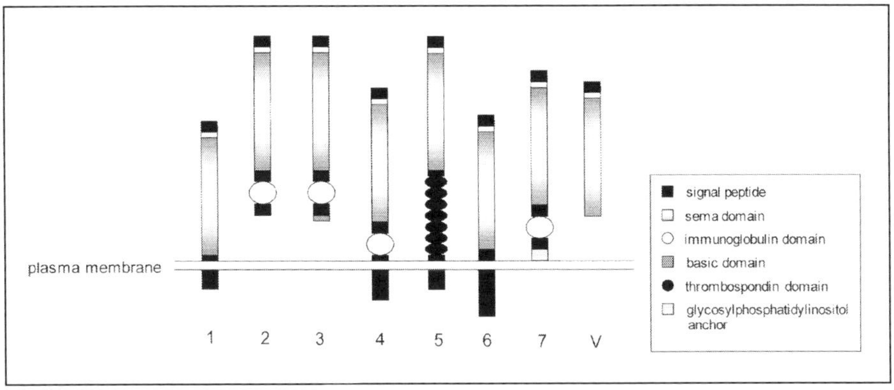

Figure 1. Semaphorin family (adapted from Unified nomenclature of semaphorins, Cell 1999).[1]

Strikingly some members have a growth promoting effect on specific neuronal subpopulations. This is the case of Sema3C that promotes the growth of cortical axons[7] and Sema3F that promotes the growth of olfactory bulb axons.[8] In vivo experiments conducted in the zebrafish showed that Sema3D triggered attraction or repulsion depending on a differential recruitment of receptor subunits.[9] In this study, a receptor complex composed of neuropilin-1 (NRP1) induced repulsion of commissural axons whereas a heterodimer composed of the two class III semaphorin receptors (neuropilin-1 and -2) triggered attraction of these axons. Thus, the temporal regulation of semaphorins and/or their receptors expression is of crucial importance to determine their functions. For example, Sema3A is expressed in the olfactory bulb from E5 to E7 in the chick when axons begin to invade the telencephalum. At E9, when axons enter the olfactory bulb, this expression is reduced.[10] Similarly, during the development of the limb innervation by the peripheral nervous system, the target limb expresses Sema3A at early time points (E10.5 in mice) thereby preventing growth cones of neuropilin-expressing motor neurons to enter this region. At E12.5 this inhibitory barrier disappears and the limb becomes permissive for the axons destined to synapse there. Again, a differential expression of semaphorin receptors contributes to the segregation of these axons since the NRP2-expressing growth cones

Table 1. Guidance effect of class III semaphorins in the nervous system

Semaphorin	Axon	Effect
Sema3A	DRG axons	repellent
	cortical axons	repellent, induces fasciculation
	sympathetic axons	repellent
	hippocampal axons	repellent
	sensitive axons	repellent
	cerebellar mossy fibers	repellent
	motoneuron axons	repellent
Sema3C	cortical axons	attractant
Sema3D	commissural axons	repellent through NRP1
		attractant through NRP1-NRP2
Sema3F	olfactory bulb axons	attractant
	motoneuron axons	repellent

are repelled by the Sema3F-containing dorsal limb region and redirected towards the ventral part of the limb while growth cones lacking NRP2 innervate the dorsal part of the limb.[11]

The genetic analysis of semaphorin function revealed several defects such as abnormal projections of sensory axons, abnormal cortical neurites orientation[12] or distorted odor map[13] in Sema3A-deficient mice. In many cases, the most severe phenotype was the defasciculation of axonal tracts in absence of Sema3A signaling.[14,15] These results are consistent with in vitro studies that have demonstrated how the inhibitory environment produced by Sema3A along cortical efferent and afferent pathways forces the axons to fasciculate.[16] Several defects in projections in the hippocampus, mid brain, forebrain and in the PNS of Sema3F deficient-mice have also been described.[17] The diversity of the guidance effects triggered by semaphorins is therefore consistent with their role in the complex wiring of various brain regions. In the cortex, a combination of Sema3A (acting as a repellent for axons and attractant for dendrites) and Sema3C (acting as a chemoattractant) is thought to control the establishment of the cortical efferent projections[7,18] and apical dendrites development.[19] Multiple combinations of semaphorins participate in the construction of axonal projections in the hippocampus,[20-22] the olfactory bulb,[8] the thalamus,[16] the spinal cord[11,23] or in the peripheral system.[14] Recently, it has been shown that not only neurons are sensitive to semaphorins in the nervous system, but also glial cells and particularly oligodendrocytes, which express semaphorin receptors. In vitro experiments showed that class 3 semaphorins control oligodendrocytes outgrowth[24,25] and are able to induce the collapse of their growth cones.[26] Oligodendrocytes migration is also controlled by class 3 semaphorin[27] and a comparable function in cell migration has been described for Sema3A and Sema3F that have been shown to differentially guide subtypes of GABAergic neurons to their appropriate target in the cortical plate or intermediate zone.[28] While the strongest expression of semaphorin is observed during development, some regions such as the hippocampus or the olfactory bulb continue to express semaphorins in adulthood.[29,30] This expression in regions presenting high level of network remodeling is consistent with evidence implicating class 3 semaphorins in the modulation of the synaptic function in the hippocampus.[31,32] These regions are also known to maintain a neurogenic potential[33-35] thereby suggesting that the role of the semaphorins in the nervous system is certainly more complex than initially thought. Indeed, some of the semaphorins have been shown to induce cell death of dopaminergic and sensory neurons[36] as well as neural precursors.[37] Hence, there is also increasing evidence for a potential role of semaphorin signaling in different pathologies of the nervous system. For example, Sema3A is over-expressed in the cerebellum of schizophrenic patients.[38] Sema3A is also accumulated in the hippocampus during Alzheimer disease.[39] In a rat model of temporal lobe epilepsy, Sema3A is down-regulated thereby permitting mossy fibers sprouting and subsequent hyper excitability of the hippocampal formation.[40] Finally, the role of class 3 semaphorins in the context of nerve lesion has also been largely documented.[41,42]

From these results, it appears that semaphorins have multiple roles ranging from axon guidance (attraction or repulsion) to cell migration or cell death. This functional diversity must be ensured by a complex signaling mechanism recruiting various receptors and co receptors coupled to specific intracellular pathways. In the following section, we will present the different families of semaphorin receptors and their interactions (Fig. 3).

Receptors of Secreted-Semaphorins: The Neuropilins

The molecular nature of semaphorin receptors remained elusive for many years until the identification of neuropilin-1 as a membrane receptor for the secreted Sema3A.[43,44] Neuropilin-1 was initially described by Fujisawa and colleagues as an orphan receptor expressed in the tadpole neuropil.[45,46] During a search for other semaphorin receptors, a neuropilin-1-related molecule, neuropilin-2, was identified.[44,47] Neuropilin-2 is 44% identical to neuropilin-1. Neuropilins are cell surface glycoproteins of about 130 kD (Fig. 2). They are composed of a large extracellular part, a unique transmembrane domain and a small 39 amino acid cytoplasmic tail. The extracellular moiety of neuropilins contains three domains with homology to several functionally diverse proteins. The N-terminus domain is referred to as the a1/a2 or CUB domain with homology to complement components C1r and C1s, this domain is followed by two coagulation-factor-homology domains (CF V/VIII, also called b1/b2) and a juxtamembrane MAM (meprin/A5/μ-phosphatase) domain also called c domain. The C-terminal part of neuropilins contains a three amino acids sequence which might bind to PDZ domain proteins such as NIP (Neuropilin Interacting Protein).[48] NRP1 or NRP2 have to dimerize to form a functional receptor complex. They can both form homo- or heterodimers[44] involving the MAM domains.[29,49,50] Our recent data suggest that the transmembrane domain of NRP1 contains a specific motif ensuring NRP1 dimerization (unpublished data, Roth and Bagnard). The constitution of homo- or heterodimers is considered to be a prerequisite for semaphorin binding. Indeed, binding experiments revealed that Sema3A may preferentially bind to NRP1 dimers, Sema3F to NRP2 dimers while Sema3C would bind to NRP1 as well as to homo or heterodimers of NRP2. Dimerization is also required at the level of ligand since to be functional, class III semaphorins dimerize through a disulfide bound located between the immunoglobulin (Ig) domain and their basic tail.[51,52] Overall, a model for the semaphorin/neuropilin complex has been proposed by several groups.[43,53-55] In this model, a semaphorin dimer binds to a neuropilin dimer on the cell surface. More precisely, the binding interface is composed of the sema domain's N-terminal part and Ig domain of semaphorins interacting with the neuropilin CUB domain while the basic C-terminal part of semaphorins bind to the coagulation factor domains of neuropilins. High affinity binding of the semaphorin Ig-basic region to the coagulation factor domains is itself not sufficient for signal transduction. Rather, it is considered to facilitate binding of the sema domain to CUB and coagulation factor domains. Interestingly, the MAM domain is not involved in the semaphorin/neuropilin interaction. The most striking feature of semaphorin/neuropilin interaction is the lack of transduction capacity supposedly due to the short intracellular tail of NRP1. The role of NRP1 and NRP2 for semaphorin signaling has been clearly shown in experiments mutating the corresponding genes[14,56,57] suggesting that the semaphorin/neuropilin interaction is the initial step for the assembly of a receptor complex recruiting transducing elements. Among the multiple possible partners that have been identified recently, members of the plexin family are crucial components of the receptor complex. Interestingly, the plexins are the binding receptor of transmembrane semaphorins.

Receptors of Transmembrane-Semaphorins: The Plexin Family

To date, nine plexins divided into four subfamilies (A-D) have been identified.[58] Plexins are single-membrane-spanning protein of approximately 240 kDa that possess a sema domain near the N-terminal part, followed by cysteine-rich motifs, Met-related sequences (MRS)[59] and glycine-proline-rich repeats (Fig. 2). The cytoplasmic part of plexins possesses a characteristic domain of 600 amino acids called the SP domain (Sex Plexins). This highly conserved domain within and across species contains some phosphorylation sites but is devoid of characteristic tyrosine kinase catalytic site.[58] Similarly to neuropilins, plexins can form homodimers[60] and heterodimers.[61,62] In 2001, Takahashi and Strittmatter have proposed that Plexin-A1 sema domain binds to the remainder of the extracellular domain of the protein thereby inducing Plexin-A1 auto-inhibition.[63] This auto-inhibition is released upon ligand binding to NRP1. An alternative view proposed by Turner et al[60] suggests that plexin dimers could auto inhibit through reciprocal interactions of their PH1 and PH2 domains.

Plexins are considered to be the primary binding sites for semaphorins that do not bind to neuropilins. Hence, with the exception of Sema3E,[64] class III semaphorins are unable to bind directly to plexins (see ref. 65 for review). Rather, plexin-neuropilin complexes are required as high-affinity receptors for secreted semaphorins, neuropilin acting as the ligand binding subunit while the plexin subunit ensures signal transduction.[66] Several neuropilin/plexin couples have been identified to transduce the diverse biological effects of semaphorins.[66-69] In any case, at least one plexin has been identified for each semaphorin class as the signaling partner. The exact nature of the domains controlling receptor complex formation and/or binding is not

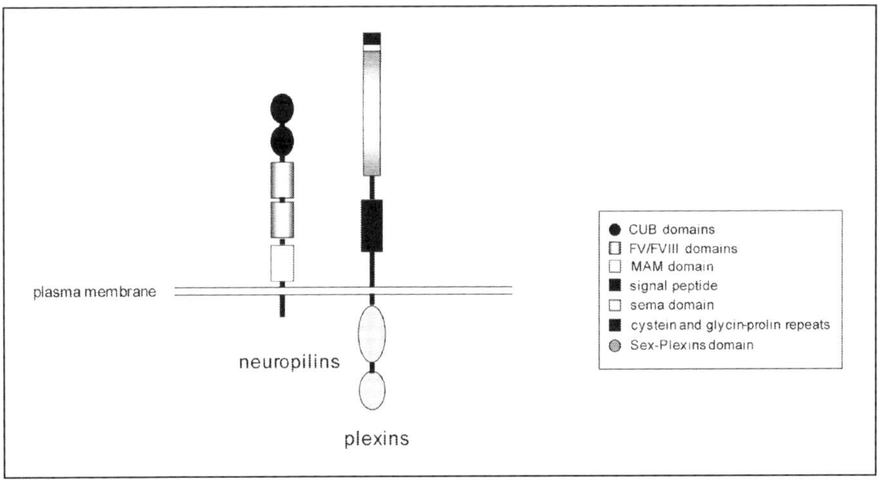

Figure 2. Characteristic features of the two major semaphorin receptors.

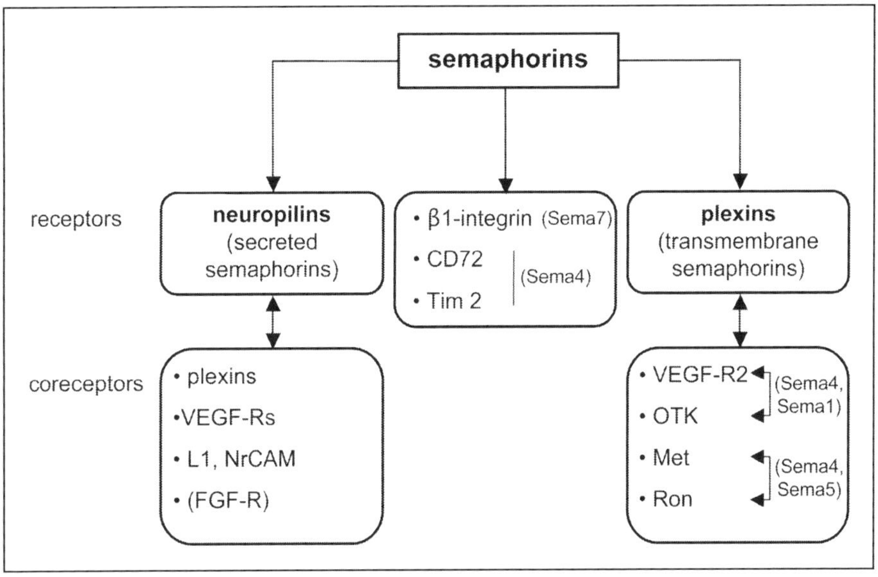

Figure 3. Molecular diversity of semaphorin receptor complexes.

known. However, structural analysis and crystal structure characterization strongly support the crucial role of the sema domain contained both in semaphorins and plexins.[70]

Other Receptors Associated with Semaphorin Signaling

Mounting evidence is now consistent with a strong molecular diversity of the receptors and coreceptors of semaphorins. This section briefly reviews the different components which have been clearly showed to control semaphorin signaling or binding.

VEGFRs

Vascular Endothelial Growth Factor (VEGF) family is a major regulator of angiogenesis[71] whose members bind to three tyrosine kinase receptors: VEGFR1, VEGFR2 and VEGFR3. Soker and collaborators discovered in 1998[72] that NRP1 is a fourth receptor for VEGF165 (a splice variant of the VEGF-A isoform), suggesting an interplay between VEGF and semaphorin signaling. Indeed, VEGF165 and Sema3A bind to NRP1 with the same affinity and compete for NRP1 binding. VEGF165 binds to NRP1 via interactions between the heparin binding domain of VEGF and the b1 domain of NRP1,[55,73] which also interacts with the basic domain of semaphorins. Soker and collaborators also demonstrated that binding of VEGF165 to NRP1 enhances affinity of VEGF for VEGFR2 and that NRP1 expression potentiates VEGF165 chemotactic effect.[72] Finally, NRP1 is also a receptor for other members of the VEGF family (VEGF-B and VEGF-E forms) and placental-growth factor-2 (PlGF-2). VEGF165 and PlGF-2, but also VEGF145 and VEGF-C bind to NRP2.[74] Gu and collaborators suggested that the NRP1-enhanced affinity of VEGF165 for VEGFR-2 may reveal a receptor complex composed of the two proteins,[55] but the domains required for the interaction between the two receptors have not been yet characterized. Nevertheless, NRP1 can bind VEGFR-1 with high affinity, and this interaction inhibits the binding of NRP1 to VEGF165. VEGFR-1 may then function as a negative regulator of angiogenesis by competing with NRP1.[73] In contrast, repulsion by Sema3A depends on both NRP1 and VEGFR-1 in neural precursor cells.[37] Thus, VEGFR-1 might serve as a co receptor for NRP1 in the transduction of Sema3A signaling. This is further supported by the Sema3A-dependent selective recruitment of MAP kinases by a receptor complex involving VEGFR-1 activation.[75] Finally, it has been shown that Plexin-A1 can form a functional complex with VEGFR-2 in the cardiac tube.[76] This complex has been reported to be involved in Sema6D signaling during cardiac morphogenesis.

L1-CAM and Nr-CAM

L1-CAM belongs to the immunoglobulin superfamily of adhesion molecules. This glycoprotein contains 6 immunoglobulin-like domains and 5 fibronectin-like domains. In addition to its classical cell adhesion function, L1 has now been implicated in Sema3A signaling. Indeed, L1-deficient axons do not respond to Sema3A. Coimmunoprecipitation and binding assays showed that L1 and NRP1, but not NRP2, interact with each other through their extracellular domains to form a stable complex. Moreover, soluble L1 converts Sema3A-induced axonal repulsion into attraction.[77] Recently, Castellani and collaborators showed that upon binding to NRP1, L1 and NRP1 are cointernalized through a clathrin-dependent mechanism mediated by L1.[78] Hence, this group has also demonstrated that NrCAM, a member of the immunoglobulin superfamily adhesion molecule of the L1 subfamily, associates with neuropilin-2 and is a component of a receptor complex for Sema3B and Sema3F.[79]

Heparin Binding Domains

A quantitative optical biosensor-based binding assay revealed that NRP1 interacts with a subset of heparin-binding proteins, notably fibroblast growth factor FGF-1, FGF-2, FGF-4, FGF-7, FGF receptor and hepatocyte growth factor/scatter factor (HGF/SF). These results suggest that NRP1 possesses a "heparin" mimetic site that is able to interact at least in part through ionic bonding with the heparin binding site of several proteins.[80] The biological relevance of these interactions has to be characterized.

Integrins

Integrins are functional heterodimers composed of α and β subunits.[81] There are 18 α and 8 β subunits that associate to form 24 integrin receptors with different ligand specificities.[82] Pasterkamp and collaborators showed that Sema7A has a pronounced effect on axon outgrowth, and that this activity is dependent on β-subunit-containing integrin receptor but is plexin-independent.[83] Several studies have also outlined the potential role of integrins in semaphorin signaling in other systems. Serini and collaborators for example showed that autocrine loops of class 3 chemorepellent semaphorins exert an essential permissive role in vasculature remodeling by inhibiting integrin-mediated adhesion of endothelial cells to the extracellular matrix, allowing the necessary de-adhesion for vascular remodeling.[84] Finally, the poxvirus A39R, member of the semaphorin family, induces actin cytoskeleton rearrangement and inhibits integrin-mediated adhesion.[85]

CD72, Tim2

CD72 is a 45-kDa type II transmembrane protein belonging to the C-type lectin family. Sema4D, the first semaphorin shown to be expressed in the immune system,[86] specifically binds to CD72 on B cell surface when Plexin-B1 is not expressed.[87,88] In that particular case, Sema4D may enhance B cells response by inhibiting the CD72 negative effect.

Tim 2 belongs to the Tim protein family characterized by expression on T cells and the presence of conserved immunoglobulin and mucin domains. Sema4A, expressed on dendritic cell surface, enhances T cells activation through binding to Tim 2.[89] Given the crucial roles of Sema4A and Sema4D signaling during the immune response, these two signals make of class IV semaphorins as a new family of immunoregulatory molecules[90] using specific receptors.

Scatter Factor Receptor

Recent evidence proposes that various tyrosine kinase receptors can be associated to the semaphorin receptor complex. Met and Ron tyrosine kinases can form disulfide-linked heterodimers and are members of the Scatter Factor receptor family containing a sema domain and a MRS sequence in their extracellular part. Giordano and collaborators showed that in cancer cells expressing the endogenous proteins, Plexin-B1 and Met associate in a complex to elicit invasive growth in response to Sema4D.[91] The same group also demonstrated that Met and Ron receptors can specifically interact with each of the three members of class B plexins. Interestingly, the role of Met/Ron interaction with plexins could be the fine-tuning of the invasive growth program of tumor cells.[92] Finally, Sema5A can trigger the intracellular signaling of Met via a receptor complex including Plexin-B3.[93]

Off-Track

Off-track (OTK) belongs to the neurotrophin receptor family. It is a glycoprotein of 160 kDa whose extracellular domain contains six immunoglobulin repeats.[94] In Drosophila, OTK and class A plexins can associate as components of a receptor complex mediating the repulsive signal in response to Sema1a.[95] Nevertheless, despite its homology with tyrosine kinase receptors, OTK itself is probably not an active tyrosine kinase and it should be determined if OTK recruits another protein for plexin phosphorylation. Moreover, Toyofuku and collaborators recently showed that Sema6D exerts expansion or narrowing of the ventricular chamber during cardiac morphogenesis through region-specific association of Plexin-A1 with off-track or VEGFR-2.[76]

Intracellular Signal Transduction

As described so far, the diversity of the semaphorin functions is related to the assembly of specific receptor complexes composed of various proteins. This diversity, not always demonstrated in neuronal cells, is the source of a wide range of possible intracellular pathways. Here, we decided to present the signaling cascades intimately linked to the signaling properties of the major transducer elements of the semaphorin receptor complex (Fig. 4).

Role of Rho GTPases

Consistent with the considerable amount of data collected on Sema3A, most of the described signaling pathways relate to this protein and its association to the neuropilin/plexin complex. As previously mentioned, NRP1 is not able to trigger any intracellular signal. Thus, the whole signaling cascade depends on the pathways recruited upon co receptors activation. Before the identification of plexins as the main signaling receptors of semaphorins, two studies by the groups of Strittmatter and Bamburg suggested the involvement of members of the Rho-GTPases family during Sema3A growth cone collapse.[96,97] These studies revealed a complex and somehow controversial role of Rac1 by transfection into neurons of dominant negative form or constitutively active form of the protein. The mechanism of action has been recently elucidated by different studies showing that Rac1 is sequestered away from its effector PAK thereby favoring actin depolymerisation pathways.[98-100] Subsequent studies also revealed the direct interaction of the cytoplamic tails of plexins with RhoGTPases. The group of Püschel for example determined that Plexin-A1 is able to bind to the small G proteins Rnd-1 and Rho-D.[101] Interestingly, these two GTPases compete for binding to the same site thereby controlling Plexin-A1 activation or inactivation respectively. Moreover, Plexin B1-Rac interaction is able to modulate the binding of Sema4D to Plexin B1.[102] Intriguingly, the Plexin-B1 possesses an intrinsic guanine triphosphate (GTP)ase activating protein activity for R-Ras, a member of Ras family of small GTPases. This particular activity has been shown to be required in promoting cell adhesion and neurite outgrowth through integrin activation.[103] Several studies also demonstrated that Rho-specific GEFs (activators of Rho-GTPases) such as PDZ-Rho-GEF and LARG ensure coupling of RhoA to Plexin B.[104-107] RhoA has been shown to mediate Sema4D-induced growth cone collapse of hippocampal neurons.[108] Finally, (PDZ)-RhoGEF

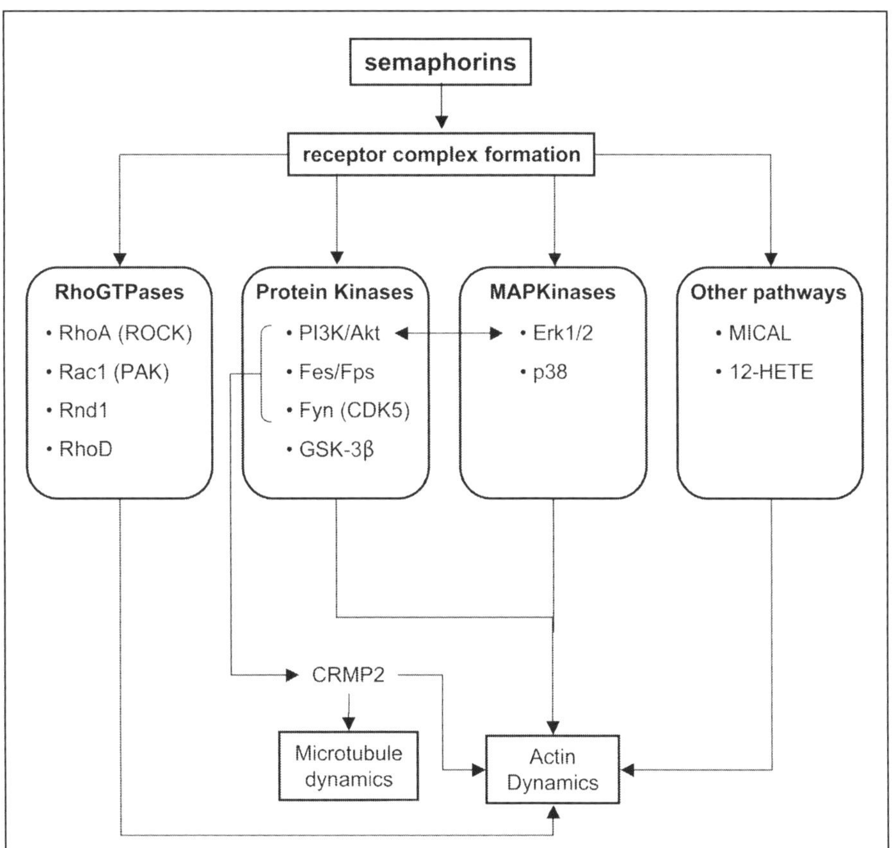

Figure 4. Representative diagram of semaphorin intracellular pathways.

and LARG can also bind to Plexin-B2 and Plexin-B3.[107] As expected from the classical description of the interplay between Rho-GTPases,[109] a concomitant inactivation of Rac1 and activation of RhoA is necessary to trigger growth cone collapse. Although differences exist between RhoGTPases/plexin interactions in vertebrates and invertebrates, the emerging scheme is the existence of a functional balance between Rho-GTPases that controls axon growth or inhibition in response to semaphorins. Ultimately, such a balance leads the modulation of actin cytoskeleton reorganization (see Nothias chapter).

The Role of Cytoskeleton Dynamics Regulators

The actin cytoskeleton is precisely regulated by multiple factors that are very similar in all cell types. Like in other cells, the Rho-GEFs and Rho-GAPs are the upstream regulators which control activation or inhibition of Rho-GTPases in neurons. Actin dynamic is thus regulated by Rho-GTPases at the level of filament nucleation and branching (role of ARP2/3 complex), filament extension (role of capping proteins) and actin recycling (role of cofilin).[110,111] It has been shown that Sema3A-induced growth cone collapse of mouse DRG neurons is correlated with a rapid increase of the phosphorylation state of cofilin, an actin binding protein supporting depolymerisation when phosphorylated.[112] This phosphorylation of cofilin is triggered by the Ser/Thr kinase LIM-kinase. Many additional kinases have been identified in the semaphorin signaling cascade.

Among them, the Fes/Fps tyrosine kinase has been implicated in Sema3A-induced growth cone collapse.[113] Fes/Fps is a nonreceptor-type tyrosine kinase shown to directly bind to the cytoplasmic region of Plexin-A1. Strikingly, the binding of Fes/Fps to Plexin-A1 is prevented by the association of NRP1 to Plexin-A1. The Fes/Fps tyrosine kinase also phosphorylates the CRMP2 (Collapsin Response Mediator Protein 2)/CRAM (CRMP-associated-molecules) complex.[113] This complex was originally identified as a crucial element of the Sema3A intracellular signaling during growth cone collapse.[114,115] In addition to Fes/Fps, the Fyn tyrosine kinase associates with and phosphorylates Plexin-A2 in response to Sema3A.[116] Furthermore, Fyn can phosphorylate the serine/threonine kinase Cdk5 which in turn can phosphorylate CRMP2.[117]

Another Ser/Thr kinase, GSK3 β, is also able to phosphorylate CRMP2. Interestingly, the phosphorylation of CRMP2 by both Cdk5 and GSK3β is essential for Sema3A-induced growth cone collapse.[118,119] CRMP2 binds to tubulin heterodimers and is supposed to control microtubules assembly.[120] Hence, Cdk5 and GSK3β may ensure the regulation of microtubules dynamics by phosphorylation of CRMP2 in response to semaphorins.

The continuing identification of novel intracellular molecular partners of plexins is demonstrated by the recent implication of MICAL, a flavoprotein oxidoreductase that binds to the C2 domain of Plexin-A in *Drosophila* to ensure Sema1a/Plexin-A-mediated repulsive axon guidance.[121] Hence, it has been shown that Sema3A stimulates the synthesis of 12(S)-hydroxyeicosatetraenoic acid (HETE) to induce DRG growth cone collapse.[122]

The Role of the MAP Kinase Pathway

The MAP kinases ERK1/2 and p38 are important signaling components often associated to cell proliferation, cell migration or cell death. Indeed, the MAP kinase pathway has been shown to regulate signaling of guidance molecules of the Netrin or Ephrin families.[123-126] As expected, the MAP kinase pathway is also recruited by semaphorins. This has been demonstrated by the activation of ERK1/2 during Sema3A-induced retinal growth cone collapse.[127] Moreover, the neurotrophic effect of Sema4D in PC12 cells also requires ERK1/2 activation.[128] The recruitment of MAP kinases during semaphorin signaling is receptor dependent. As shown by Pasterkamp and collaborators,[83] the Sema7A growth promoting effect requires an integrin-dependent MAP kinase signaling. Moreover, our work demonstrated the selective recruitment of ERK1/2 during Sema3A/VEGFR-1-mediated neural precursor cells repulsion and p38 activation to trigger cell death.[75] The integrity of the MAP kinase cascade has been shown to be important to ensure outgrowth promoting effect of L1 and other adhesion molecules such as NCAM.[129] It would therefore not be surprising that L1-mediated semaphorin effects also require activation of MAP kinases. Many arguments including the involvement of MAP kinases in the signaling of VEGF,[130] the activation of MAP kinases by small GTPases[131] and the multiplicity of the semaphorin receptors linked to MAP kinases strongly support this pathway as a point of convergence in semaphorin signaling. Hence the complexity of the interactions between intracellular pathways is illustrated by the ability of Sema3F to suppress NGF-dependent activation of the PI3K/Akt and ERK1/2 pathways,[132] two pathways involved in cancer cells to trigger EGF-induced NRP1 expression.[133]

Conclusion

Extensive work has been conducted over the past decade to elucidate the biological functions of semaphorins. A particular effort has been done to understand the complex signaling pathways recruited by semaphorins to exert their various roles. While the inhibitory pathways have been well dissected, semaphorin-triggered growth promoting pathways remain obscure. Nevertheless, compelling data being fully detailed in the chapter by Holt and collaborators (see Holt chapter, this book) identified cGMP as a key regulator of semaphorin function.[134,135] Future studies will have to investigate this fascinating question making things much more complex since inhibition and growth promotion can take place in the same cell in the presence of a single semaphorin.[19] Moreover, the driving force of guidance cues is intimately linked to the existence of gradients.[136] In the case of class 3 semaphorins, we have shown that growth cones have stereotyped responses to semaphorin gradients and do not require precise dimensioned gradients to be attracted or repulsed.[137] The exact molecular mechanism allowing growth cones to read and integrate semaphorin gradients remain obscure. Hence, the molecular hierarchy of the signaling pathways will have to be determined when cells are exposed to semaphorin combinations and, to address the biological reality, when neuronal growth cones are exposed to multiple families of guidance cues that may have converging or diverging signaling pathways.

Acknowledgement

The lab of DB is supported by grants from the FRM (#INE20011109006), BQR-ULP (#BE/CH/2001-225V) and ACI JC (#5327) and APETREIMC. BG is supported by Neurex. We thank AW Püschel for critical reading of the manuscript.

References

1. Semaphorin Nomenclature Committee. Unified nomenclature for the semaphorins/collapsins. Cell 1999; 97:551-552.
2. Kolodkin AL, Matthes DJ, Goodman CS. The semaphorin genes encode a family of transmembrane and secreted growth cone guidance molecules. Cell 1993; 75:1389-1399.
3. Hinck L. The versatile roles of "axon guidance" cues in tissue morphogenesis. Dev Cell 2004; 7:783-793.
4. Chédotal A, Kerjan G, Moreau-Fauvarque C. The brain within the tumor: New roles for axon guidance molecules in cancers. Cell Death Differ 2005; 12:1044-1056.
5. Luo Y, Raible D, Raper JA. Collapsin: A protein in brain that induces the collapse and paralysis of neuronal growth cones. Cell 1993; 75:217-227.
6. Raper JA. Semaphorins and their receptors in vertebrates and invertebrates. Curr Opin Neurobiol 2000; 10:88-94.
7. Bagnard D, Lohrum M, Uziel D et al. Semaphorins act as attractive and repulsive guidance signals during the development of cortical projections. Development 1998; 125:5043-5053.
8. de Castro F, Hu L, Drabkin H et al. Chemoattraction and chemorepulsion of olfactory bulb axons by different secreted semaphorins. J Neurosci 1999; 19:4428-4436.
9. Wolman MA, Liu Y, Tawarayama H et al. Repulsion and attraction of axons by semaphorin 3D are mediated by different neuropilins in vivo. J Neurosci 2004; 24:8428-8435.
10. Renzi MJ, Wexler TL, Raper JA. Olfactory sensory axons expressing a dominant-negative semaphorin receptor enter the CNS early and overshoot their target. Neuron 2000; 28:437-447.
11. Huber AB, Kania A, Tran TS et al. Distinct roles for secreted semaphorin signaling in spinal motor axon guidance. Neuron 2005; 48:949-964.

12. Behar O, Golden JA, Mashimo H et al. Semaphorin III is needed for normal patterning and growth of nerves, bones and heart. Nature 1996; 383:525-528.
13. Taniguchi M, Nagao H, Takahashi YK et al. Distorted odor maps in the olfactory bulb of semaphorin 3A-deficient mice. J Neurosci 2003; 23:1390-1397.
14. Kitsukawa T, Shimizu M, Sanbo M et al. Neuropilin-semaphorin III/D-mediated chemorepulsive signals play a crucial role in peripheral nerve projection in mice. Neuron 1997; 19:995-1005.
15. Taniguchi M, Yuasa S, Fujisawa H et al. Disruption of semaphorin III/D gene causes severe abnormality in peripheral nerve projection. Neuron 1997; 19:519-530.
16. Bagnard D, Chounlamountri N, Püschel AW et al. Axonal surface molecules act in combination with semaphorin 3a during the establishment of corticothalamic projections. Cereb. Cortex 2001; 11:278-285.
17. Sahay A, Molliver ME, Ginty DD et al. Semaphorin 3F is critical for development of limbic system circuitry and is required in neurons for selective CNS axon guidance events. J Neurosci 2003; 23:6671-6680.
18. Polleux F, Giger RJ, Ginty DD et al. Patterning of cortical efferent projections by semaphorin-neuropilin interactions. Science 1998; 282:1904-1906.
19. Polleux F, Morrow T, Ghosh A. Semaphorin 3A is a chemoattractant for cortical apical dendrites. Nature 2000; 404:567-573.
20. Chédotal A, Del Rio JA, Ruiz M et al. Semaphorins III and IV repel hippocampal axons via two distinct receptors. Development 1998; 125:4313-4323.
21. Steup A, Ninnemann O, Savaskan NE et al. Semaphorin D acts as a repulsive factor for entorhinal and hippocampal neurons. Eur J Neurosci 1999; 11:729-734.
22. Steup A, Lohrum M, Hamscho N et al. Sema3C and netrin-1 differentially affect axon growth in the hippocampal formation. Mol Cell Neurosci 2000; 15:141-155.
23. Cohen S, Funkelstein L, Livet J et al. A semaphorin code defines subpopulations of spinal motor neurons during mouse development. Eur J Neurosci 2005; 21:1767-1776.
24. Ricard D, Stankoff B, Bagnard D et al. Differential expression of collapsin response mediator proteins (CRMP/ULIP) in subsets of oligodendrocytes in the postnatal rodent brain. Mol Cell Neurosci 2000; 16:324-337.
25. Ricard D, Rogemond V, Charrier E et al. Isolation and expression pattern of human Unc-33-like phosphoprotein 6/collapsin response mediator protein 5 (Ulip6/CRMP5): Coexistence with Ulip2/CRMP2 in Sema3a- sensitive oligodendrocytes. J Neurosci 2001; 21:7203-7214.
26. Cohen RI, Rottkamp DM, Maric D et al. A role for semaphorins and neuropilins in oligodendrocyte guidance. J Neurochem 2003; 85:1262-1278.
27. Spassky N, de Castro F, Le Bras B et al. Directional guidance of oligodendroglial migration by class 3 semaphorins and netrin-1. J Neurosci 2002; 22:5992-6004.
28. Marín O, Yaron A, Bagri A et al. Sorting of striatal and cortical interneurons regulated by semaphorin-neuropilin interactions. Science 2001; 293:872-875.
29. Giger RJ, Pasterkamp RJ, Heijnen S et al. Anatomical distribution of the chemorepellent semaphorin III/collapsin-1 in the adult rat and human brain: Predominant expression in structures of the olfactory-hippocampal pathway and the motor system. J Neurosci Res 1998; 52:27-42.
30. Hirsch E, Hu LJ, Prigent A et al. Distribution of semaphorin IV in adult human brain. Brain Res 1999; 823:67-79.
31. Sahay A, Kim CH, Sepkuty JP et al. Secreted semaphorins modulate synaptic transmission in the adult hippocampus. J Neurosci 2005; 25:3613-3620.
32. Bouzioukh F, Daoudal G, Falk J et al. Semaphorin3A regulates synaptic function of differentiated hippocampal neurons. Eur J Neurosci 2006; 23:2247-2254.
33. Altman J, Das GD. Autoradiographic and histological evidence of postnatal hippocampal neurogenesis in rats. J Comp Neurol 1965; 124:319-335.
34. Luskin MB. Restricted proliferation and migration of postnatally generated neurons derived from the forebrain subventricular zone. Neuron 1993; 11:173-189.
35. Taupin P, Gage FH. Adult neurogenesis and neural stem cells of the central nervous system in mammals. J Neurosci Res 2002; 69:745-749.
36. Shirvan A, Ziv I, Fleminger G et al. Semaphorins as mediators of neuronal apoptosis. J Neurochem 1999; 73:961-971.
37. Bagnard D, Vaillant C, Khuth ST et al. Semaphorin 3A-vascular endothelial growth factor-165 balance mediates migration and apoptosis of neural progenitor cells by the recruitment of shared receptor. J Neurosci 2001; 21:3332-3341.
38. Eastwood SL, Law AJ, Everall IP et al. The axonal chemorepellant semaphorin 3A is increased in the cerebellum in schizophrenia and may contribute to its synaptic pathology. Mol Psychiatry 2003; 8:148-155.

39. Good PF, Alapat D, Hsu A et al. A role for semaphorin 3A signaling in the degeneration of hippocampal neurons during Alzheimer's disease. J Neurochem 2004; 91:716-736.
40. Holtmaat AJ, Gorter JA, De Wit J et al. Transient downregulation of Sema3A mRNA in a rat model for temporal lobe epilepsy. A novel molecular event potentially contributing to mossy fiber sprouting. Exp Neurol 2003; 182:142-150.
41. Pasterkamp RJ, Verhaagen J. Emerging roles for semaphorins in neural regeneration. Brain Res Brain Res Rev 2001; 35:36-54.
42. Niclou SP, Ehlert EM, Verhaagen J. Chemorepellent axon guidance molecules in spinal cord injury. J Neurotrauma 2006; 23:409-421.
43. He Z, Tessier-Lavigne M. Neuropilin is a receptor for the axonal chemorepellent Semaphorin III. Cell 1997; 90:739-751.
44. Kolodkin AL, Levengood DV, Rowe EG et al. Neuropilin is a semaphorin III receptor. Cell 1997; 90:753-762.
45. Takagi S, Tsuji T, Amagai T et al. Specific cell surface labels in the visual centers of Xenopus laevis tadpole identified using monoclonal antibodies. Dev Biol 1987; 122:90-100.
46. Fujisawa H. From the discovery of neuropilin to the determination of its adhesion sites. In: Bagnard D, ed. Neuropilin from Nervous system to vascular and tumor biology. Adv Exp Med Biol 2002; 515:1-12.
47. Chen H, Chédotal A, He Z et al. Neuropilin-2, a novel member of the neuropilin family, is a high affinity receptor for the semaphorins Sema E and Sema IV but not Sema III. Neuron 1997; 19:547-559.
48. Cai H, Reed RR. Cloning and characterization of neuropilin-1-interacting protein: A PSD-95/Dlg/ZO-1 domain-containing protein that interacts with the cytoplasmic domain of neuropilin-1. J Neurosci 1999; 19:6519-6527.
49. Nakamura F, Tanaka M, Takahashi T et al. Neuropilin-1 extracellular domains mediate semaphorin D/III-induced growth cone collapse. Neuron 1998; 21:1093-1100.
50. Chen H, He Z, Bagri A et al. Semaphorin-neuropilin interactions underlying sympathetic axon responses to class III semaphorins. Neuron 1998; 21:1283-1290.
51. Klostermann A, Lohrum M, Adams RH et al. The chemorepulsive activity of the axonal guidance signal semaphorin D requires dimerization. J Biol Chem 1998; 273:7326-7331.
52. Koppel AM, Raper JA. Collapsin-1 covalently dimerizes, and dimerization is necessary for collapsing activity. J Biol Chem 1998; 273:15708-15713.
53. Feiner L, Koppel AM, Kobayashi H et al. Secreted chick semaphorins bind recombinant neuropilin with similar affinities but bind different subsets of neurons in situ. Neuron 1997; 19:539-545.
54. Giger RJ, Urquhart ER, Gillespie SK et al. Neuropilin-2 is a receptor for semaphorin IV: Insight into the structural basis of receptor function and specificity. Neuron 1998; 21:1079-1092.
55. Gu C, Limberg BJ, Whitaker GB et al. Characterization of neuropilin-1 structural features that confer binding to semaphorin 3A and vascular endothelial growth factor 165. J Biol Chem 2002; 277:18069-18076.
56. Giger RJ, Cloutier JF, Sahay A et al. Neuropilin-2 is required in vivo for selective axon guidance responses to secreted semaphorins. Neuron 2000; 25:29-41.
57. Walz A, Rodriguez I, Mombaerts P. Aberrant sensory innervation of the olfactory bulb in neuropilin-2 mutant mice. J Neurosci 2002; 22:4025-4035.
58. Tamagnone L, Artigiani S, Chen H et al. Plexins are a large family of receptors for transmembrane, secreted, and GPI-anchored semaphorins in vertebrates. Cell 1999; 99:71-80.
59. Tamagnone L, Comoglio PM. Control of invasive growth by hepatocyte growth factor (HGF) and related scatter factors. Cytokine Growth Factor Rev 1997; 8:129-142.
60. Turner LJ, Nicholls S, Hall A. The activity of the plexin-A1 receptor is regulated by Rac. J Biol Chem 2004; 279:33199-33205.
61. Artigiani S, Barberis D, Fazzari P et al. Functional regulation of semaphorin receptors by proprotein convertases. J Biol Chem 2003; 278:10094-10101.
62. Usui H, Taniguchi M, Yokomizo T et al. Plexin-A1 and plexin-B1 specifically interact at their cytoplasmic domains. Biochem Biophys Res Commun 2003; 300:927-931.
63. Takahashi T, Strittmatter SM. Plexina1 autoinhibition by the plexin sema domain. Neuron 2001; 29:429-439.
64. Gu C, Yoshida Y, Livet J et al. Semaphorin 3E and plexin-D1 control vascular pattern independently of neuropilins. Science 2005; 307:265-268.
65. Potiron V, Roche J. Class 3 semaphorin signaling: The end of a dogma. Sci STKE 2005; 285:pe24.
66. Takahashi T, Fournier A, Nakamura F et al. Plexin-neuropilin-1 complexes form functional semaphorin-3A receptors. Cell 1999; 99:59-69.

67. Rohm B, Ottemeyer A, Lohrum M et al. Plexin/neuropilin complexes mediate repulsion by the axonal guidance signal semaphorin 3A. Mech Dev 2000; 93:95-104.
68. Cheng HJ, Bagri A, Yaron A et al. Plexin-A3 mediates semaphorin signaling and regulates the development of hippocampal axonal projections. Neuron 2001; 32:249-263.
69. Suto F, Murakami Y, Nakamura F et al. Identification and characterization of a novel mouse plexin, plexin-A4. Mech Dev 2003; 120:385-396.
70. Gherardi E, Love CA, Esnouf RM et al. The sema domain. Curr Opin Struct Biol 2004; 14:669-678.
71. Cleaver O, Melton DA. Endothelial signaling during development. Nat Med 2003; 9:661-668.
72. Soker S, Takashima S, Miao HQ et al. Neuropilin-1 is expressed by endothelial and tumor cells as an isoform-specific receptor for vascular endothelial growth factor. Cell 1998; 92:735-745.
73. Fuh G, Garcia KC, de Vos AM. The interaction of neuropilin-1 with vascular endothelial growth factor and its receptor flt-1. J Biol Chem 2000; 275:26690-26695.
74. Gluzman-Poltorak Z, Cohen T, Herzog Y et al. Neuropilin-2 is a receptor for the vascular endothelial growth factor (VEGF) forms VEGF-145 and VEGF-165 [corrected]. J Biol Chem 2000; 275:18040-18045.
75. Bagnard D, Sainturet N, Meyronet D et al. Differential MAP kinases activation during semaphorin3A-induced repulsion or apoptosis of neural progenitor cells. Mol Cell Neurosci 2004; 25:722-731.
76. Toyofuku T, Zhang H, Kumanogoh A et al. Dual roles of Sema6D in cardiac morphogenesis through region-specific association of its receptor, Plexin-A1, with off-track and vascular endothelial growth factor receptor type 2. Genes Dev 2004; 18:435-447.
77. Castellani V, Chédotal A, Schachner M et al. Analysis of the L1-deficient mouse phenotype reveals cross-talk between Sema3A and L1 signaling pathways in axonal guidance. Neuron 2000; 27:237-249.
78. Castellani V, Falk J, Rougon G. Semaphorin3A-induced receptor endocytosis during axon guidance responses is mediated by L1 CAM. Mol Cell Neurosci 2004; 26:89-100.
79. Falk J, Bechara A, Fiore R et al. Dual functional activity of semaphorin 3B is required for positioning the anterior commissure. Neuron 2005; 48:63-75.
80. West DC, Rees CG, Duchesne L et al. Interactions of multiple heparin binding growth factors with neuropilin-1 and potentiation of the activity of fibroblast growth factor-2. J Biol Chem 2005; 280:13457-13464.
81. Hynes RO. Integrins: A family of cell surface receptors. Cell 1987; 48:549-554.
82. Hynes RO. Integrins: Bidirectional, allosteric signaling machines. Cell 2002; 110:673-687.
83. Pasterkamp RJ, Peschon JJ, Spriggs MK et al. Semaphorin 7A promotes axon outgrowth through integrins and MAPKs. Nature 2003; 424:398-405.
84. Serini G, Valdembri D, Zanivan S et al. Class 3 semaphorins control vascular morphogenesis by inhibiting integrin function. Nature 2003; 424:391-397.
85. Walzer T, Galibert L, Comeau MR et al. Plexin C1 engagement on mouse dendritic cells by viral semaphorin A39R induces actin cytoskeleton rearrangement and inhibits integrin-mediated adhesion and chemokine-induced migration. J Immunol 2005; 174:51-59.
86. Delaire S, Elhabazi A, Bensussan A et al. CD100 is a leukocyte semaphorin. Cell Mol Life Sci 1998; 54:1265-1276.
87. Kumanogoh A, Watanabe C, Lee I et al. Identification of CD72 as a lymphocyte receptor for the class IV semaphorin CD100: A novel mechanism for regulating B cell signaling. Immunity 2000; 13:621-631.
88. Ishida I, Kumanogoh A, Suzuki K et al. Involvement of CD100, a lymphocyte semaphorin, in the activation of the human immune system via CD72: Implications for the regulation of immune and inflammatory responses. Int Immunol 2003; 15:1027-1034.
89. Kumanogoh A, Marukawa S, Suzuki K et al. Class IV semaphorin Sema4A enhances T-cell activation and interacts with Tim-2. Nature 2002; 419:629-633.
90. Takegahara N, Kumanogoh A, Kikutani H. Semaphorins: A new class of immunoregulatory molecules. Philos. Trans R Soc Lond B Biol Sci 2005; 360:1673-1680.
91. Giordano S, Corso S, Conrotto P et al. The semaphorin 4D receptor controls invasive growth by coupling with Met. Nat Cell Biol 2002; 4:720-724.
92. Conrotto P, Corso S, Gamberini S et al. Interplay between scatter factor receptors and B plexins controls invasive growth. Oncogene 2004; 23:5131-5137.
93. Artigiani S, Conrotto P, Fazzari P et al. Plexin-B3 is a functional receptor for semaphorin 5A. EMBO Rep 2004; 5:710-714.
94. Pulido D, Campuzano S, Koda T et al. Dtrk, a Drosophila gene related to the trk family of neurotrophin receptors, encodes a novel class of neural cell adhesion molecule. EMBO J 1992; 11:391-404.

95. Winberg ML, Tamagnone L, Bai J et al. The transmembrane protein Off-track associates with Plexins and functions downstream of Semaphorin signaling during axon guidance. Neuron 2001; 32:53-62.
96. Jin Z, Strittmatter SM. Rac1 mediates collapsin-1-induced growth cone collapse. J Neurosci 1997; 17:6256-6263.
97. Kuhn TB, Brown MD, Wilcox CL et al. Myelin and collapsin-1 induce motor neuron growth cone collapse through different pathways: Inhibition of collapse by opposing mutants of rac1. J Neurosci 1999; 19:1965-1975.
98. Driessens MH, Hu H, Nobes CD et al. Plexin-B semaphorin receptors interact directly with active Rac and regulate the actin cytoskeleton by activating Rho. Curr Biol 2001; 11:339-344.
99. Rohm B, Rahim B, Kleiber B et al. The semaphorin 3A receptor may directly regulate the activity of small GTPases. FEBS Lett 2000; 486:68-72.
100. Vikis HG, Li W, He Z et al. The semaphorin receptor plexin-B1 specifically interacts with active Rac in a ligand-dependent manner. Proc Natl Acad Sci USA 2000; 97:12457-12462.
101. Zanata SM, Hovatta I, Rohm B et al. Antagonistic effects of Rnd1 and RhoD GTPases regulate receptor activity in Semaphorin 3A-induced cytoskeletal collapse. J Neurosci 2002; 22:471-477.
102. Vikis HG, Li W, Guan KL. The plexin-B1/Rac interaction inhibits PAK activation and enhances Sema4D ligand binding. Genes Dev 2002; 16:836-845.
103. Negishi M, Oinuma I, Katoh H. R-ras as a key player for signaling pathway of plexins. Mol Neurobiol 2005; 32:217-222.
104. Aurandt J, Vikis HG, Gutkind JS et al. The semaphorin receptor plexin-B1 signals through a direct interaction with the Rho-specific nucleotide exchange factor, LARG. Proc Natl Acad Sci USA 2002; 99:12085-12090.
105. Chikumi H, Fukuhara S, Gutkind JS. Regulation of G protein-linked guanine nucleotide exchange factors for Rho, PDZ-RhoGEF, and LARG by tyrosine phosphorylation: Evidence of a role for focal adhesion kinase. J Biol Chem 2002; 277:12463-12473.
106. Hirotani M, Ohoka Y, Yamamoto T et al. Interaction of plexin-B1 with PDZ domain-containing Rho guanine nucleotide exchange factors. Biochem Biophys Res Commun 2002; 297:32-37.
107. Perrot V, Vazquez-Prado J, Gutkind JS. Plexin B regulates Rho through the guanine nucleotide exchange factors leukemia-associated Rho GEF (LARG) and PDZ-RhoGEF. J Biol Chem 2002; 277:43115-43120.
108. Swiercz JM, Kuner R, Behrens J et al. Plexin-B1 directly interacts with PDZ-RhoGEF/LARG to regulate RhoA and growth cone morphology. Neuron 2002; 35:51-63.
109. Hall A. Rho GTPases and the actin cytoskeleton. Science 1998; 279:509-514.
110. Wear MA, Schafer DA, Cooper JA. Actin dynamics: Assembly and disassembly of actin networks. Curr Biol 2000; 10:R891-895.
111. Dickson BJ. Rho GTPases in growth cone guidance. Curr Opin Neurobiol 2001; 11:103-110.
112. Aizawa H, Wakatsuki S, Ishii A et al. Phosphorylation of cofilin by LIM-kinase is necessary for semaphorin 3A-induced growth cone collapse. Nat Neurosci 2001; 4:367-373.
113. Mitsui N, Inatome R, Takahashi S et al. Involvement of Fes/Fps tyrosine kinase in semaphorin3A signaling. EMBO J 2002; 21:3274-3285.
114. Goshima Y, Nakamura F, Strittmatter P et al. Collapsin-induced growth cone collapse mediated by an intracellular protein related to UNC-33. Nature 1995; 376:509-514.
115. Inatome R, Tsujimura T, Hitomi T et al. Identification of CRAM, a novel unc-33 gene family protein that associates with CRMP3 and protein-tyrosine kinase(s) in the developing rat brain. J Biol Chem 2000; 275:27291-27302.
116. Sasaki Y, Cheng C, Uchida Y et al. Fyn and Cdk5 mediate semaphorin-3A signaling, which is involved in regulation of dendrite orientation in cerebral cortex. Neuron 2002; 35:907-920.
117. Uchida Y, Oshima T, Sasaki et al. Semaphorin3A signalling is mediated via sequential Cdk5 and GSK3beta phosphorylation of CRMP2: implication of common phosphorylating mechanism underlying axon guidance and Alzheimer's disease. Genes Cells 2005; 10:165-79.
118. Brown M, Jacobs T, Eickholt B et al. Alpha2-chimaerin, cyclin-dependent Kinase 5/p35, and its target collapsin response mediator protein-2 are essential components in semaphorin 3A-induced growth-cone collapse. J Neurosci 2004; 24:8994-9004.
119. Eickholt BJ, Walsh FS, Doherty P. An inactive pool of GSK-3 at the leading edge of growth cones is implicated in Semaphorin 3A signaling. J Cell Biol 2002; 157:211-217.
120. Fukata Y, Itoh TJ, Kimura T et al. CRMP-2 binds to tubulin heterodimers to promote microtubule assembly. Nat Cell Biol 2002; 4:583-591.
121. Terman JR, Mao T, Pasterkamp RJ et al. MICALs, a family of conserved flavoprotein oxidoreductases, function in plexin-mediated axonal repulsion. Cell 2002; 109:887-900.

122. Mikule K, Gatlin JC, de la Houssaye BA et al. Growth cone collapse induced by semaphorin 3A requires 12/15-lipoxygenase. J Neurosci 2002; 22:4932-4941.
123. Forcet C, Stein E, Pays L et al. Netrin-1-mediated axon outgrowth requires deleted in colorectal cancer-dependent MAPK activation. Nature 2002; 417:443-447.
124. Miao H, Wei BR, Peehl DM et al. Activation of EphA receptor tyrosine kinase inhibits the Ras/MAPK pathway. Nat Cell Biol 2001; 3:527-530.
125. Kim I, Ryu YS, Kwak HJ et al. EphB ligand, ephrinB2, suppresses the VEGF- and angiopoietin 1-induced Ras/mitogen-activated protein kinase pathway in venous endothelial cells. FASEB J 2002; 16:1126-1128.
126. Elowe S, Holland SJ, Kulkarni S et al. Downregulation of the Ras-mitogen-activated protein kinase pathway by the EphB2 receptor tyrosine kinase is required for ephrin-induced neurite retraction. Mol Cell Biol 2001; 21:7429-7441.
127. Campbell DS, Holt CE. Apoptotic pathway and MAPKs differentially regulate chemotropic responses of retinal growth cones. Neuron 2003; 37:939-952.
128. Fujioka S, Masuda K, Toguchi M et al. Neurotrophic effect of Semaphorin 4D in PC12 cells. Biochem Biophys Res Commun 2003; 301:304-310.
129. Doherty P, Williams G, Williams EJ. CAMs and axonal growth: A critical evaluation of the role of calcium and the MAPK cascade. Mol Cell Neurosci 2000; 16:283-295.
130. Zachary I, Gliki G. Signaling transduction mechanisms mediating biological actions of the vascular endothelial growth factor family. Cardiovasc Res 2001; 49:568-581.
131. Vojtek AB, Cooper JA. Rho family members: Activators of MAP kinase cascades. Cell 1995; 82:527-529.
132. Atwal JK, Singh KK, Tessier-Lavigne M et al. Semaphorin 3F antagonizes neurotrophin-induced phosphatidylinositol 3-kinase and mitogen-activated protein kinase kinase signaling: A mechanism for growth cone collapse. J Neurosci 2003; 23:7602-7609.
133. Parikh AA, Liu WB, Fan F et al. Expression and regulation of the novel vascular endothelial growth factor receptor neuropilin-1 by epidermal growth factor in human pancreatic carcinoma. Cancer 2003; 98:720-729.
134. Song H, Ming G, He Z et al. Conversion of neuronal growth cone responses from repulsion to attraction by cyclic nucleotides. Science 1998; 281:1515-1518.
135. Campbell DS, Regan AG, Lopez JS et al. Semaphorin 3A elicits stage-dependent collapse, turning, and branching in Xenopus retinal growth cones. J Neurosci 2001; 21:8538-8547.
136. Goodhill GJ. Mathematical guidance for axons. Trends Neurosci 1998; 21:226-231.
137. Bagnard D, Thomasset N, Lohrum M et al. Spatial distributions of guidance molecules regulate chemorepulsion and chemoattraction of growth cones. J Neurosci 2000; 20:1030-1035.

Chapter 5

Slits and Their Receptors

Alain Chédotal*

Abstract

Slit was identified in *Drosophila* embryo as a gene involved in the patterning of larval cuticle.[1] It was later shown that Slit is synthesized in the fly central nervous system by midline glia cells.[2-5] Slit homologues have since been found in *C. elegans*[6] and many vertebrate species, from amphibians,[7] fishes,[8] birds[9-11] to mammals.[7,12-14] A single *slit* was isolated in invertebrates, whereas there are three *slit* genes (*slit1-slit3*) in mammals, that have around 60% homology.[12] All encodes large ECM glycoproteins of about 200 kDa[15,16] (Fig. 1A), comprising, from their N terminus to their C terminus, a long stretch of four leucine rich repeats (LRR) connected by disulphide bonds, seven to nine EGF repeats, a domain, named ALPS (Agrin, Perlecan, Laminin, Slit) or laminin G-like module (see ref. 17), and a cystein knot (Fig. 1A). Alternative spliced transcripts have been reported for *Drosophila* Slit[2], human Slit2 and Slit3,[14] and Slit1.[18,19] Moreover, two Slit1 isoforms exist in zebrafish as a consequence of gene duplication.[20] Last, in mammals, two Slit2 isoforms can be purified from brain extracts, a long 200 kDa one[15,16] and a shorter 150 kDa form (Slit2-N) that was shown to result from the proteolytic processing of full-length Slit2.[21] Human Slit1 and Slit3 and Drosophila Slit are also cleaved by an unknown protease in a large N-terminal fragment and a shorter C-terminal fragment, suggesting conserved mechanisms for Slit cleavage across species.[12,21-23] Moreover, Slit fragments have different cell association characteristics in cell culture suggesting that they may also have different extents of diffusion, different binding properties, and, hence, different functional activities in vivo. This conclusion is supported by in vitro data showing that full-length Slit2 functions as an antagonist of Slit2-N in the DRG branching assay, and that Slit2-N, not full-length Slit2, causes collapse of OB growth cones.[24] In addition, Slit1-N and full-length Slit1 can induce branching of cortical neurons (see below), but only full-length Slit1 repels cortical axons.[23]

Structure-function analysis in vertebrates and *Drosophila* demonstrated that the LRRs of Slits are required and sufficient to mediate their repulsive activities in neurons.[24-26] More recent detailed structure function analysis of the LRR domains of *Drosophila* Slit,[27] revealed that the active site of Slit (at least regarding its pro-angiogenic activity) is located on the second of the fourth LRR (LRR2), which is highly conserved between Slits. Slit can also dimerize through the LRR4 domain and the cystein knot.[18] However, a Slit1 spliced-variant that lacks the cysteine knot and does not dimerize is still able to repel OB axons.[18]

Introduction

The first *roundabout* gene, *robo*, was identified in *Drosophila* during a comprehensive screen for genes regulating midline crossing in the CNS.[28] If SAX-3 is the unique *robo* ortholog in

*Alain Chédotal—CNRS UMR 7102, Université Paris 6, Bâtiment B, pièce 611, Case 12, 9 Quai Saint Bernard, 75005 Paris, France. Email: chedotal@infobiogen.fr

Axon Growth and Guidance, edited by Dominique Bagnard.

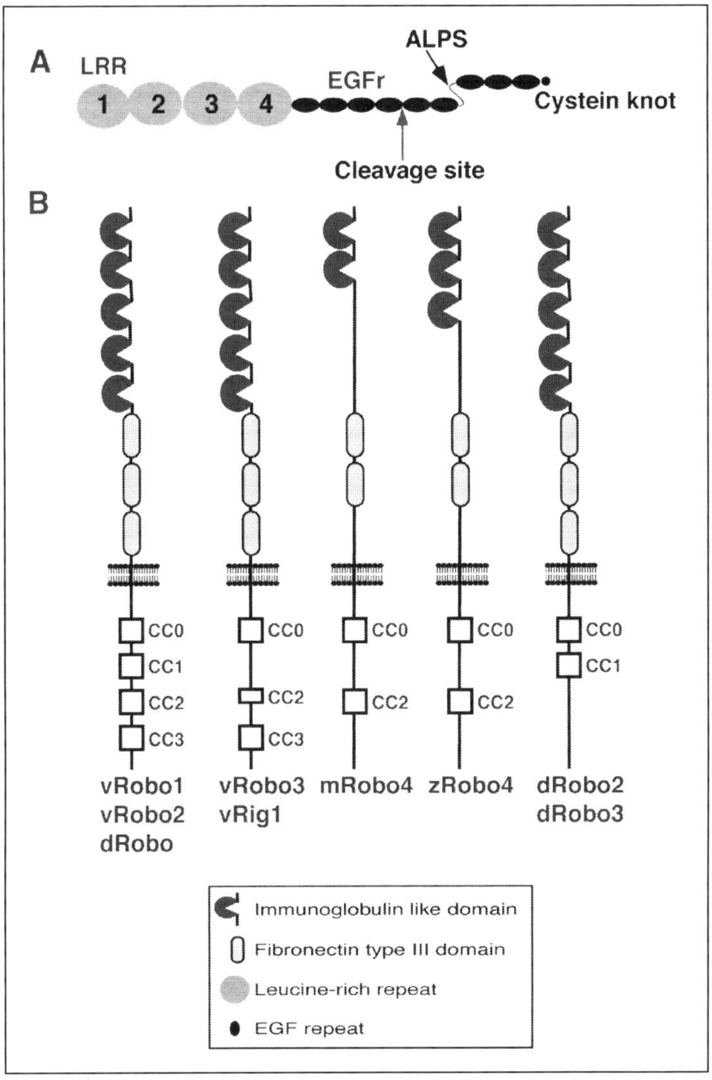

Figure 1. A) Structure of the slit proteins. B) Diagrammatic comparison of the structure of Robo receptors in invertebrate and vertebrate species. v: vertebrate; d: Drosophila; z: zebrafish. ALPS: domain found in Agrin, Laminin, Perlecan and Slit.

C.elegans,[6] three *robo* genes have been found in Drosophila,[5,29,30] zebrafish,[31,32] chick[9] (Chedotal unpublished data) and mammals.[7,12,33] Robo proteins belong to the immunoglobulin (Ig) superfamily and have five Ig-like domains followed by three fibronectin type III (FNIII) repeats, a transmembrane portion and a long intracellular tail containing up to four conserved cytoplasmic motifs, CC0-CC3, with no obvious catalytic domains (Fig. 1B). The first two Ig domains are the most highly conserved portion and are also found in another protein called Robo4 or magic roundabout that is only expressed by endothelial cells and plays a role in angiogenesis.[34,35] However, Robo4 lacks the last three Ig domains, some FNIII domains and the CC1 and CC3 motifs found in other Robo proteins. Moreover, its capacity to bind Slits is still debated.[36,37]

CC0 has no known function but is a site of tyrosine phosphorylation.[38] CC1 also contains tyrosine residues that can be phosphorylated and was shown to bind to the P3 domain of the netrin-1 receptor, deleted in colorectal cancer (DCC; see below).

CC2 is a proline-rich sequence that matches the consensus binding site for *Drosophila* Enabled (Ena; see below), CC3 is also a polyproline stretch.[29] Drosophila Robo2 and Robo3 lack the CC2 and CC3 domains[5,30] and the second half of CC2 is also not conserved in mouse and zebrafish Robo3. Furthermore, mouse Robo3/Rig-1 (Rig-1 for retinoblastoma inhibiting gene 1[39]) lacks the CC1 motif[33,39] but zebrafish Robo3 has it.[32] In addition, some spliced variants of mouse Robo3/Rig-1, including a secreted form, may exist.[39] Last, Robo 1 can be cleaved in transfected cells.[26]

In *Drosophila*, genetic and biochemical evidence demonstrated that Slit is a ligand of the Robo1-Robo3 receptors.[4,5,29,30,40] Likewise, mammalian Slits can bind to all Robo receptors with comparable affinity.[7,12,41] Slit cleavage fragments appear to have different cell association characteristics, with the smaller C-terminal fragment being more diffusible and the larger N-terminal and full length fragments being more tightly cell-associated.[21] In addition, the C-fragment does not bind to Robo.[24] More recent studies have shown that in *Drosophila* all three Robo receptors compete for a single active binding site in the second LRR of Slit[27] and that neither the FNIII domains nor Robo dimerization are required for Slit binding. The major Robo1-3 binding site of Slit is in the second of the four LRRs, is evolutionary conserved and has a similar affinity for all Robos. However, Slit affinity is higher when all LRRs are present, probably due to its dimerization. On the receptor side, several results suggest that the first two Ig domains of Robos are required for Slit binding. First, the genetic deletion of Ig1 and Ig2 results in abnormal lung development.[42] Second, antibodies against Robo Ig1 inhibit tumor growth in mice[43] and neurite outgrowth in vitro.[44] Third, Robo1 Ig1-2 are important for Slit binding and function in vitro.[45]

Several studies suggest that Slit can bind to other proteins than Robo, in particular heparan sulfate glycosaminoglycans that are negatively charged carbohydrates found on the cell surface. Slit1 and Slit2 were shown to bind to heparin column[21,46] and to Glypican-1,[46] a glycosyl phosphatidyl inositol (GPI)-anchored heparan sulfate proteoglycan known to interact with positively charged molecules. Biochemical data suggest that Slit binds to glypican-1 through its C-terminus.[47] Moreover, heparinase III treatment reduces Slit2 activity and binding to Robo1.[48] In *Drosophila*, expression of the transmembrane heparan sulfate proteoglycan syndecan in target cells appears to be required for Slit signaling.[49] There is also genetic evidence in mouse supporting interaction between Slit and heparan sulfates in vivo.[50] Heparan sulfates could help stabilizing the Robo/Slit complex or function as coreceptors presenting Slits to Robos or to alternative receptors.

Robo Partners

The analysis of Frazzled-Robo chimeric proteins in Drosophila, first revealed that the cytoplasmic domain of Robo is required to control the lateral positioning of post-crossing axons.[51] Genetic and biochemical studies have then led to the identification of a number of transmembrane and cytoplasmic proteins that may participate or modulate Slit signaling through Robo receptors (Fig. 2). However, only a few of these proteins have been shown to directly participate to Robo signaling upon binding Robo CC domains.

Abelson Tyrosine Kinase

In *Drosophila*, Robo was found to be a substrate for the cytoplasmic tyrosine kinase Abelson (Abl). Abl is able to phosphorylate Robo's CC0 and CC1 leading to Robo inactivation.[38] In the *Drosophila* visual system, Abl was also found to interact with Robo2 and Robo3. An Abl substrate, the actin binding protein Enabled (Ena), is also involved in Robo repulsion in *Drosophila*[38] and *C. elegans*.[52] Ena was shown to bind Robo's CC2 motif, therefore participating to Robo signaling. However, other studies showed that Abl rather than inactivating Robo, could

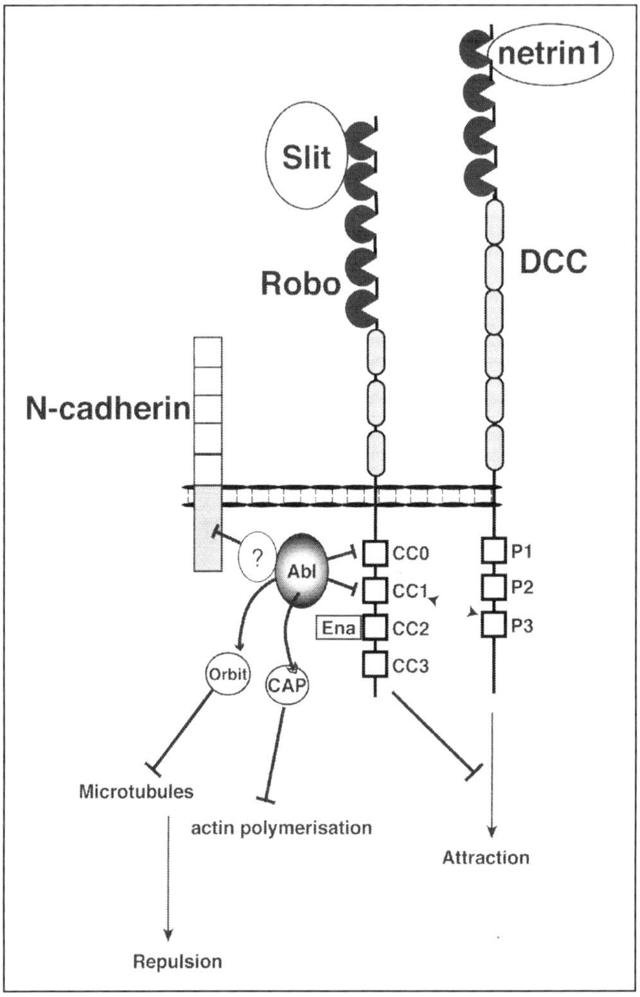

Figure 2. A central role for Abelson tyrosine kinase in Robo repulsion. Abl can inactivate Robo signaling through phosphorylation of CC0 and CC1. Other data suggest that Abl is recruited following Robo activation. Abl interacts with multiple effectors such as Enabled (Ena), capulet (CAP) and Orbit/MAST/CLASP that control cytoskeletal dynamics. Slit binding to Robo was also shown to inactivate the cell adhesion molecule N-cadherin. Abl also mediates interaction between Robo and N-cadherin, most likely through an unknown partner. Robo can also bind to the netrin-1 receptor DCC upon slit binding. This leads to the inhibition of netrin-1 attractive activity.

promote repulsion downstream of Robo.[53] The adenylyl cyclase associated proteins (CAP) regulate actin polymerization and bind to the SH3 domain of Abl. Interestingly, axon guidance defects at the midline were observed in the Drosophila CAP homolog *capulet* (*capt*), or in *capt-slit* or *capt-robo-robo2* transheterozygotes. In this system, Abl and Capt are recruited by Robo activation and inhibit actin polymerization, therefore acting positively in the Slit pathway. Thus, Abl may play both positive and negative roles in Slit signaling.[54] Slit binding to Robo was also shown to inactivate the cell adhesion molecule N-cadherin that mediates homophilic binding. Interestingly, Abl is required for Robo binding to N-cadherin. Robo is thought to inhibit N-cadherin function by interfering with its cytoplasmic domain and by inducing a decrease in N-cadherin-mediated adhesion. Slit binding to Robo increases the phosphorylation of β-catenin and thus its ability to bind to N-cadherin. This induces the binding of Robo to N-cadherin resulting in its inhibition.[55]

Abl function downstream of Robo may involve microtubule associated proteins (MAP) in addition to actin binding proteins.[56] The MAP *Orbit/MAST*, ortholog of the vertebrate cytoplasmic linker protein (CLIP)-associated proteins (CLASP) are microtubule-associated plus end tracking proteins that seem to reduce microtubule stability. Genetic evidence supports a role for Orbit/MAST downstream of Abl in the Slit repellent pathway.

Although all these studies suggest that Abl plays a pivotal role in mediating Robo signaling in Drosophila, it remains to determine if Abl function is conserved in vertebrates.

Rho Family of Small GTPases

Rho family of small GTP-binding proteins (Rho GTPases) are major modulators of the actin cytoskeleton and play a central role in axonal growth and cell migration.[57] Rho GTPases are activated upon GTP binding and inactive when bound to GDP. The switch from their active to their inactive state is controlled by two families of proteins: the guanine nucleotide exchange factors (GEFs) and the GTPase activating proteins (GAPs). GEFs activate Rho GTPAses while GAPs inactivate them by inducing GTP hydrolysis.[57]

Many studies have shown that Rho GTPases play an important role in the modulation of Slit function (Fig. 3). First, activation of RhoA in Drosophila causes axons to cross the midline.[58] Second, In this system, Rac1 inactivation or Cdc42 activation can overcome the effect of constitutively active Robo suggesting that Cdc42 and Rac1 function downstream of Robo.[59] Likewise, in vertebrate neurons, a constitutively active Cdc42 blocks the repulsive effect of Slit.[60] Biochemical studies have shown that the SH3-SH2 adaptator protein Dock, can bind directly to Robo and that this interaction requires the SH3 domain of Dock and the CC2 and CC3 motifs of Robo.[61,62] Dock is known to interact with key regulators of the actin cytoskeleton such as p21-activated protein kinase (Pak), a serine-threonine kinase which in turn can interact with RhoGTPases such as Rac1 and Cdc42. Slit binding to Robo increases the level of Dock-Robo association, recruits Pak and stimulates Rac (in particular Rac1). Thus Slit regulates the assembly of a multiproteic complex composed of Dock, Pak and Rac that couples Robo receptor activation to the regulation of the actin cytoskeleton.

Robo also controls the activity of Rho GTPases through a family of Slit/Robo specific GAPs, SrGAP1-3, that were identified using yeast two hybrid and Robo1 CC3 domain as a bait. SrGAPs consist of a RhoGAP domain, a SH3 domain and a Fes/CIP4 (FCH) homology domain. SrGAP1 and srGAP3 bind to the CC3 domain of Robo1 through their SH3 domain. Slit2 increases Robo1 binding to srGAP1 and its activity and this regulation requires CC3. In turn, SrGAP1 inactivates Cdc42 and activates RhoA but not Rac1.[60] Contrary to srGAP1, srGAP3 is mainly a repressor of Rac1.[63] Recently, it was also shown that in *Drosophila*, Slit/Robo signaling could also control Rac activity upon binding the GAP Vilse/CrossGAP, that is conserved in vertebrates.[64,65] Genetic evidence showed that CrossGAP may be involved in Robo-dependent axonal repulsion and tracheal cell migration and that it specifically inactivates Rac. Moreover, CrossGAP WW domains can directly bind to the CC2 domain of Robo and thus may not function downstream of other Robo receptors that lack the CC2 domain.[64] Moreover, the two proteins can be coimmunoprecipitated from brain extracts.[65] Although it inactivates Rac, Vilse seems to have a positive role in Robo repulsion.[64]

The Netrin Receptor DCC

Deleted in colorectal cancer (DCC) is a transmembrane receptor for the secreted protein netrin-1 (see Chapter by Moore et al). Robo1 can bind DCC and this results in the inhibition of netrin-1 attraction.[66] This silencing activity requires the binding of the CC1 domain of Robo 1 and to the P3 domain of DCC. The *C. elegans* homolog of DCC, UNC-40, can also

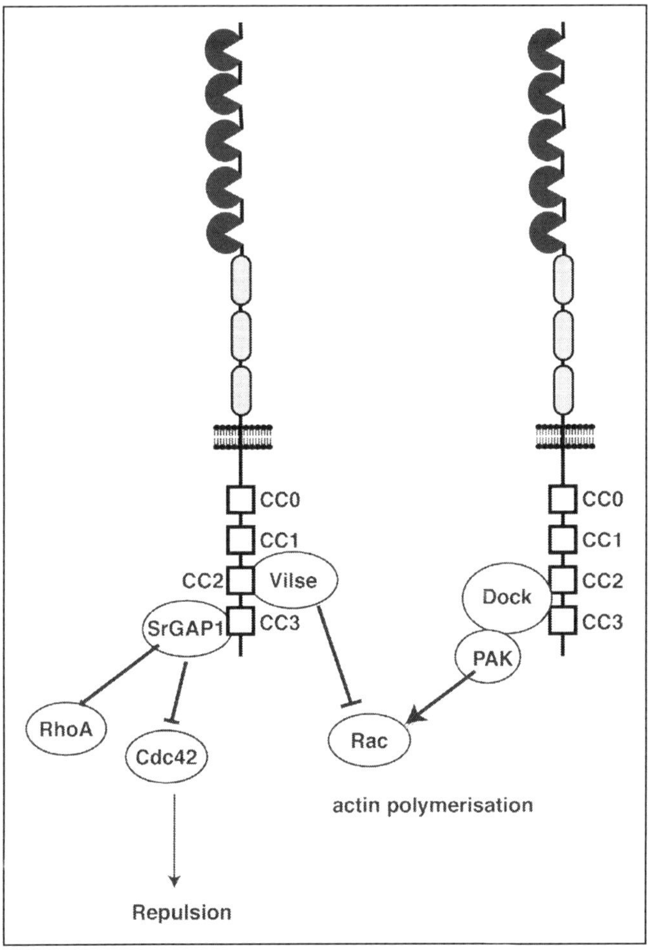

Figure 3. Rho GTPases in Slit/Robo signaling. Rho family of small GTP-binding proteins (Rho GTPases) are activated upon GTP binding and inactive when bound to GDP. Rho GTPases play an important role in the modulation of Slit function. The SH3-SH2 adaptator protein Dock, can bind directly to Robo. Dock interacts with key regulators of the actin cytoskeleton such as p21-activated protein kinase (Pak), which in turn can interact with RhoGTPases such as Rac1 and Cdc42. Slit binding to Robo increases the level of Dock-Robo association, recruits Pak and stimulates Rac. Robo also controls the activity of Rho GTPases through a family of Slit/Robo specific GAPs. SrGAP1 inactivates Cdc42 and activates RhoA but not Rac1. Slit/Robo signaling could also control Rac activity upon binding the GAP Vilse/CrossGAP. CrossGAP WW domains can directly bind to the CC2 domain of Robo. Although it inactivates Rac, Vilse seems to have a positive role in Robo repulsion.

bind SAX-3.[52] Interestingly, Slit also binds netrin-1,[12] but the functional consequence of this interaction is unknown.

Other Modulators of Slit/Robo Function

ECM molecules, specially laminin-1 have been shown to influence the response of retinal axons to netrin-1.[67] Thus, exposure of *Xenopus* retinal axons to laminin converts their response to netrin-1 from attraction to repulsion, apparently by lowering cAMP levels in the growth cones.[67] The level of cyclic nucleotides in the growth cone in part determines the action of many guidance cues.[68] It was found that Slit2-N growth-promoting action could be converted into an inhibition by lowering cGMP levels. Along this line, the activation of CXCR4 by the chemokine SDF-1 reduces the repulsive activity of Slit2 on retinal axons indirectly, by stimulating PKA[69] and increasing cAMP levels. Other second messengers such as calcium may also participate to Slit signaling.[70] Interestingly, as for netrin-1, a laminin-1 peptide was able to convert Slit2-N activity and this may also involve integrins.[71] Accordingly, there is genetic evidence in flies for a regulation of Slit action by integrins.[62]

Last, there is also genetic evidence suggesting that receptor-linked tyrosine phosphatases,[72] Calmodulin and the Ras/Rho GEF Son of Sevenless (SOS) critical for Ras activation[73] may participate to Slit/Robo signaling. Both Sos and CaM signaling pathways are required to prevent certain axons from crossing the midline. However, the link with the transduction of Robo signaling is unclear and these proteins may just function in parallel pathways.

Molecular Control of Slit and Robo Expression

Transcriptional Regulation

In *Drosophila*, Slit function is controlled by the BTB transcription factor Lola.[74] In addition, several transcription factors were shown to control Slit expression in fly embryo[75] such as the PAS bHLH single minded.[76] Slit promoter region also contains binding sites for the SOX HMG domain protein Fish-Hook and the POU domain protein Drifter.[75] Likewise, in the chick retina optic layer, the Irx homeobox gene family member, *Irx4*, negatively controls Slit1 expression.[77] Last, Islet-2, a LIM/homeodomain-type transcription factor of the Islet-1 family was also proposed to control in zebrafish sensory neurons the expression of some factors important for Slit signaling.[78] One of these factors may be the semaphorin receptor plexin-A4.[79]

Similarly, *Robos* could be subject to transcriptional regulation.[74,80] For instance, in fly embryo, Robo2 expression in the mesoderm is likely to be controlled by homeotic genes such as homothorax[81] and there is a hox binding site in the *robo2* gene.

Post Transcriptional and Post-Translational Regulation Commissureless

In *Drosophila* and rodents, Robo expression is regionally restricted[29,41,82] to longitudinal axons and absent from commissures. In fly, this localization of Robo to the post-crossing segment of commissural axons is controlled by the transmembrane protein commissureless (Comm).[82,83]

Comm was initially proposed to be expressed and required at the midline for appropriate midline crossing and to triggers Robo internalization.[84] However, more recent studies have shown that Comm is expressed by commissural axons and acts autonomously in commissural neurons.[85] Moreover, there is no need for Comm at the midline for restoring midline crossing in *comm* mutants.[86] Comm protein appears to be prevented from reaching the contralateral portion of commissural axons and accumulates at the midline. In turn, Comm prevents the delivery of Robo at the growth cone, by recruiting it to late endosomes.[86] It is still unclear why Robo is present on the post-crossing segment.

Comm is a predicted transmembrane protein of 370 amino acids with no known domains. Structure-function analysis revealed that the N-terminal and transmembrane domains of Comm are required to downregulate Robo.[87] The intracellular portion of Comm is also essential for its function and contains an endosomal sorting domain that is required, together with the membrane proximal region of comm (108-131) to relocalize Robo in transfected cells.[86] It also includes a binding site for the ubiquitine ligase dNedd4 and interaction with Nedd4 is required for Comm to localize within vesicles in transfected S2 cells.[88] In yeast two hybrid, Nedd4 was shown to bind Robo.[88] However, more recent studies have shown that comm ubiquitination is not required for its function and that Nedd4 does not influence midline guidance in vivo.[86] Robo2 and Robo3 expression can also be negatively regulated by Comm when the protein is overexpressed but this probably does not occur in vivo.[4,5,30] In normal condition, the restricted expression of Robo2 and Robo3 expression may be controlled by other Comm proteins, but may also be transcriptionally regulated.[5]

Despite its major role in Drosophila, so far no *commissureless* homolog has been found in vertebrates suggesting the existence of additional regulatory mechanisms and modulators one of which could be Robo3/Rig1.

In the mouse spinal cord, commissural axons become responsive to midline repellents, including Slit2, after crossing.[89] Moreover, in mouse spinal cord, Robo1 and Robo2 expression is upregulated after crossing.[90] This suggests that the expression and function of vertebrate Robo is also precisely controlled at the midline. Surprisingly, this regulation seems to involve the receptor Robo3/Rig1. Rig1 expression overlaps with Robo1 in dorsal spinal cord[41] and is downregulated in post-crossing axons and neurons.[41,91] In addition, axons from Rig1 knockout exhibit a premature response to Slit. In the spinal cord and hindbrain,[91] Rig1 seems to function as an inhibitor of Slit signaling in precrossing axons. Accordingly, there is a significant rescue of midline crossing by commissural axons in *rig1/slit2* and *rig1/robo1* double mutants and *rig1/slit1/slit2* triple mutants.[41,90] Rig1 exact function is unknown. It may sequester Slit, or interfere with Robo1 signaling, but there is still no evidence for direct Robo/Rig1 interaction.[41]

Other Regulators

As mentioned above, all Slits, and possibly some Robo receptors can be proteolytically processed into shorter fragments. The enzymes regulating the cleavage of these proteins are unknown, although there is some evidence[92] for a role of the metalloprotease of the ADAM family kuzbanian in *Drosophila*. There is also some data supporting a postranscriptional modulation of Slit function by the Arf6-GEF, Schizo, through a regulation of endocytosis or membrane dynamics.[93]

Multiple Functions for Slit/Robo in the Nervous System

Slits play a major role in axon guidance in many systems and animal species. In most cases Slits act as repellents but there is some evidence that they may act positively on some axons.[71,77]

Midline Crossing

Slit and Robo are primarily known for their function in regulating midline crossing in the nervous system. In *Drosophila robo* mutants, many axons abnormally cross the CNS midline and some multiple times.[29] In *Drosophila*, Robo also controls midline crossing in the olfactory system.[94] In the CNS of *slit* mutant, axons converge to the midline and remains there. Slit was later shown to be a repellent for noncrossing axons and for commissural axons once they have crossed the midline. Biochemical and genetic studies showed that Slit is produced by midline glia cells and that its binding to Robo triggers axonal repulsion. The different midline phenotype between *robo* and *slit* mutants suggested that additional Slit receptors may be present on commissural axons. Accordingly, Robo2 was shown to act redundantly with Robo to control midline crossing in *Drosophila*.[4,5] However, each receptor has a unique role and their function in controlling midline crossing is only partially redundant. In contrast, Robo3 does not seem to play a role in midline crossing in fly.[5]

Interestingly, this essential function of Slit/Robo at the CNS midline is evolutionary conserved from *C. elegans* to humans.[33] In all these species, Slits are expressed at or near the midline, such as the floor plate and septum in vertebrates, or are expressed around decussating axons, canalizing them as they approach the midline. Thus, in vertebrates, Slit/Robo were shown to govern midline crossing by retinal axons[95,96] commissural axons in the spinal cord,[41,90] olfactory bulb axons,[97] cortical axons,[98,99] precerebellar axons.[91] They were also shown to control midline crossing by migrating neurons in the hindbrain.[91]

In the vertebrate visual system both ipsilaterally and controlaterally projecting axons respond to Slits and in their absence, pathfinding errors are observed prior to crossing. In this system, Slit expressing cells surround retinal axons, channeling the axons before and after the chiasm up to the diencephalon.[96,100] The same occurs in the neocortex where Slit2 expression in the glial wedge and *induseum griseum* prevent callosal axons from entering the septum.[98,99] In vertebrates, the function of the three *slit* genes, that are often totally or partially coexpressed[101] appears largely redundant. Thus, axonal tracts are only slightly perturbed in mice deficient for a single *slit* gene and sometimes for two *slit* genes.[90,95,97,98] This redundancy may explain why some major commissures such as the anterior commissure and the hippocampal commissure are normal in mice deficient for both Slit1 and Slit2. Accordingly, it is only in the spinal cord of triple Slit1/2/3 knockouts[90] that many commissural axons stay at the midline and recross it, a phenotype reminiscent of the *Drosophila slit* mutant. The organization of the brain of *slit1/slit2/slit3* triple knockouts will have to be fully studied to determine if Slit/Robo controls the development of all commissural tracts in vertebrates.

Projection Map Formation

In many systems, in particular those conveying sensory informations, axonal projections are topographically ordered in the target territory. Slit and Robo seem to play an important role in regulating axonal targeting in vertebrates and invertebrates. Thus, in the *Drosophila* visual system, Slit and Robo control the segregation of lamina cells (that express Slit) and lobula cells (that express all Robo receptors) by preventing cell mixing.[102] Likewise, in the visual system of zebrafish, Robo2 (*Astray*) in addition to control axon guidance at the chiasm regulates pathfinding within the tectum.[96,100]

In the *Drosophila* olfactory system, distinct subtypes of olfactory axons express various combinations of Robo receptors and Robo controls axonal positioning in the olfactory lobes.[94] In rodents, the projection from the vomeronasal organ (VNO) to the accessory olfactory bulb (AOB) is topographically organized. Neurons in the apical part of the VNO send axons to glomeruli in the anterior half of the AOB and VNO neurons in the basal part project to the posterior AOB. All VNO axons were shown to express *robo1* mRNA during development, while *robo2* is present only in basal ones.[101,103] Slit1 and Slit3 are also expressed in the VNO (preferentially in the apical part) and the anterior AOB and VNO axons are repelled by Slit in collagen gel.[101,103,104] The important role of Slit1 in VNO axon targeting was recently confirmed in vivo using *slit1*-deficient mice.[104] In zebrafish, Robo2 controls the development of olfactory projections from to the olfactory bulb, in particular the establishment of the glomerular map.[105]

Branching

In vertebrate, Slit2 was originally purified as a factor able to stimulate the formation of axon collateral branches by NGF-responsive neurons of the dorsal root ganglia (DRG).[21] It was also shown that only the N-terminal fragment of Slit2, but not the full length protein is capable of stimulating DRG elongation and branching.[21,24] Moreover, full-length Slit2 can antagonize the effect of Slit2-N.[24] Slit2 also controls the branching/arborization of central trigeminal sensory axons in the brainstem of rodents[106] and in zebrafish.[78] In this later case, the branching activity of Slit2 is modulated by the semaphorin receptor plexin-A4.[79] Last, although DRG express Robo2,[21] and trigeminal axons express both Robo1 and Robo2, the axonal receptor mediating Slit branching activity is unknown.

Interestingly, Slit/Robo not only influence axonal branching but also dendritic branching. First, in Drosophila, the directionality of dendritic outgrowth at the midline is controlled cell-autonomously by Robo.[107] In *robo* mutant, dendrites of motor neurons grow abnormally toward the midline while no phenotype was observed in *robo2* and *robo3* mutants. Likewise, Slit1 has a dual activity on rat cortical neurons,[108] as it repels their axons but induces dendritic growth and branching. This effect appears to involve Robo signaling.

Longitudinal Tract Formation

In *Drosophila*, Robo, Robo2 and Robo3 are expressed in overlapping domains within longitudinal tracts of the CNS, and this combination of Robo receptors is thought to control the lateral position of longitudinal axons. Thus, genetic alterations of the Robo code displace longitudinal axons along the mediolateral axis. However, it is not known if these changes involves Slit signaling. As Robos are immunoglobulins and able to mediate homophilic and heterophilic binding, it is possible that the control of lateral positioning by Robo involves Robo-Robo interactions and selective axonal fasciculation.

In mouse, there is also some data supporting a differential expression of Robo receptors by longitudinal axons that project at distinct ventro-dorsal position in the spinal cord.[41]

Control of Cell Differentiation

In fly, serotoninergic neurons are bilaterally organized and must cross the midline to achieve their differentiation. Robo2 and Robo3 were shown to regulate the expression of the serotonin transporter (SerT) as many serotoninergic neurons fail to express SerT in *robo2* and *robo3* mutants. Moreover, Robo2 and Robo3 are required for eagle expression, a transcription factor controlling serotoninergic differentiation.[109] Interestingly, SerT activity is normal in *slit* mutants suggesting that Robo2/3 function in serotoninergic differentiation is Slit independent. In *Drosophila*, Slit also promotes the terminal asymmetric division of ganglion mother cells by regulating the asymmetric distribution of Inscuteable and by downregulating the expression of POU genes.[110] In vertebrates,[111] Robo1 may also control cell differentiation as its overexpression in Xenopus leads to ectopic neuronal differentiation. Last, during kidney development in mouse, *slit2* and *robo2* inactivation leads to supernumerary ureteric buds, possibly through postranscriptional effect on other developmental genes.[112]

Cell Migration

Another important function for Slits and Robos is the control of cell migration in the nervous system (both neurons and glia) and in several other tissues. As for axons, Slits were found to be important regulators of the behavior of migrating cells at the midline. But in contrast with axons, migrating cells can either be attracted or repelled by Slits.

In *Drosophila*, longitudinal glia is generated from glioblasts that migrate ventrally to contact pioneer neurons at a distance from the midline. These cells express Robo1 and in *robo* mutant, glial cells migrate over the midline[113] suggesting that Slit is repulsive. Likewise, Muscle precursors in Drosophila embryos[114] fail to migrate away from the midline in *slit* mutant. In this system also, Slit produced by midline glia acts as a repellent. However, at later stages Slit expressed at muscle attachment sites attracts muscle precursors that express both Robo and Robo2. Interestingly, Comm also cooperates with Robo and Robo2 to control muscle precursors migration.[30] The mechanism responsible for the switch from repulsion to attraction is still unknown but may involve signaling through different Robo receptors as suggested in other systems. Hence, Robo2 was proposed to mediate the long-range attraction of tracheal cell into the CNS[115] while Robo may mediate a repulsive action of Slit on tracheal cells. This different activity of the two receptors may rely on differences in their cytoplasmic domains. In *C. elegans*, Slit was also shown to be positive regulator of neuronal migration *C.elegans* along the anterior posterior axis.[6]

In vertebrates, Slits and Robo participate to the migration of many neurons in the CNS and PNS but so far there is only evidence for a repulsive activity. Moreover, whereas Slit and Robo were shown to guide tangentially migrating neurons, they do not seem to participate to radial migration. During development and throughout adulthood, several types of olfactory bulb (OB) interneurons (the granule cells and the tufted cells) are generated from progenitors located in the so-called subventricular zone (SVZ) that surrounds the lateral ventricles[116] and migrate to the OB via the rostral migratory stream (RMS). The rostral migration of SVZ-derived neuroblasts was shown to involve chemorepellents secreted by the septum.[117-119] Biochemical and in vitro studies have since demonstrated that Slit1 and Slit2 are mediating this repulsive activity.[15] Accordingly, some SVZ-derived neuroblasts showed abnormal migration pattern in *slit1* deficient-mice.[119] However, those cells were also shown to express Slit1[119] that may act cell autonomously. Migrating OB neuroblasts express *robo2* and *robo3* mRNAs[101] and srGAP1.[60] Although dominant-negative srGAP1 blocks Slit repulsion of SVZ cells,[60] the contribution of Robo signaling in this system is largely unknown. Elegant in vitro assay also showed that Slit repels migrating svz cells without blocking their migration. Thus, Slit may just control the directionality of the migration without affecting cell motility,[120] although this issue is still controversial.[121] In chick embryo, Slit2 repels the migration of trunkal but not vagal neural crest cells[122] that take different migratory pathways during development. As observed in the SVZ system, Slit2 appears to enhance cell motility: neural crest cells migrate further in the presence of soluble Slit2.

In the telencephalon, Slit1 repels in vitro the migration of GABAergic interneurons from the ganglionic eminence.[123] However, the phenotypic analysis of *slit1/slit2* deficient mice revealed that they are not necessary for tangential migration of GABAergic interneurons to the cortex in vivo.[124] On the other hand, this study revealed that Slits influence the migration of cholinergic neurons of the basal magnocellular complex.

In the hindbrain, Slit and Robo participate to the migration of rhombic lip derivatives both in chick and rodents.[11,70,91] Although Slits primarily act as repellents for rhombic lip-derived cells, they were also proposed to antagonize the attractive activity of floor plate-derived netrin-1.[125]

Last, Slits and Robo also influence the migration of other vertebrate cells; either negatively as shown for leukocytes,[126] but sometimes positively as shown for endothelial cells[27,127] In this later case, Slit attractive activity involves Robo1 signaling.

A Role for Slit and Robo in Neurological Disorders?

There is relatively little direct evidence so far indicating that Slit and Robo may be involved in pathological processes except in cancers.[128] However, several patients suffering from a rare congenital syndrome named Horizontal gaze palsy with progressive scoliosis and hindbrain dysplasia (HGPPS) were recently shown to bear mutations in the ROBO3 gene. In these patients, the pyramidal tract and the dorsal column-medial lemniscus are uncrossed.[33] Moreover, there is a reduced pontine nucleus and abducens nuclei. As shown in *robo3* knockout mice the pontine nucleus defect is probably caused by an abnormal migration during development.[91] It is still unknown if the other human brain defects are also present in mice lacking Robo3 and as those die at birth, behavioral analysis are not possible.

One of the Slit/Robo GAP, SrGAP3 has a putative role in idiopathic mental retardation[63] as it is mutated in patients with X chromosome-linked MR. However, the cellular basis for these defects and the normal function of SrGAP3 in the CNS are unknown.

The incapacity of adult axons to regenerate in the CNS of mammals is known to rely for a large extent on the existence of inhibitors of axonal growth expressed either in the glial scar or in myelin. Several studies have shown that repulsive axon guidance molecules may be responsible for some of the inhibition.[129] Slit an Robo expression after injury has not been extensively studied so far. However, in a model of cryoinjury, Slit2 was found to be expressed in reactive astrocytes together with glypican-1.[130] In addition, adult DRG neurons also express mRNAs for Robo2 and Slit1 but their expression does not change after sciatic nerve transection or dorsal column lesion in the spinal cord.[131] In contrast, Glypican-1 and SrGAP2 expression are upregulated after such lesions.[131,132] Thus, it is still unclear if Slit and Robo play any role in preventing axonal regeneration.

Perspectives

There is mounting evidence that Slits regulate a large range of biological functions, from axon guidance, neuronal migration, immune response[133] to cell differentiation most likely through Robo signaling. However there are still many open questions.

First, could Robo functions independently of Slit in particular through Robo/Robo interaction and what are the signaling pathways involved. Thus, in the Drosophila PNS, Robo2 expressed on visceral mesoderm binds Slit and present it to Robo expressing chordotonal sensory neurons. This may also involve Robo-Robo2 direct interaction.[81,134] In vitro experiments also showed that the growth of retinal and olfactory Robo expressing axons is stimulated on Robo expressing cells, suggesting that Robo might work as cell-adhesion molecule to regulate outgrowth.[44] In

Drosophila, Robo and Robo2 can dimerize in vitro[40] and ectopic expression of low level of Robo2 causes a Robo-like phenotype suggesting that Robo2 could interfere with Robo function.[5]

The function of Slits and Robos to in the normal adult brain and in pathological condition also remains to be clarified. Many data support a role for these molecules in tumorigenesis, in particular in gliomas[128] but this needs to be further demonstrated. As all Slits and Robos are expressed in adult neurons it is likely that they modulate synaptic transmission as shown recently for other secreted axon guidance molecules of the semaphorin family.[135] Many answers to these questions should come from the analysis of mouse deficient for one or several of these proteins.

References

1. Nusslein-Volhard C, Wiechaus E, Kluding H. Mutations affecting the pattern of the larval cuticle in Drosophila melanogaster. I. Zygotic loci on the second chromosome. Roux Arch Dev Biol 1984; 193:267-283.
2. Rothberg JM, Artavanis-Tsakonas S. Modularity of the Slit protein characterization of a conserved carboxy-terminal sequence in secreted proteins and a motif implicated in extracellular protein interactions. J Mol Biol 1992; 227:367-370.
3. Kidd T, Bland KS, Goodman CS. Slit is the midline repellent for the robo receptor in drosophila. Cell 1999; 96:785-794.
4. Rajagopalan S, Nicolas E, Vivancos V et al. Crossing the midline: Roles and regulation of robo receptors. Neuron 2000; 28:767-777.
5. Simpson JH, Kidd T, Bland KS et al. Short-range and long-range guidance by slit and its robo receptors: Robo and Robo2 play distinct roles in midline guidance. Neuron 2000; 28:753-766.
6. Hao JC, Yu TW, Fujisawa K et al. C. elegans Slit Acts in midline, dorsal-ventral, and anterior-posterior guidance via the SAX-3/Robo receptor. Neuron 2001; 32(1):25-38.
7. Li HS, Chen JH, Wu W et al. Vertebrate Slit, a secreted ligand for the transmembrane protein roudabout, is a repellent for olfactory bulb axons. Cell 1999; 96:807-818.
8. Yee KT, Simon HH, Tessier-Lavigne M et al. Extension of long leading processes and neuronal migration in the mammalian brain directed by the chemoattractant netrin-1. Neuron 1999; 24(3):607-622.
9. Vargesson N, Luria V, Messina I et al. Expression patterns of Slit and Robo family members during vertebrate limb development. Mech Dev 2001; 106(1-2):175-180.
10. Holmes G, Niswander L. Expression of slit-2 and slit-3 during chick development. Dev Dyn 2001; 222(2):301-307.
11. Gilthorpe JD, Papantoniou EK, Chedotal A et al. The migration of cerebellar rhombic lip derivatives. Development 2002; 129(20):4719-4728.
12. Brose K, Bland KS, Wang KH et al. Slit proteins bind Robo receptors and have an evolutionarily conserved role in repulsive axon guidance. Cell 1999; 96:795-806.
13. Holmes GP, Negus K, Burridge L et al. Distinct but overlapping expression patterns of two vertebrate slit homologs implies functional roles in CNS development and organogenesis. Mech Dev 1998; 79:57-72.
14. Itoh A, Miyabayashi T, Ohno M et al. Cloning and expressions of three mammalian homologues of Drosophila slit suggest possible roles for slit in the formation and maintenance of the nervous system. Mol Brain Res 1998; 62:175-186.
15. Hu H. Chemorepulsion of neuronal migration by slit2 in the developing mammalian forebrain. Neuron 1999; 23:703-711.
16. Niclou SP, Jia L, Raper JA. Slit2 is a repellent for retinal ganglion cell axons. J Neurosci 2000; 20(13):4962-4974.
17. Nguyen-Ba-Charvet KT, Chédotal A. Role of Slit proteins in the vertebrate brain. J Physiol Paris 2002; 96:91-98.
18. Tanno T, Takenaka S, Tsuyama S. Expression and function of Slit1{alpha}, a novel alternative splicing product for Slit1. J Biochem 2004; 136(5):575-581.
19. Little M, Rumballe B, Georgas K et al. Conserved modularity and potential for alternate splicing in mouse and human Slit genes. Int J Dev Biol 2002; 46(4):385-391.
20. Hutson LD, Jurynec MJ, Yeo SY et al. Two divergent slit1 genes in zebrafish. Dev Dyn 2003; 228(3):358-369.
21. Wang KH, Brose K, Arnott D et al. Biochemical purification of a mammalian Slit protein as a positive regulator of sensory axon elongation and branching. Cell 1999; 96:771-784.
22. Patel K, Nash JAB, Itoh A et al. Slit proteins are dominant chemorepellents for olfactory tract and spinal motor axons. Development 2001; 128:5031-5037.

23. Whitford KL, Dijkhuizen P, Polleux F et al. Molecular control of cortical dendrite development. Annu Rev Neurosci 2002; 25:127-149.
24. Nguyen Ba-Charvet KT, Brose K, Ma L et al. Diversity and specificity of actions of Slit2 proteolytic fragments in axon guidance. J Neurosci 2001; 21:4281-4289.
25. Battye R, Stevens A, Jacobs JR. Axon repulsion from the midline of the Drosophila CNS requires slit function. Development 1999; 126:2475-2481.
26. Chen JH, Wen L, Dupuis S et al. The N-terminal leucine-rich regions in slit are sufficient to repel olfactory bulb axons and subventricular zone neurons. J Neurosci 2001; 21:1548-1556.
27. Howitt JA, Clout NJ, Hohenester E. Binding site for Robo receptors revealed by dissection of the leucine-rich repeat region of Slit. EMBO J 2004; 23(22):4406-4412.
28. Seeger M, Tear G, Ferres-Marco D et al. Mutations affecting growth cone guidance in Drosophila: Genes necessary for guidance toward or away from the midline. Neuron 1993; 10(3):409-426.
29. Kidd T, Brose K, Mitchell KJ et al. Roundabout controls axon crossing of the CNS midline and defines a novel subfamily of evolutionarily conserved guidance receptors. Cell 1998; 92:205-215.
30. Rajagopalan S, Vivancos V, Nicolas E et al. Selecting a longitudinal pathway: Robo receptors specify the lateral position of axons in the Drosophila CNS. Cell 2000; 103:1033-1045.
31. Yuan W, Zhou L, Chen JH et al. The mouse SLIT family: Secreted ligands for ROBO expressed in patterns that suggest a role in morphogenesis and axon guidance. Dev Biol 1999; 212:290-306.
32. Lee JS, Ray R, Chien CB. Cloning and expression of three zebrafish roundabout homologs suggest roles in axon guidance and cell migration. Dev Dyn 2001; 221(2):216-230.
33. Jen JC, Chan WM, Bosley TM et al. Mutations in a human ROBO gene disrupt hindbrain axon pathway crossing and morphogenesis. Science 2004; 304(5676):1509-1513.
34. Huminiecki L, Gorn M, Suchting S et al. Magic roundabout is a new member of the roundabout receptor family that is endothelial specific and expressed at sites of active angiogenesis. Genomics 2002; 79(4):547-552.
35. Bedell VM, Yeo SY, Park KW et al. Roundabout4 is essential for angiogenesis in vivo. Proc Natl Acad Sci USA 2005; 102(18):6373-6378.
36. Park KW, Morrison CM, Sorensen LK et al. Robo4 is a vascular-specific receptor that inhibits endothelial migration. Dev Biol 2003; 261:251-267.
37. Suchting S, Heal P, Tahtis K et al. Soluble Robo4 receptor inhibits in vivo angiogenesis and endothelial cell migration. FASEB J 2005; 19(1):121-123.
38. Bashaw GJ, Kidd T, Murray D et al. Repulsive axon guidance: Abelson and enabled play opposing roles downstream of the roundabout receptor. Cell 2000; 101:703-715.
39. Yuan S-SF, Cox LA, Dasika GK et al. Cloning and functional studies of a novel gene aberrantly expressed in RB-deficient embryos. Dev Biol 1999; 207:62-75.
40. Simpson JH, Bland KS, Fetter RD et al. Short-range and long-range guidance by slit and its robo receptors: A combinatorial code of Robo receptors controls lateral position. Cell 2000; 103:1019-1032.
41. Sabatier C, Plump AS, Le M et al. The divergent robo family protein rig-1/robo3 is a negative regulator of slit responsiveness required for midline crossing by commissural axons. Cell 2004; 117(2):157-169.
42. Xian J, Clark KJ, Fordham R et al. Inadequate lung development and bronchial hyperplasia in mice with a targeted deletion in the Dutt1/Robo1 gene. Proc Natl Acad Sci USA 2001; 4:4.
43. Xian J, Aitchison A, Bobrow L et al. Targeted disruption of the 3p12 gene, Dutt1/Robo1, predisposes mice to lung adenocarcinomas and lymphomas with methylation of the gene promoter. Cancer Res 2004; 64(18):6432-6437.
44. Hivert B, Liu Z, Chuang JY et al. Robo1 and Robo2 are hompohilic binding molecules that promote axonal growth. Mol Cell Neurosci 2002; 21:534-545.
45. Liu Z, Patel K, Schmidt H et al. Extracellular Ig domains 1 and 2 of Robo are important for ligand (Slit) binding. Mol Cell Neurosci 2004; 26(2):232-240.
46. Liang Y, Annan RS, Carr SA et al. Mammalian homologues of the Drosophila slit protein are ligands of the heparan sulfate proteoglycan glypican-1 in brain. J Biol Chem 1999; 274(25):17885-17892.
47. Ronca F, Andersen JS, Paech V et al. Characterization of Slit protein interactions with glypican-1. J Biol Chem 2001; 276:29141-29147.
48. Hu H. Cell-surface heparan sulfate is involved in the repulsive guidance activities of Slit2 protein. Nature Neurosci 2001; 4:695-701.
49. Steigemann P, Molitor A, Fellert S et al. Heparan sulfate proteoglycan syndecan promotes axonal and myotube guidance by Slit/Robo signaling. Curr Biol 2004; 14:225-230.
50. Inatani M, Irie F, Plump AS et al. Mammalian brain morphogenesis and midline axon guidance require heparan sulfate. Science 2003; 302(5647):1044-1046.
51. Bashaw GJ, Goodman CS. Chimeric axon guidance receptors: The cytoplasmic domains of slit and netrin receptors specificity attraction versus repulsion. Neuron 1999; 97:917-926.

52. Yu TW, Hao JC, Lim W et al. Shared receptors in axon guidance: SAX-3/Robo signals via UNC-34/ Enabled and a Netrin-independent UNC-40/DCC function. Nat Neurosci 2002; 5(11):1147-1154.
53. Wills Z, Emerson M, Rusch J et al. A Drosophila homolog of cyclase-associated proteins collaborates with the Abl tyrosine kinase to control midline axon pathfinding. Neuron 2002; 36(4):611-622.
54. Hsouna A, Kim YS, VanBerkum MF. Abelson tyrosine kinase is required to transduce midline repulsive cues. J Neurobiol 2003; 57(1):15-30.
55. Rhee J, Mahfooz NS, Arregui C et al. Activation of the repulsive receptor Roundabout inhibits N-cadherin-mediated cell adhesion. Nat Cell Biol 2002; 4(10):798-805.
56. Lee H, Engel U, Rusch J et al. The microtubule plus end tracking protein Orbit/MAST/CLASP acts downstream of the tyrosine kinase Abl in mediating axon guidance. Neuron 2004; 42(6):913-926.
57. Huber AB, Kolodkin AL, Ginty DD et al. Signaling at the growth cone: Ligand-receptor complexes and the control of axon growth and guidance. Annu Rev Neurosci 2003; 26:509-563.
58. Bashaw GJ, Hu H, Nobes CD et al. A novel Dbl family RhoGEF promotes Rho-dependent axon attractin to the central nervous system midline in Drosophila and overcomes Robo repulsion. JCB 2001; 155:1117-1122.
59. Matsuura R, Tanaka H, Go MJ. Distinct functions of Rac1 and Cdc42 during axon guidance and growth cone morphogenesis in Drosophila. Eur J Neurosci 2004; 19(1):21-31.
60. Wong K, Ren XR, Huang YZ et al. Signal transduction in neuronal migration. roles of gtpase activating proteins and the small gtpase cdc42 in the slit-robo pathway. Cell 2001; 107(2):209-221.
61. Fan X, Labrador JP, Hing H et al. Slit stimulation recruits Dock and Pak to the roundabout receptor and increases Rac activity to regulate axon repulsion at the CNS midline. Neuron 2003; 40(1):113-127.
62. Stevens A, Jacobs JR. Integrins regulate responsiveness to slit repellent signals. J Neurosci 2002; 22(11):4448-4455.
63. Endris V, Wogatzky B, Leimer U et al. The novel Rho-GTPase activating gene MEGAP/ srGAP3 has a putative role in severe mental retardation. Proc Natl Acad Sci USA 2002; 99(18):11754-11759.
64. Lundstrom A, Gallio M, Englund C et al. Vilse, a conserved Rac/Cdc42 GAP mediating Robo repulsion in tracheal cells and axons. Genes Dev 2004; 18(17):2161-2171.
65. Hu H, Li M, Labrador JP et al. Cross GTPase-activating protein (CrossGAP)/Vilse links the Roundabout receptor to Rac to regulate midline repulsion. Proc Natl Acad Sci USA 2005; 102(12):4613-4618.
66. Stein E, Tessier-Lavigne M. Hierarchical organization of guidance receptors: Silencing of netrin attraction by slit through a Robo/DCC receptor complex. Science 2001; 291(5510):1928-1938.
67. Höpker VH, Shewan D, Tessier-Lavigne M et al. Growth-cone attraction to netrin-1 is converted to repulsion by laminin-1. Nature 1999; 401:69-73.
68. Song H, Ming G, He Z et al. Conversion of neuronal growth cone responses from repulsion to attraction by cyclic nucleotides. Science 1998; 281(5382):1515-1518.
69. Chalasani SH, Sabelko KA, Sunshine MJ et al. A chemokine, SDF-1, reduces the effectiveness of multiple axonal repellents and is required for normal axon pathfinding. J Neurosci 2003; 23(4):1360-1371.
70. Xu HT, Yuan XB, Guan CB et al. Calcium signaling in chemorepellant Slit2-dependent regulation of neuronal migration. Proc Natl Acad Sci USA 2004; 101(12):4296-4301.
71. Nguyen-Ba-Charvet KT, Brose K, Marillat V et al. Sensory axons response to substrate-bound Slit2 is modulated by laminin and cyclicGMP. Mol Cell Neurosci 2001; 17:1048-1058.
72. Sun Q, Bahri S, Schmid A et al. Receptor tyrosine phosphatases regulate axon guidance across the midline of the drosophila embryo. Development 2000; 127:801-812.
73. Fritz JL, VanBerkum MFA. Calmodulin and son of sevenless dependent signaling pathways regulate midline crossing of axons in the Drosophila CNS. Development 2000; 127:1991-2000.
74. Crowner D, Madden K, Goeke S et al. Lola regulates midline crossing of CNS axons in Drosophila. Development 2002; 129(6):1317-1325.
75. Ma Y, Certel K, Gao Y et al. Functional interactions between Drosophila bHLH/PAS, Sox, and POU transcription factors regulate CNS midline expression of the slit gene. J Neurosci 2000; 20(12):4596-4605.
76. Wharton Jr K, Franks RG, Kasai Y et al. Control of CNS midline transcription by asymetric E-box-like elements: Similarity to xenobiotic responsive regulation. Development 1994; 120(12):3563-3569.
77. Jin Z, Zhang J, Klar A et al. Irx4-mediated regulation of Slit1 expression contributes to the definition of early axonal paths inside the retina. Development 2003; 130:1037-1048.
78. Yeo SY, Miyashita T, Fricke C et al. Involvement of Islet-2 in the Slit signaling for axonal branching and defasciculation of the sensory neurons in embryonic zebrafish. Mech Dev 2004; 121(4):315-324.

79. Miyashita T, Yeo SY, Hirate Y et al. PlexinA4 is necessary as a downstream target of Islet2 to mediate Slit signaling for promotion of sensory axon branching. Development 2004; 131(15):3705-3715.
80. Zlatic M, Landgraf M, Bate M. Genetic specification of axonal arbors. Atonal regulates robo3 to position terminal branches in the Drosophila nervous system. Neuron 2003; 37(1):41-51.
81. Kraut R, Zinn K. Roundabout 2 regulates migration of sensory neurons by signaling in trans. Curr Biol 2004; 14(15):1319-1329.
82. Kidd T, Russell C, Goodman CS et al. Dosage-sensitive and complementary functions of roudabout and commissureless control axon crossing of the CNS midline. Neuron 1998; 20:25-33.
83. Tear G, Harris R, Sutaria S et al. Commissureless controls growth cone guidance across the CNS midline in Drosophila and encodes a novel membrane protein. Neuron 1996; 16(3):501-514.
84. Georgiou M, Tear G. Commissureless is required both in commissural neurones and midline cells for axon guidance across the midline. Development 2002; 129(12):2947-2956.
85. Keleman K, Rajagopalan S, Cleppien D et al. Comm sorts robo to control axon guidance at the Drosophila midline. Cell 2002; 110(4):415.
86. Keleman K, Ribeiro C, Dickson BJ. Comm function in commissural axon guidance: Cell-autonomous sorting of Robo in vivo. Nat Neurosci 2005; 8(2):156-163.
87. Georgiou M, Tear G. The N-terminal and transmembrane domains of Commissureless are necessary for its function and trafficking within neurons. Mech Dev 2003; 120(9):1009-1019.
88. Myat A, Henry P, McCabe V et al. Drosophila Nedd4, a ubiquitin ligase, is recruited by Commissureless to control cell surface levels of the roundabout receptor. Neuron 2002; 35(3):447-459.
89. Zou Y, Stoeckli E, Chen H et al. Squeezing axons out of the gray matter: A role for slit and semaphorin proteins from the midline and ventral spinal cord. 2000; 102:363-375.
90. Long H, Sabatier C, Le M et al. Conserved roles for slit and robo proteins in midline commissural axon guidance. Neuron 2004; 42(2):213-223.
91. Marillat V, Sabatier C, Failli V et al. The Slit receptor Rig-1/Robo3 controls midline crossing by hindbrain precerebellar neurons and axons. Neuron 2004; 43:1-20.
92. Schimmelpfeng K, Gogel S, Klambt C. The function of leak and kuzbanian during growth cone and cell migration. Mech Dev 2001; 106(1-2):25-36.
93. Onel S, Bolke L, Klambt C. The Drosophila ARF6-GEF Schizo controls commissure formation by regulating Slit. Development 2004; 131(11):2587-2594.
94. Jhaveri D, Saharan S, Sen A et al. Positioning sensory terminals in the olfactory lobe of Drosophila by Robo signaling. Development 2004; 131(9):1903-1912.
95. Plump AS, Erskine L, Sabatier C et al. Slit1 and Slit2 cooperate to prevent premature midline crossing of retinal axons in the mouse visual system. Neuron 2002; 33:219-232.
96. Hutson LD, Chien CB. Pathfinding and error correction by retinal axons: The role of astray/robo2. Neuron 2002; 33(2):205-217.
97. Nguyen-Ba-Charvet KT, Plump AS, Tessier-Lavigne M et al. Slit1 and Slit2 proteins control the development of the lateral olfactory tract. J Neurosci 2002; 22:5473-5480.
98. Bagri A, Marin O, Plump AS et al. Slit proteins prevent midline crossing and determine the dorsoventral position of major axonal pathways in the mammalian forebrain. Neuron 2002; 33:233-248.
99. Shu T, Sundaresan V, McCarthy MM et al. Slit2 guides both precrossing and postcrossing callosal axons at the midline in vivo. J Neurosci 2003; 23(22):8176-8184.
100. Fricke C, Lee J-S, Geiger-Rudolph S et al. Astray, a zebrafish roundabout homolog required for retinal axon guidance. Science 2001; 292:507-510.
101. Marillat V, Cases O, Nguyen-Ba-Charvet KT et al. Spatiotemporal expression patterns of slit and robo genes in the rat brain. J Comp Neurol 2002; 442:130-155.
102. Tayler TD, Robichaux MB, Garrity PA. Compartmentalization of visual centers in the Drosophila brain requires Slit and Robo proteins. Development 2004; 131(23):5935-5945.
103. Knoll B, Schmidt H, Andrews W et al. On the topographic targeting of basal vomeronasal axons through Slit-mediated chemorepulsion. Development 2003; 130:5073-5082.
104. Cloutier JF, Sahay A, Chang EC et al. Differential requirements for semaphorin 3F and Slit-1 in axonal targeting, fasciculation, and segregation of olfactory sensory neuron projections. J Neurosci 2004; 24(41):9087-9096.
105. Miyasaka N, Sato Y, Yeo SY et al. Robo2 is required for establishment of a precise glomerular map in the zebrafish olfactory system. Development 2005; 132:1283-1293.
106. Ozdinler PH, Erzurumlu RS. Slit2, a branching-arborization factor for sensory axons in the Mammalian CNS. J Neurosci 2002; 22(11):4540-4549.
107. Furrer MP, Kim S, Wolf B et al. Robo and Frazzled/DCC mediate dendritic guidance at the CNS midline. Nat Neurosci 2003; 6(3):223-230.
108. Whitford KL, Marillat V, Stein E et al. Regulation of cortical dendrite development by Slit-Robo interactions. Neuron 2002; 33:47-61.

109. Couch JA, Chen J, Rieff HI et al. Robo2 and robo3 interact with eagle to regulate serotonergic neuron differentiation. Development 2004; 131(5):997-1006.
110. Mehta B, Bhat KM. Slit signaling promotes the terminal asymmetric division of neural precursor cells in the Drosophila CNS. Development 2001; 128(16):3161-3168.
111. Connor RM, Key B. Expression and role of Roundabout-1 in embryonic Xenopus forebrain. Dev Dyn 2002; 225(1):22-34.
112. Grieshammer U, Ma L, Plump AS et al. SLIT2-mediated ROBO2 signaling restricts kidney induction to a single site. Dev Cell 2004; 6(5):709-717.
113. Kinrade EFV, Bartes T, Tear G et al. Roundabout signalling, cell contact and trophic support confine longitudinal glia and axons in the Drosophila CNS. Development 2001; 128:207-216.
114. Kramer SG, Kidd T, Simpson JH et al. Switching repulsion to attraction: Changing responses to slit during transition in mesoderm migration. Science 2001; 292:737-740.
115. Englund C, Steneberg P, Falileeva L et al. Attractive and repulsive functions of Slit are mediated by different receptors in the Drosophila trachea. Development 2002; 129(21):4941-4951.
116. Doetsch F. The glial identity of neural stem cells. Nat Neurosci 2003; 6(11):1127-1134.
117. Hu H, Rutishauser U. A septum-derived chemorepulsive factor for migrating olfactory interneuron precursors. Neuron 1996; 16:933-940.
118. Wu W, Wong K, Chen JH et al. Directional guidance of neuronal migration in the olfactory system by the protein Slit. Nature 1999; 400:331-336.
119. Nguyen-Ba-Charvet KT, Picard-Riera N, Tessier-Lavigne M et al. Multiple roles for slits in the control of cell migration in the rostral migratory stream. J Neurosci 2004; 24(6):1497-1506.
120. Ward M, McCann C, DeWulf M et al. Distinguishing between directional guidance and motility regulation in neuronal migration. J Neurosci 2003; 23(12):5170-5177.
121. Mason HA, Ito S, Corfas G. Extracellular signals that regulate the tangential migration of olfactory bulb neuronal precursors: Inducers, inhibitors, and repellents. J Neurosci 2001; 21:7654-7663.
122. De Bellard ME, Rao Y, Bronner-Fraser M. Dual function of Slit2 in repulsion and enhanced migration of trunk, but not vagal, neural crest cells. J Cell Biol 2003; 162(2):269-279.
123. Zhu Y, Li HS, Zhou L et al. Cellular and molecular guidance of GABAergic neuronal migration from an extracortical origin to the neocortex. Neuron 1999; 23:473-485.
124. Marin O, Plump AS, Flames N et al. Directional guidance of interneuron migration to the cerebral cortex relies on subcortical Slit1/2-independent repulsion and cortical attraction. Development 2003; 130:1889-1901.
125. Causeret F, Danne F, Ezan F et al. Slit antagonizes netrin-1 attractive effects during the migration of inferior olivary neurons. Dev Biol 2002; 246(2):429-440.
126. Wu JY, Feng L, Park H-T et al. The neuronal repellent Slit inhibits leukocyte chemotaxis induced by chemotactic factors. Nature 2001; 410:948-952.
127. Wang B, Xiao Y, Ding BB et al. Induction of tumor angiogenesis by Slit-Robo signaling and inhibition of cancer growth by blocking Robo activity. Cancer Cell 2003; 4(1):19-29.
128. Chédotal A, Kerjan G, Moreau-Fauvarque C. The brain within the tumor: New role for axon guidance molecules in cancer. Cell Death Differ 2005; 12(8):1044-1056.
129. Filbin MT. Myelin-associated inhibitors of axonal regeneration in the adult mammalian CNS. Nat Rev Neurosci 2003; 4(9):703-713.
130. Hagino S, Iseki K, Mori T et al. Slit and glypican-1 mRNAs are coexpressed in the reactive astrocytes of the injured adult brain. Glia 2003; 42(2):130-138.
131. Bloechlinger S, Karchewski LA, Woolf CJ. Dynamic changes in glypican-1 expression in dorsal root ganglion neurons after peripheral and central axonal injury. Eur J Neurosci 2004; 19(5):1119-1132.
132. Madura T, Yamashita T, Kubo T et al. Changes in mRNA of Slit-Robo GTPase-activating protein 2 following facial nerve transection. Brain Res Mol Brain Res 2004; 123(1-2):76-80.

Chapter 6

Neurotrophic Factors:
Are They Axon Guidance Molecules?

Sarah Guthrie*

Introduction

Neurotrophic factors play a multiplicity of roles during development. They can promote axonal growth in vitro and are capable of orienting axons. Recently, molecular evidence suggests that neurotrophic factors and axon guidance molecules regulate the same signalling pathways in neurons. Yet only a few critical studies have tested a possible role for neurotrophic factors as guidance molecules in vivo. In this chapter I review the evidence for and against neurotrophic factors as axon guidance molecules.

Evaluation of Neurotrophic Factors as Axon Guidance Molecules

In the past two decades, the Semaphorin, Ephrin, Netrin and Slit molecules have been found to exert positive (growth promoting/chemoattractant) and/or negative (growth inhibitory/chemorepellent) effects on axon guidance.[1] Neurotrophic factors are known to positively regulate neuronal survival and axon growth, but their status as guidance molecules is much less certain. What criteria should therefore be used in evaluating the ability of neurotrophic factors to guide axons? Firstly, it is important to distinguish between 'trophic' effects—promotion of neuronal survival and axon outgrowth, and 'tropic' effects—axonal chemoattraction. In vitro studies should therefore test not only the ability of neurotrophic factors to increase axonal growth when applied globally, but also their capacity to chemoattract axons when applied focally. Secondly, in vivo gain and loss of function approaches should cause ectopic axon projections or pathfinding defects respectively, or alterations in branching and terminal arborisation. Thirdly, evidence of common signalling pathways might provide a circumstantial link for a role of neurotrophic factors as guidance molecules.

The Discovery of the Neurotrophins

The ground-breaking work of Rita Levi-Montalcini and Viktor Hamburger in the 1940s first revealed the vital role played by neurotrophic factors in the development of the nervous system. They discovered that neuronal cell death occurs in normal embryos, and that the size of the surviving neuronal population depends on the size of the target field they innervate. The full implications of these findings were not clear at the time, but the work paved the way for the development of the idea that neuronal survival depends on limiting amounts of factors taken up by neurons from their targets.[2,3] Experiments in which tumour tissues were implanted into mice showed that diffusible factors from this source potently promoted the survival and outgrowth of sensory and sympathetic neurons. The active component proved to be Nerve Growth Factor, the founder member of the family of neurotrophins and of the

*Sarah Guthrie—MRC Centre for Developmental Neurobiology, 4th Floor New Hunt's House, King's College London, Guy's Campus, London SE1 1UL, U.K. Email: sarah.guthrie@kcl.ac.uk

Axon Growth and Guidance, edited by Dominique Bagnard. ©2007 Landes Bioscience and Springer Science+Business Media.

extended family of neurotrophic factors.[4] When anti-NGF antiserum was injected into embryonic or neonatal mice and rats, there was extensive atrophy of the sympathetic and sensory ganglia, showing a physiological requirement for NGF in neuronal survival.[5] Injection of NGF into the brains of neonatal rats resulted in abundant ingrowth of sympathetic axons, which entered the CNS via the dorsal and ventral roots, apparently growing towards the source of NGF.[6] Moreover, exposure of embryonic sympathetic or sensory ganglia to NGF in vitro resulted in a massive enhancement of axon outgrowth.[2] Subsequent experiments used subdivided tissue culture dishes to demonstrate that NGF-dependent neuronal survival was dependent upon uptake of NGF from nerve terminals and retrograde transport to the cell body.[7] Together, these findings established NGF as a target-derived factor with potent effects on neuronal survival and axon outgrowth.

Whilst it was identified by its actions on sensory and sympathetic neurons, NGF was later found to exert biological effects on defined populations of both peripheral and central neuronal types; in the latter category for example are cholinergic neurons of the basal forebrain.[8] This theme was developed and extended, with the purification of the related molecules, brain-derived neurotrophic factor (BDNF), neurotrophin-3 (NT-3), and neurotrophin-4/5 (NT-4/5). Together with NGF, these molecules form the neurotrophin family. Correspondingly, the Trk family of tyrosine kinase receptors, TrkA, TrkB and TrkC bind preferentially to NGF, BDNF/NT4/5 and NT-3 respectively In addition, all neurotrophins bind with low affinity to the common p75 neurotrophin receptor (p75NTR). The different neurotrophins are capable of supporting the survival of specific and (sometimes) overlapping groups of neurons.[9-11] Mice lacking the function of individual neurotrophins have deficits in the production of particular groups of neurons, and this is largely mirrored in the deficits in mice lacking the corresponding receptors. So, for example, mice mutant for *NGF/TrkA* display loss of sympathetic and nociceptive sensory neurons, whilst mutants for *BDNF/TrkB* lose vestibular and other subsets of sensory neurons, and *NT-3/TrkC* mutants lose proprioceptive sensory neurons. *TrkB* and *TrkC* mutants show an additional deficit in a subset of motor neurons.[9,12] A key aspect of neurotrophin function is thus to promote neuronal survival, regulating programmed cell death and matching neuronal numbers to target structures. However, a much broader range of activities have subsequently been revealed; neurotrophins are capable of regulating cell proliferation, differentiation, axon extension and branching, as well as influencing the regrowth and synaptic plasticity of adult neurons.[11,13]

The Extended Family of Neurotrophic Factors

At the time of the identification of NGF it could hardly have been suspected that the neurotrophins form a subset within a huge and diverse family of neurotrophic factors, possessing a multiplicity of influences in neural development. These include the glial cell line-derived neurotrophic factor (GDNF) family, the interleukin-6 (IL-6) family of cytokines, and hepatocyte growth factor (HGF). Although other molecules might be considered to belong under the general umbrella of neurotrophic factors, I shall restrict my discussion to these families together with the neurotrophins. GDNF was originally characterised as a trophic factor for midbrain dopaminergic neurons,[14] and was subsequently found to have wider effects on the survival of motor neurons,[15] peripheral sensory neurons, and autonomic and noradrenergic neurons.[16] The related molecules neurturin (NTN) and artemin (ART) also have survival effects on sympathetic and sensory neurons and dopaminergic neurons, while persephin (PSP) supports motor and dopaminergic neurons but not peripheral neurons.[17] GDNF family members signal via a receptor complex consisting of the Ret tyrosine kinase receptor and a glycosyl phosphatidylinositol (GPI)- anchored ligand binding-component (GFRα). GDNF, NRTN, ART and PSP bind preferentially to GFRα1, GFRα2, GFRα3 and GFRα4 respectively, although promiscuity of binding to alternative receptors can also occur. In concordance with the in vitro data on the actions of GDNF family members, there is evidence that mice deficient in GDNF ligands or receptors lose specific neuronal subtypes, namely peripheral sensory and autonomic neurons and motor neurons; for example, defects in the parasympathetic nervous system have been found in mice lacking either NTN or GFRα2.[18,19]

The IL-6 family of cytokines comprises ciliary neurotrophic factor (CNTF), leukaemia inhibitory factor (LIF), cardiotrophin-1 (CT-1) and interleukin-6 (IL-6); some members of the family act via a common receptor complex including the transmembrane proteins gp130 and LIFRβ. The CNTF and CT-1 receptors contain an additional, GPI-anchored subunit.[20] CNTF was originally named based on its survival effects on ciliary ganglion neurons, but can also support motor neurons in vitro.[21] However, CNTF mutant mice failed to show a significant loss of motor neurons, in contrast to a 50% loss of motor neurons in mice mutant for CNTFαR or LIFRβ.[22-24] CT-1 is the most potent motor neuron trophic factor in vitro[25] as well as having actions on other central and peripheral neuronal types. Mice lacking CT-1 show loss of spinal and subsets of cranial motor neurons.[26]

HGF was originally identified as a mitogen and a regulator of cell motility, which acts via the c-MET tyrosine kinase receptor. HGF is capable of inducing axon growth of embryonic motor, sensory, sympathetic and cortical neurons.[27-31]

Neurotrophic Factors Can Promote Axon Growth and Orient Axons in Vitro

There is a large body of evidence showing that neurotrophic factors applied in the medium can promote axon outgrowth; e.g., the effects of BDNF and NT-4 on retinal ganglion neurons.[32] However, the role of neurotrophic factors in maintaining neuronal survival presents a problem experimentally, since these effects may not be readily separable from those on axon growth and guidance. Since some types of primary neurons are initially neurotrophic factor-independent, it is also possible to test effects of these factors on axon outgrowth at early developmental timepoints. For example, young cranial motor neurons show an increase in axon outgrowth in response to HGF, BDNF, CNTF and CT-1 at times before standard survival assays are carried out.[33] Another strategy to separate survival and outgrowth effects in vitro is to overexpress the anti-apoptotic protein Bcl-2 in neurons, leading to neuronal survival in the absence of neurotrophic support. Retinal ganglion neurons which expressed Bcl-2 failed to extend axons unless BDNF or CNTF were not supplied,[34] thus implying that these factors also regulate axon extension. NTN and most strikingly ART (but not GDNF) promote axon outgrowth from embryonic sympathetic ganglia in vitro, and ART is also capable of chemoattracting these axons.[35] GDNF, and to a lesser effect NTN exerts a potent effect on the outgrowth of motor axons in postnatal rat spinal cords in vitro,[36] and GDNF supports survival of oculomotor neurons while PSP increases their axon outgrowth.[37]

The first evidence of NGF's chemoattractant role came from experiments in which chick dorsal root ganglion neurons turned and grew towards NGF applied focally from a pipette.[38,39] Presentation of NGF via diffusion from polystyrene beads also chemoattracts growth cones, in a manner which is dependent on the TrkA receptor.[40] Moreover, BDNF and NT-3 can both induce a chemoattractant response of *Xenopus* spinal neurons.[41] Rat cranial motor axons are chemoattracted by HGF presented on beads or transfected into cell clusters, and assumed to diffuse to form a gradient.[28]

Neurotrophic Factors and Axon Guidance Molecules Use Common Signalling Pathways

Dissection of the pathways by which neurotrophic factors influence neuronal growth has identified sets of molecules which overlap with those which mediate responses to axon guidance cues (see Fig. 1).[13,42] Therefore at the level of signal integration at least, neurotrophic factors have the potential to influence axon guidance. Key structures involved in growth and guidance are the actin filaments in the growth cone which interact dynamically with the ends of microtubules located in the axon shaft, and splaying out into the periphery of the growth

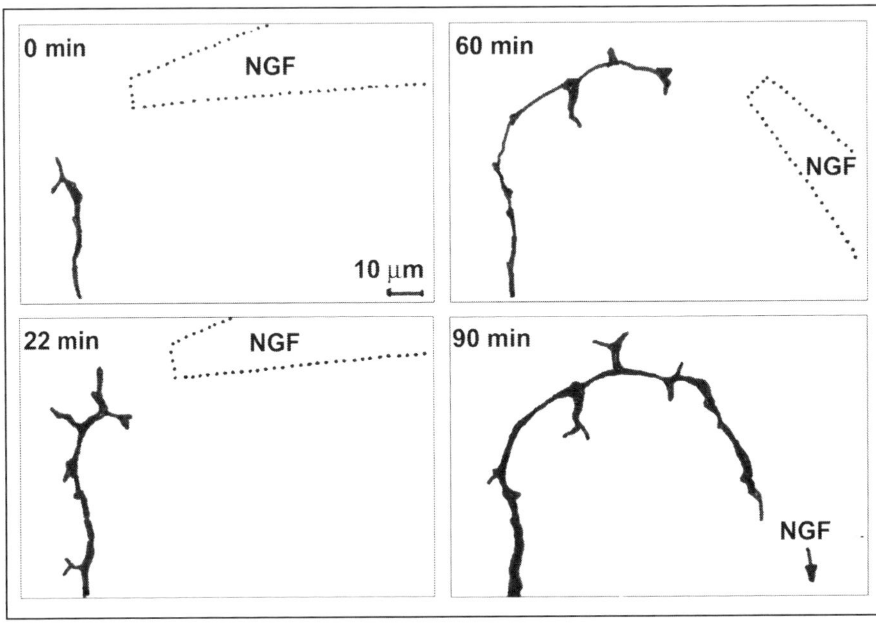

Figure 1. Chemoattractive turning of chick dorsal root ganglion axon in response to focal application of NGF. Time after application of NGF is indicated at top left; position of NGF pipette shown by dotted line. Adapted from Gundersen and Barrett (1979).[39]

cone. Extension, remodelling and turning of growth cones depends on dynamic instability of the cytoskeleton, including polymerisation/depolymerisation of actin, actomyosin contractility and microtubule polymerisation in the peripheral region of the growth cone, retrograde flow and actin depolymerisation in the central region of the growth cone, and actomyosin contractility. This 'treadmilling' determines the rate of protrusion or retraction of the growth cone leading edge, and is in turn dependent on the action of proteins such as profilin and cofilin, actin-binding proteins which control actin polymerisation. Local variations in these processes across the growth cone in turn determine turning, i.e., chemoattraction/chemorepulsion.

The signalling cascades by which neurotrophic factors regulate neuronal survival, differentiation and growth are best understood for the neurotrophins. Binding of neurotrophin ligands to Trks triggers receptor tyrosine phosphorylation at several locations, and docking of proteins containing SH2 or PTB domains to the receptor cytoplasmic domains, leading to the formation of signalling complexes. Among the downstream consequences of this process is the activation of three key pathways; the Ras-Raf-Mek-Erk cascade (also called the MAP kinase pathway), phosphatidylinositol-3-kinase (PI3K) pathway and phospholipase C gamma (PLC-γ) pathway.

The MAP kinases are known to have different downstream targets which mediate transcriptional and differentiation effects in cells. For example, Erk1, Erk2 and Erk5 phosphorylate the Rsk family, which together with MAPK-activated protein kinase 2 phosphorylate CREB, leading to survival and differentiation.[43,44] However, this pathway also appears to be involved in the axon-growth-promoting effects of the neurotrophins. For example, the effects of BDNF and NGF in promoting axon growth in sensory neurons have been shown to require the Ras-MAPK pathway.[45,46] The PI3K pathway regulates the protein kinase Akt, which is involved in cell survival pathways, but also activates other effectors including those which regulate the RhoA GTPase family, GSK3β and microtubule dynamics.[47] PI3K together with the

PTEN lipid phosphatase regulate levels of 3-phosphoinositides such PIP_2 and PIP_3 which interact with PH-domain containing proteins and in turn regulate further elements in downstream signalling cascades.[48] The PI3K pathway is involved in neurotrophin-dependent axon outgrowth, since sprouting of dorsal root ganglion neurons induced by NGF-coated beads is blocked by application of an inhibitor of PI3K, wortmannin.[49] Ligand binding to Trk receptors also activates PLC-γ, which converts PIP_2 to diacylglycerol (DAG) and inositol tri-phosphate (IP_3), leading to release of calcium from internal stores, with activation of Ca^{2+}-regulated isoforms of protein kinase C, and Ca^{2+} -calmodulin-regulated protein kinases including PKC-δ. The latter has been shown to be required for NGF-promoted neurite outgrowth from PC12 cells.[50]

Evidence is also accumulating that the neurotrophins regulate the activity of the Rho family of small GTPases, which are targets of axon guidance molecules and play a pivotal role in actin cytoskeleton rearrangements. Whilst Rac and Cdc42 are generally positive regulators of axon extension, RhoA is a negative regulator which causes collapse of growth cones; some exceptions to this have been reported however.[51] The chain of events leading from receptor activation to the Rho GTPases is not completely elucidated. Possible targets of Trk activation are RhoG, Rac and Cdc42, while RhoA is inactivated.[1,52] For example, BDNF activates Rac1 and Cdc42 in *Xenopus* spinal neurons.[47] In a differentiation model in PC12 cells, NGF signals through the PI3K pathway to activate Rac and to inactivate RhoA.[53] Another possible link is that activation of Trk receptors causes an elevation in cAMP, which increases PKA activity, inactivating RhoA, so that Rac1 and Cdc42 predominate, thereby favouring increased axonal growth and chemoattraction.[11]

There is currently less detailed information concerning the signalling pathways downstream of neurotrophic factors other than the neurotrophins. However, in general it appears that similar signalling cascades are activated. One recent study shows that mutation of two separate residues within RET linked respectively to the P13K and Rac1 pathways both inhibit lamellipodia formation in response to GDNF.[54] Receptor activation by IL-6 cytokines results in activation of signalling pathways including those involving PI3, MAPK and PLC-γ.[55] Activation of the gp130/LIFRβ receptor activation also leads to phosphorylation of Jak (Janus kinase) tyrosine kinases which are then docking sites for members of the STAT (signal transducer and activator of transcription) family of transcription factors.[20] HGF signalling depends on the presence within the Met receptor cytoplasmic domain of a multifunctional docking site for various SH2-domain containing signal transducers, including PI3K and PLC-γ.[56] Survival and growth effects of HGF on sympathetic neurons has been shown to depend on PI3K and MAP-kinase pathways.[57]

The Role of Signalling Pathways in Axonal Chemoattraction by Neurotrophins and Netrin-1

There are striking similarities in the signalling pathways mediating the effects of the axon guidance molecule Netrin-1 and neurotrophins. For example, as mentioned for BDNF, Netrin-1 can activate Rac1 and Cdc42 in cell lines,[58] and can stimulate the MAP kinase pathway in cell lines and during chemoattraction of commissural neurons.[59] In a turning assay utilising *Xenopus* spinal neurons to analyse the role of particular second messenger signalling pathways, the PI3K pathway and PLCγ were implicated in chemoattractive responses to both Netrin-1 and neurotrophins. Expression in *Xenopus* spinal neurons of separate TrkA forms which contained mutated tyrosines unable to mediate effects via PI3K or PLCγ abolished the turning response to NGF.[60] Netrin-1-induced turning was also abolished in normal spinal neurons in the presence of PI3K inhibitors, and activation of either the PI3K pathway or the PLCγ pathway alone via focal exposure to NGF was sufficient to abolish turning in the presence of Netrin-1.[60] Similarly, there was a dose-dependent cross-desensitisation of growth cones to Netrin-1 in response to BDNF, and activation of the MAPK pathway in response to Netrin-1 or BDNF was required for resensitisation to guidance cues.[61] These observations on the mode of action of Netrin-1 and the neurotrophins imply that the latter could be chemoattractant in vivo, but do not provide direct evidence for such a role.

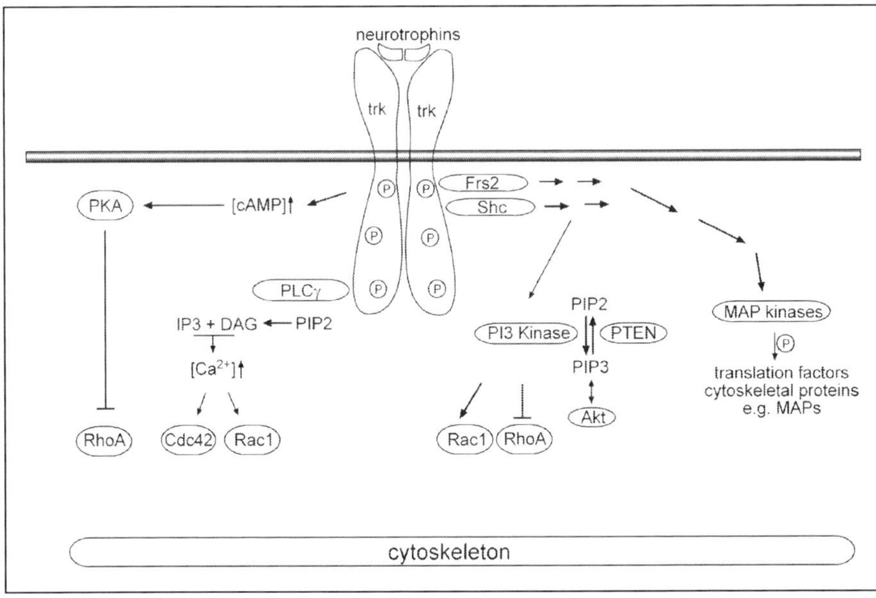

Figure 2. Speculative scheme of major neurotrophin-mediated signalling pathways.

Role of cAMP and Calcium Concentrations in Chemoattraction

Another compelling parallel between the operation of neurotrophins and more established families of axon guidance molecules is the apparent importance of ambient levels of cyclic nucleotides and calcium ions in determining a chemoattractant or chemorepellent response. NGF can cause an increase in intracellular cAMP[62] and the application of compounds which increase cAMP effected an attractive turning response of dorsal root ganglion axons.[63] Conversely, attractive responses to NGF, BDNF and Netrin-1 can be converted to repulsion by decreasing levels of cAMP.[64,65] Modulating levels of cyclic GMP can alter responses to NT-3 and Semaphorins.[65,66] Calcium levels within the growth cone also profoundly influence guidance behaviour, and are implicated in growth cone turning in response to local application of both Netrin-1 and BDNF.[67,68] Indeed, creation of an intracellular gradient of calcium or localised release of calcium is sufficient to trigger growth cone turning in vitro.[67,69] For netrin-mediated chemoattraction, a link has been established between cAMP levels and Calcium influx since cAMP directly modulates activity of L-type Calcium channels.[70] The PLC-γ pathway is one route capable of causing calcium release intracellularly, but these initial calcium increases are known to be amplified by ion influxes across the membrane. Recently, chemoattractive growth cone turning in response to both Netrin-1 and BDNF has been shown to utilise transient receptor potential (TRP) channels, which are activated in response to intracellular calcium release and depolarise cells sufficiently to activate voltage-dependent calcium channels, together allowing sufficient local calcium influx to cause growth cone turning.[71,72] There are multiple targets of calcium in growth cones which lead to chemoattractant turning (see previous section).[13] Cdc42 and Rac1 may also be upregulated in response to elevated calcium during chemoattractive growth cone turning.[68]

In view of these multiple levels of similarity between the action of the neurotrophins and other diffusible guidance cues, it appears that neurotrophic factors modulate responses to a number of axon guidance cues. For example, embryonic dorsal root ganglion neurons grown in the presence of BDNF show increased sensitivity to Sema3A compared with those grown in NGF.[73] NGF provides a dose-dependent protection against Sema3A-mediated growth cone collapse in embryonic sensory neurons.[74] Similarly, cross-talk occurs between positive and negative cues in sympathetic neurons in vitro, with Sema3F-induced growth cone collapse being attenuated by NGF-stimulation of PI3K and MAP kinase pathways.[75]

In Vivo Tests of Neurotrophic Factors in Axon Guidance

Despite the abundant evidence that neurotrophic factors guide axons in vitro, critical in vivo tests are required. These experiments are not so numerous, and I have also included some using explant cultures. As in vitro, an important constraint of 'knocking out' neurotrophic factors in vivo is that it leads to neuronal death, but this problem can be circumvented by eliminating components of the genetic pathways required for neuronal apoptosis. Thus, in *NGF* or *TrkA* mutant mice crossed with mice mutant for the *Bax* pro-apoptotic gene, sensory neurons (which would otherwise die) survive, and their peripheral (though not central) axonal processes are missing.[76] Separate studies showed that in these animals the majority of sympathetic target organs showed a reduced innervation, implying a role for NGF in this process.[77] In *BDNF/Bax* double mutant mice, cranial sensory neurons which were rescued from cell death navigated correctly to their targets, but showed local deficiencies in target innervation.[78] Overall, the axon growth defects in these animals were fairly local deficiencies in branching within the target region.

One interesting system is the innervation of the 'whisker field' by trigeminal sensory axons in the mouse embryo. Collagen gel cocultures, in which gradients of diffusible molecules can be established, were used to show that regions of the maxillary and mandibular primordia exerted a chemoattractant influence on embryonic trigeminal sensory neurons, producing a strong bias in the normally radial outgrowth from ganglion explants.[79,80] NGF did not appear to be responsible since anti-NGF antibodies did not affect the chemoattraction,[79] and expression studies later confirmed that NGF was not expressed in the target region until after the arrival of the sensory axons.[81] However, subsequent experiments showed the presence of BDNF and NT-3 in the target tissue at the time of axon outgrowth, while antibodies against these factors could completely eliminate the directed outgrowth of trigeminal sensory axons in vitro.[82] Maxillary tissue taken from either *BDNF* or *NT-3* mutant mice showed a reduced ability to promote directional outgrowth. However, in *BDNF/NT-3* double mutant embryos, the initial pattern of outgrowth of the trigeminal ganglion was normal, pointing to the existence of additional chemoattractants in this system.

One study which demonstrates that neurotrophic factors affect axon outgrowth and orientation utilised slice cultures of embryonic spinal cord and limb bud derived from a *tau-EGFP* reporter transgenic mouse line, in which sensory and motor projections were labelled.[83] When beads soaked in any of the neurotrophins (NGF, BDNF, NT-3 or NT4/5) were placed on slice cultures in the path of nerve growth, portions of the spinal nerves diverted their trajectories to grow towards the beads (Fig. 3). Retrograde labelling of the nerve branches at the bead labelled sensory neurons within the dorsal root ganglia, rather than motor neurons. However, the growth of mixed spinal nerves containing both sensory and motor axons was inhibited in proximity to beads carrying neutralising antibodies against the neurotrophins.[83] These blocking experiments suggest that neurotrophins play a role in nerve extension towards targets in vivo, as opposed to in the last stages of arbour formation. Expression of NGF and NT-3 in the periphery and the limb bud in particular are broadly consistent with a role for these molecules in guidance of sensory and motor branches.[84,85] The GDNF family of neurotrophic factors in particular has been implicated in growth and guidance of the autonomic nervous system.[86] In *RET* mutant embryos there are sympathetic axon outgrowth defects which may be due to the inability of axons to reach the blood vessels which act as their intermediate targets, en route to the sympathetic ganglia.[35] Mice mutant for *ART* or *GFRα3* showed defects in axon projections, while implantation of an artemin-soaked bead near to the sympathetic chain in mouse embryos followed by 24 hours in culture revealed an induction of axonal outgrowth (Fig. 4).[87] In a separate study, the projection of parasympathetic neurons of the ciliary ganglion was shown to

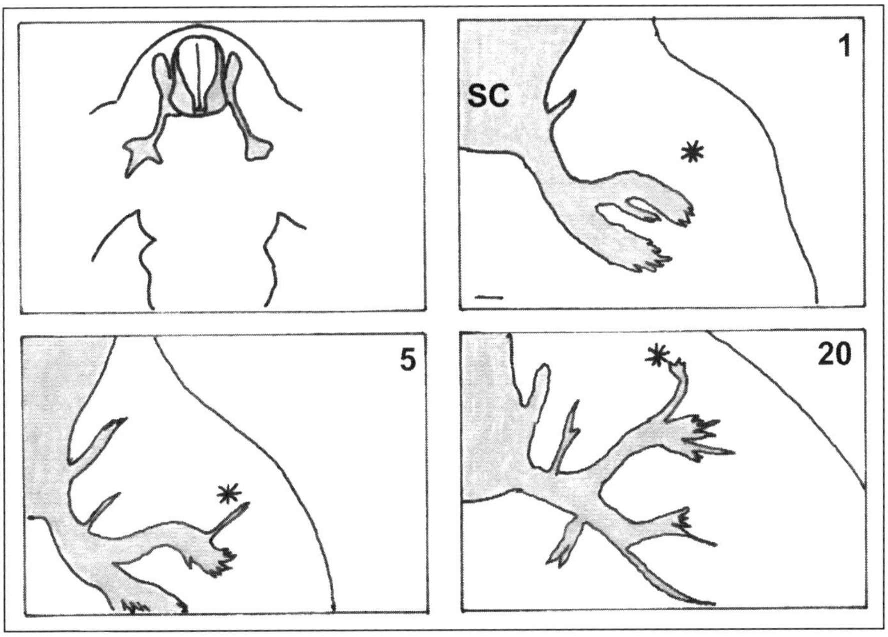

Figure 3. Growth of mouse motor and sensory axons in response to NGF-loaded beads in slice cultures. Shaded areas represent *tau-EGFP* expressing axon tracts. Top left panel shows section through mouse embryo expressing *tau-EGFP* in spinal cord and peripheral pathways. Top right panel and bottom panels show higher power views of axon growth at various timepoints after application of an NGF-loaded bead (asterisk). Adapted from Tucker et al (2001).[83]

depend on GDNF and NTN released from their target eye muscles, with blockade of GDNF using antibodies in chick embryos in vivo leading to a lack of axon projections from the ciliary ganglion.[88] In addition, there was an absence of parasympathetic nerve supply to the sublingual and lacrimal glands in mice lacking the function of *GFRα2*, implying a role for NTN in the innervation of these structures.[18]

Experiments in explant cultures and in vivo have shown that HGF acts as a chemoattractant for motor axons. Blocking experiments demonstrated that HGF is responsible for a portion of the chemoattractant activity of limb bud tissue and branchial arch tissue on spinal and cranial motor axons respectively.[28,89] Mice which were mutant for *HGF* showed defects in the branching patterns of spinal motor axons in the limbs and the hypoglossal motor nerve in the head.[28,89] The effects of HGF also extend to sensory neurons, since in mice lacking c-MET function, innervation of the skin by sensory axons is deficient and there are pathfinding errors within the hindlimb.[29]

Role of Neurotrophic Factors in Branching

Aside from a possible role in guiding axon projections, do neurotrophic factors influence the distinct phenomenon of branching? Many axons form branches, either during innervation of multiple targets, or in the final target area. In some cases, branching may be under distinct molecular control from the set of guidance cues required to reach the primary target. Although the molecular control of collateral branching is poorly understood, there is some evidence that neurotrophins play a role in this process. While ephrins and their Eph receptors are required for the correct targeting of retinal axons to particular regions of the optic tectum/superior colliculus,[90] a uniformly-expressed activity appears necessary to ensure branching occurs.

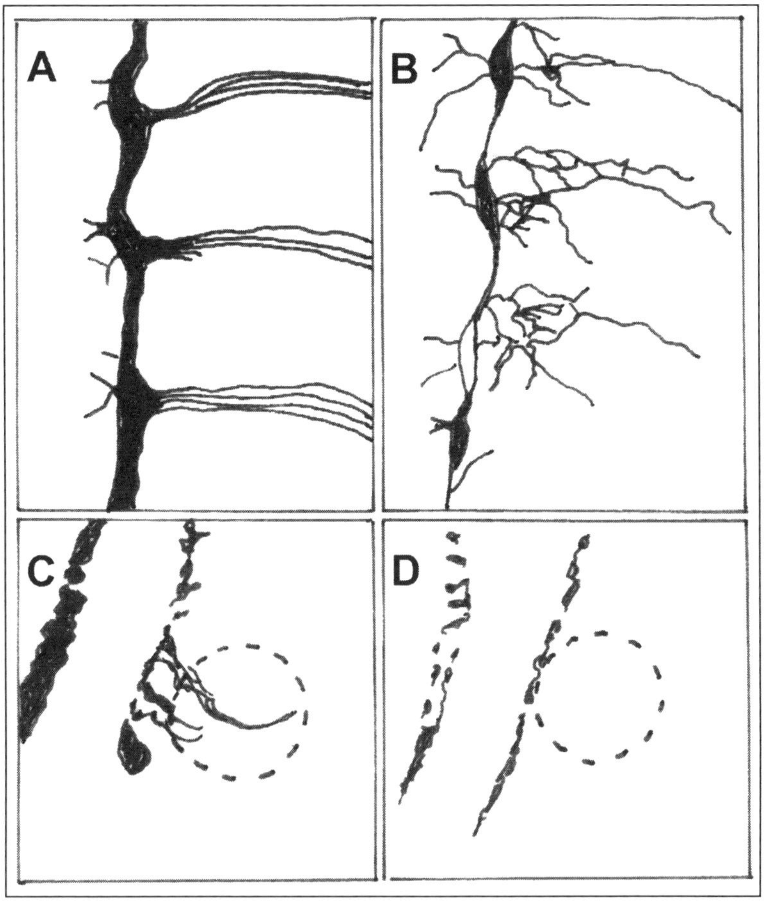

Figure 4. Growth of sympathetic axons in Artemin-deficient mice and in response to application of Artemin-loaded beads in vivo. A,B) show pathways of sympathetic axons in wild-type mice (A) and ART mutant mice (B). C,D) shows ectopic outgrowth of sympathetic axons in response to ART-loaded bead in a wild-type (C) and GFRα3-deficient mouse (D). Dotted line indicates position of the bead. Adapted from Honma et al (2002).[87]

Several studies have implicated BDNF as accounting for at least part of this branching activity. In the *Xenopus* visual system, BDNF and TrkB are expressed by the optic tectum and retinal neurons respectively, and injections of BDNF into the optic tectum in vivo increased retinal axon branching and arbour complexity, while injection of anti-BDNF antibodies had the opposite effect.[91-93] In vivo time-lapse imaging of retinal axons also demonstrated that axon arborisation was enhanced in the presence of microinjected BDNF.[94]

In vivo studies have also implicated neurotrophic factors in regulating axon branching. For example, in NT-3 deficient mice, sympathetic axons approach but fail to invade the pineal gland, but this process is rescued following infusion of NT-3.[95] NGF and GDNF influence the terminal arborisation of sensory/sympathetic axons and motor axons respectively.[96,97] In regeneration models in the adult rat following corticospinal axon transection, BDNF and NT-3 can cause enhanced collateral sprouting, either in the presence or the absence of foetal spinal cord transplants to provide permissive conditions for growth.[98,99]

In vitro, focal application of neurotrophins on beads or from pipettes has been used to show branching effects of NT3 and NGF on trigeminal and dorsal root ganglion sensory neurons.[49,100,101] BDNF application in vitro rapidly induces filopodia and lamellipodia along the axon shaft in *Xenopus* spinal axons, accompanied by increases in F-actin and local microtubule debundling.[102] The probable pathways whereby neurotrophins influence branching have been little studied, but point to the PI3K pathway, since NGF-responsive sensory neurons showed increased branching when transfected with Akt[46] and application of PI3K inhibitors blocked branching in response to NGF.[49] Akt can also increase axon diameter and cause branching, suggesting that distinct facets of neurotrophin signalling mediate different events in pathfinding and cytoskeletal remodelling.[46]

Conclusions

Are neurotrophic factors axon guidance molecules? There can be no question that neurotrophic factors are capable of guiding growth cones in vitro. They promote axon outgrowth under conditions in which their survival effects have been eliminated, and when applied focally can elicit chemoattractant growth cone turning. Neurotrophins in particular share signalling pathways with axon guidance molecules including Netrin-1, and regulate the same key intracellular molecules such as the Rho GTPases. Nevertheless, it remains to be determined whether the details of how neurotrophic factors and axon guidance molecules chemoattract axons are the same. The major weakness in the argument that neurotrophic factors are axon guidance molecules comes in discussing the in vivo data. A strong case is emerging that the GDNF family regulates axon growth/guidance of sympathetic and parasympathetic neurons, while NGF is involved with the growth of peripheral branches of sensory neurons. However, the majority of the evidence, arguably, points to predominantly short-range effects of neurotrophic factors in the later stages of axon branching and arborisation at the target. As yet, there is little evidence for long-range action, and in studies such as that on the trigeminal/branchial arch system, neurotrophic factors do not appear to be the primary guidance cues. Nor is there evidence that neurotrophic factors form gradients in vivo, but this has been extremely difficult to demonstrate for even the most intensively-researched guidance molecules. There are also difficulties in extrapolating from in vitro to in vivo systems, and demonstrations that axons turn in a gradient of NGF in vitro are hard to interpret in in vivo terms. Interestingly, recent studies which modelled axon behaviour in a strictly controlled gradient of NGF demonstrated that axons 10 μm wide could orient within a gradient of 0.1% at 1nM, implying much higher sensitivity to gradient steepness than has hitherto been suspected.[103] In summary, current evidence suggests that neurotrophic factors may act at shorter range than axon guidance molecules and may collaborate with them to modify axonal behaviour. Since it is highly likely that axons are exposed to established axon guidance cues in conjunction with neurotrophic factors, it will be important but extremely challenging to evaluate the range of interactions which occur.

Acknowledgements

Many thanks to Uwe Drescher for extensive discussions and for making (Fig. 2). Thanks also to Britta Eickholt and Andrew Lumsden for critical comments on the manuscript.

References

1. Huber AB, Kolodkin AL, Ginty DD et al. Signalling at the growth cone: Ligand-receptor complexes and the control of axon growth and guidance. Annu Rev Neurosci 2003; 26:509-563.
2. Levi-Montalcini R. The nerve growth factor 35 years later. Science 1987; 237:1154-1162.
3. Hamburger V. The history of the discovery of the nerve growth factor. J Neurobiol 1993; 24:893-897.
4. Cowan WM. Viktor Hamburger and Rita Levi-Montalcini: The path to the discovery of nerve growth factor. Annu Rev Neurosci 2001; 24:551-600.
5. Levi-Montalcini R. The morphological effects of immunosympathectomy. In: Steiner G, Schonbaum E, eds. Immunosympathectomy. Amsterdam: Elsevier, 1972:55-78.

6. Menesini Chen MG, Chen JS, Levi-Montalcini R. Sympathetic nerve fibers ingrowth in the central nervous system of neonatal rodent upon intracerebral NGF injections. J Cell Biol 1978; 87:546-554.
7. Campenot RB. Regeneration of neurites in long-term cultures of sympathetic neurons deprived of nerve growth factor. Science 1981; 214:579-581.
8. Hefti F. Nerve growth factor promotes survival of septal cholinergic neurons after fimbrial transactions. J Neurosci 1986; 6:2155-2162.
9. Klein R. Role of neurotrophins in mouse neuronal development. FASEB J 1994; 8:738-744.
10. Henderson CE. Role of neurotrophic factor in neuronal development. Curr Opin Neurobiol 1996; 6:64-70.
11. Gillespie LN. Regulation of axonal growth and guidance by the neurotrophin family of neurotrophic factors. Clin and Exp Pharmacol and Physiol 2003; 30:724-733.
12. Snider WD. Function of the neurotrophins during nervous system development: What the knockouts are teaching us. Cell 1994; 77:627-638.
13. Huang EJ, Reichardt LF. Trk receptors: Roles in neuronal signal transduction. Annu Rev Biochem 2003; 72:609-642.
14. Lin LF, Doherty DH, Lile JD et al. GDNF: A glial line-derived neurotrophic factor for midbrain dopaminergic neurons. Science 1993; 260:1130-1132.
15. Henderson CE, Phillips HS, Pollock RA et al. GDNF: A potent survival factor for motoneurons present in peripheral nerve and muscle. Science 1994; 266:1062-4.
16. Airaksinen MS, Saarma M. The GDNF family: Signalling, biological functions and therapeutic value. Nat Rev Neurosci 2002; 3:383-394.
17. Sariola H, Saarma M. Novel functions and signalling pathways for GDNF. J Cell Sci 2003; 116:3855-3862.
18. Rossi J, Luukko K, Poteryaev D et al. Retarded growth and deficits in the enteric and parasympathetic nervous system in mice lacking GFR alpha2, a functional neurturin receptor. Neuron 1999; 22:243-252.
19. Heuckeroth RO, Enomoto H, Grider JR et al. Gene targeting reveals a critical role for neurturin in the development and maintenance of enteric, sensory, and parasympathetic neurons. Neuron 1999; 22:253-263.
20. Stahl N, Yancopoulos GD. The tripartite CNTF receptor complex: Activation and signalling involves components shared with other cytokines. J Neurobiol 1994; 25:1454-1466.
21. Arakawa Y, Sendtner M, Thoenen H. Survival effect of ciliary neurotrophic factor (CNTF) on chick embryonic motoneurons in culture: Comparison with other neurotrophic factors and cytokines. J Neurosci 1990; 11:3507-3515.
22. Masu Y, Wolf E, Holtmann B et al. Disruption of the CNTF gene results in motor neuron degeneration. Nature 1993; 365:27-32.
23. DeChiara TM, Vejsada R, Poueymirou WT et al. Mice lacking the CNTF receptor, unlike mice lacking CNTF, exhibit profound motor neuron deficits at birth. Cell 1995; 83:313-322.
24. Li M, Sendtner M, Smith A. Essential function of LIF receptor in motor neurons. Nature 1995; 378:724-727.
25. Pennica D, Arce V, Swanson TA et al. Cardiotrophin-1, a cytokine present in embryonic muscle, supports long-term survival of spinal motorneurons. Neuron 1996; 17:63-74.
26. Oppenheim RW, Wiese S, Prevette D et al. Cardiotrophin-1, a muscle-derived cytokine, is required for the survival of subpopulations of developoing motorneurons. J Neurosci 2001; 21:1283-1291.
27. Yamamoto Y, Livet J, Pollock RA et al. Hepatocyte growth factor (HGF/SF) is a muscle-derived survival factor for a subpopulation of embryonic motoneurons. Development 1997; 124:2903-2913.
28. Caton A, Hacker A, Naeem A et al. The branchial arches and HGF are growth-promoting and chemoattractant for cranial motor axons. Development 2000; 127:1751-1766.
29. Maina F, Hilton MC, Ponzetto C et al. Met receptor signalling is required for sensory nerve development and HGF promotes axonal growth and survival of sensory neurons. Genes Dev 1997; 11:3341-3350.
30. Maina F, Hilton MC, Andres R et al. Multiple roles for hepatocyte growth factor in sympathetic neuron development. Neuron 1998; 20:835-846.
31. Hamanoue M, Takemoto N, Matsumoto K et al. Neurotrophic effect of hepatocyte growth factor on central nervous system neurons in vitro. J Neurosci Res 1996; 43:554-564.
32. Bosco A, Linden R. BDNF and NT-4 differentially modulate neurite outgrowth in developing retinal ganglion cells. J Neurosci Res 1999; 57:759-769.
33. Naeem A, Abbas L, Guthrie S. Comparison of the effects of HGF, BDNF, CT-1, CNTF, and the branchial arches on the growth of embryonic cranial motor neurons. J Neurobiol 2002; 51:101-114.

34. Goldberg JL, Espinosa JS, Xu Y et al. Retinal ganglion cells do not extend axons by default: Promotion by neurotrophic signalling and electrical activity. Neuron 2002; 33:689-702.
35. Enomoto H, Crawford PA, Gorodinsky A et al. RET signalling is essential for migration, axonal growth and axon guidance of developing sympathetic neurons. Development 2001; 128:3963-3974.
36. Ho TW, Bristol LA, Coccia C et al. TGFB trophic factors differentially modulate motor axon outgrowth and protection from excitotoxicity. Exp Neurol 2000; 161:664-675.
37. Chen J, Butowt R, Rind HB et al. GDNF increases the survival of developing oculomotor neurons through a target-derived mechanism. Mol Cell Neurosci 2003; 24:41-56.
38. Letourneau P. Chemotactic response of nerve fiber elongation to nerve growth factor. Dev Biol 1978; 66:183-196.
39. Gundersen RW, Barrett JN. Neuronal chemotaxis: Chick dorsal root axons turn towards high concentrations of nerve growth factor. Science 1979; 206:1079-1080.
40. Gallo G, Lefcort FB, Letourneau PC. The TrkA receptor mediates growth cone turning toward a localized source of nerve growth factor. J Neurosci 1997; 17:5445-5554.
41. Ming G, Lohof AM, Zheng JQ. Acute morphogenic and chemotropic effects of neurotrophins on cultured embryonic Xenopus spinal neurons. J Neurosci 1997a; 17:7860-7871.
42. Markus A, Patel TD, Snider WD. Neurotrophic factors and axonal growth. Curr Opin in Neurobiol 2002a; 12:523-531.
43. Lonze BE, Riccio A, Cohen S et al. Apoptosis, axonal growth defects, and degeneration of peripheral neurons in mice lacking CREB. Neuron 2002; 34:371-385.
44. Riccio A, Ahn S, Davenport CM et al. Mediation by a CREB family transcription factor of NGF-dependent survival of sympathetic neurons. Science 1999; 286:2358-2361.
45. Atwal JK, Massie B, Miller FD et al. The TrkB-Shc site signals neuronal survival and local axon growth via MEK and PI3-kinase. Neuron 2000; 27:265-277.
46. Markus A, Zhong J, Snider WD. Raf and Akt mediate distinct aspects of sensory axon growth. Neuron 2002b; 35:65-76.
47. Yuan X, Jin M, Xu X et al. Signalling and crosstalk of Rho GTPases in mediating axon guidance. Nature Cell Biol 2003; 5:1-8.
48. Comer FI, Parent CA. PI 3-kinases and PTEN: How opposites chemoattract. Cell 2002; 109:541-544.
49. Gallo G, Letourneau PC. Localized sources of neurotophins initiate axon collateral sprouting. J Neurosci 1998; 18:5403-5414.
50. Corbit KC, Soh JW, Yoshida K et al. Different protein kinase C isoforms determine growth factor specificity in neuronal cells. Mol Cell Biol 2000; 20:5392-5403.
51. Dickson BJ. Rho GTPases in growth and guidance. Curr Opin Neurobiol 2001; 11:103-110.
52. Mueller BK. Growth Cone Guidance: First steps to a deeper understanding. Annu Rev Neurosci 1999; 22:351-388.
53. Nusser N, Gosmanova E, Zheng Y et al. Nerve growth factor signals through TrkA, phosphatidylinositol 3-kinase, and Rac1 to inactivate RhoA during the initiation of neuronal differentiation of PC12 cells. J Biol Chem 2002; 277:35840-35846.
54. Fukuda T, Kiuchi K, Takahashi M. Novel mechanism of regulation of Rac activity and lamellipodia formation by RET tyrosine kinase. J Biol Chem 2002; 277:19114-19121.
55. Dolcet X, Soler RM, Gould TW et al. Cytokines promote motorneuron survival through the janus kinase-dependent activation of the phosphatidylinositol 3-kinase pathway. Mol Cell Neurosci 2001; 18:619-631.
56. Ponzetto C, Bardelli A, Zhen Z et al. A multifunctional docking site mediates signaling and transformation by the hepatocyte growth factor/scatter factor receptor family. Cell 1994; 77:261-271.
57. Thompson J, Dolcet X, Hilton M et al. HGF promotes survival and growth of maturing sympathetic neurons by PI-3 kinase and MAP kinase-dependent mechanisms. Mol Cell Neurosci 2004; 27:441-452.
58. Shekarabi M, Kennedy TE. The Netrin-1 receptor DCC promotes filopodia formation and cell spreading by activating Cdc42 and Rac1. Mol Cell Neurosci 2002; 19:1-17.
59. Forcet C, Stein E, Pays L et al. Netrin-1 mediated axon outgrowth requires deleted in colorectal cancer-dependent MAPK activation. Nature 2002; 417:443-447.
60. Ming G, Song H, Berninger B et al. Phospholipase C-gamma and phosphoinositide 3-kinase mediare cytoplasmic signalling in nerve growth cone guidance. Neuron 1999; 23:139-148.
61. Ming G, Wong S, Henley J et al. Adaptation in the chemotactic guidance of nerve growth cones. Nature 2002; 417:411-418.
62. Hier DB, Arnason BG, Young M. Studies on the mechanism of action of nerve growth factor. Proc Natl Acad Sci 1972; 69:2268-2272.

63. Gundersen RW, Barrett JN. Characterization of the turning response of dorsal root neurites towards nerve growth factor. J Cell Biol 1980; 86:546-554.
64. Ming G, Song H, Berninger B et al. cAMP-dependent growth cone guidance by netrin-1. Neuron 1997b; 19:1225-1235.
65. Song H, Ming G, Poo M. cAMP-induced switching in turning direction of nerve growth cones. Nature 1997; 388:275-279.
66. Song H, He Z, Lehmann M et al. Conversion of neuronal growth cone responses from repulsion to attraction by cyclic nucleotides. Science 1998; 281:1465-1466.
67. Hong K, Nishiyama M, Henley J et al. Calcium signalling in the guidance of nerve growth by netrin-1. Nature 2000; 403:93-98.
68. Jin M, Guan CB, Jiang YA et al. Ca^{2+}-dependent regulation of RhoGTPases triggers turning of nerve growth cones. J Neurosci 2005; 25:2338-2347.
69. Wen Z, Guirland C, Ming GL et al. A CaMKII/calcineurin switch controls the direction of Ca(2+)-dependent growth cone guidance. Neuron 2004; 43:760-762.
70. Nishiyama M, Hoshino A, Tsai L et al. Cyclic AMP/GMP-dependent modulation of Ca2+ channels sets the polarity of nerve growth-cone turning. Nature 2003; 423:990-995.
71. Li Y, Jia YC, Cui K et al. Essential role of TRPC chennels in the guidance of nerve growth cones by brain-derived neurotrophic factor. Nature 2005; 434:894-897.
72. Wang GX, Poo MM. Requirement of TRPC channels in netrin-1 induced chemotropic turning of nerve growth cones. Nature 2005; 434:835-904.
73. Tuttle R, O'Leary DDM. Neurotrophins rapidly modulate growth cone response to the axon guidance molecule, collapsing-1. Mol Cell Neurosci 1998; 11:1-8.
74. Dontchev VD, Letourneau PC. Nerve growth factor and semaphoring 3A signalling pathways interact in regulating sensory neuronal growth cone motility. J Neurosci 2002; 22:6659-6669.
75. Atwal JK, Singh KK, Tessier-Lavigne M et al. Semaphorin 3F anatagonizes neurotrophin-induced phosphatidylinositol 3-kinase and mitogen-activated protein kinase kinase signalling: A mechanisms for growth cone collapse. J Neurosci 2003; 23:7602-7609.
76. Patel HD, Jackman A, Rice FL et al. Development of sensory neurons in the absence of NGF/TrkA signalling in vivo. Neuron 2000; 25:345-357.
77. Glebova NO, Ginty DD. Heterogeneous requirement for NGF for sympathetic target innervation in vivo. J Neurosci 2004; 24:743-751.
78. Hellard D, Brosenitsch T, Fritzsch B et al. Cranial sensory neuron development in the absence of brain-derived neurotrophic factor in BDNF/Bax double null mice. Devl Biol 2004; 275:34-43.
79. Lumsden AGS, Davies AM. Earliest sensory nerve fibers are guided to peripheral targets by attractants other than NGF. Nature 1983; 306:786-788.
80. Lumsden AGS, Davies AM. Chemotropic effect of specific targe epithelium in the developing mammalian nervous system. Nature 1986; 323:538-539.
81. Davies AM, Bandtlow C, Heumann R et al. Timing and site of nerve growth factor synthesis in developing skin I relation to innervation and expression of the receptor. Nature 1987; 326:353-358.
82. O'Connor, Tessier-Lavigne M. Identification of maxillary factor, a maxillary process-derived chemoattractant for developing trigeminal sensory axons. Neuron 1999; 24:265-178.
83. Tucker KL, Meyer M, Barde YA. Neurotrophins are required for nerve growth during development. Nat Neurosci 2001; 4:29-37.
84. White FA, Silos-Santiago I, Molliver DC et al. Synchronous onset of NGF and TrkA survival dependence in developing dorsal root ganglia. J Neurosci 1996; 16:4662-4672.
85. Farinas I, Yoshida CK, Backus C et al. Lack of neurotrophin-3 results in death of spinal sensory neurons and premature differentiation of their precursors. Neuron 1996; 17:1065-1078.
86. Young HM, Anderson RB, Anderson CR. Guidance cues involved in the development of the peripheral autonomic nervous system. Autonomic Neuroscience: Basic and Clinical 2004; 112:1-14.
87. Honma Y, Araki T, Gianino S et al. Artemin is a vascular-derived neurotrophic factor for developing sympathetic neurons. Neuron 2002; 35:267-282.
88. Hashino E, Shero M, Junghans D et al. GDNF and neurturin are target-derived factors essential for cranial parasympathetic neuron development. Development 2001; 128:3773-3782.
89. Ebens A, Brose K, Leonardo ED et al. Hepatocyte growth factor/scatter factor is an axonal chemoattractant and a neurotrophic factor for spinal motor neurons. Cell 1996; 82:359-370.
90. Mann F, Harris WA, Holt CE. New views on retinal axon development: A navigation guide. Int J Dev Biol 2004; 48:957-64.
91. Cohen-Cory S, Escandon E, Fraser SE. The cellular patterns of BDNF and trkB expression suggest multiple roles for BDNF during Xenopus visual system development. Dev Biol 1996; 179:102-115.

92. Cohen-Cory S, Fraser SE. Effects of brain-derived neurotrophic factor on optic axon branching and remodelling in vivo. Nature 1995; 378:192-196.
93. Lom B, Cohen-Cory S. Brain-derived neurotropic factor differentially regulates retinal ganglion cell dendritic and axonal arborisation in vivo. J Neurosci 1999; 19:9928-9938.
94. Alsina B, Vu T, Cohen-Cory S. Visualizing synapse formation in arborizing optic axons in vivo: Dynamics and modulation by BDNF. Nat Neurosci 2001; 4:1093-1101.
95. El Shamy WM, Linnarsson S, Lee KF et al. Prenatal and postnatal requiremens of NT-3 for sympathetic neuroblast survival and innervation of specific targets. Development 1996; 122:491-500.
96. Kennedy TE, Tessier-Lavigne M. Guidance and induction of branch formation in developing axons by target-derived diffusible factors. Curr Opin Neurobiol 1995; 5:83-90.
97. Keller-Peck CR, Feng G, Sanes JR et al. Glial cell line-derived neurotrophic factor administration in postnatal life results in motor unit enlargement and continuous synaptic remodelling at the neuromuscular junction. J Neurosci 2001; 21:6136-6146.
98. Schnell L, Schneider R, Kolbeck R et al. Neurotrophin-3 enhances sprouting of corticospinbal tract during development and after adult spinal cord lesion. Nature 1994; 367:112-113.
99. Coumans JV, Lin TT, Dai HN et al. Axonal regeneration and functional recovery after complete spinal cord transaction in rats by delayed treatment with transplants and neurotrophins. J Neurosci 2001; 21:9334-9344.
100. Ulupinar E, Jacquin MF, Erzurumlu RS. Differential effects of NGF and NT-3 on embryonic trigeminal axon growth patterns. J Comp Neurol 2000; 425:202-218.
101. Ozdinler PH, Ulupinar E, Erzurumlu RS. Local neurotrophin effects on central trigeminal axon growth patterns. Brain Res Dev 2004; 151:55-66.
102. Gibney J, Zheng JQ. Cytoskeletal dynamics underlying collateral membrane protrusions induced by neurotrophins in cultured Xenopus embryonic neurons. J Neurobiol 2003; 54:393-405.
103. Rossoff WJ, Urbach JS, Esrick MA et al. A new chemotaxis assay shows the extreme sensitivity of axons to molecular gradients. Nat Neurosci 2004; 7:678-682.

Chapter 7

The Role of Cell Adhesion Molecules in Axon Growth and Guidance

Hiroyuki Kamiguchi*

Abstract

During development, axons elongate along the correct path toward their final targets. Growing axons maintain adhesive interactions with specific environmental cues via cell adhesion molecules (CAMs). The axon-environment adhesion must be dynamically controlled, both temporally and spatially, to enable the axons to navigate and migrate correctly. In this way, CAMs play a central role in mediating contact-dependent regulation of motile behavior of the axons. This chapter examines the mechanisms underlying how CAMs control axon growth and guidance, with a particular focus on intracellular signaling, trafficking, and interactions with the actin cytoskeleton.

Introduction

So far, three major classes of cell adhesion molecules (CAMs) have been identified in the nervous system: integrins, cadherins, and the immunoglobulin (Ig) superfamily. During development, CAMs play critical roles not only in static cell adhesion but also in dynamic cellular events such as cell migration and process outgrowth. For example, NCAM, a CAM in the Ig superfamily, mediates side-to-side adhesion of axons (fasciculation)[1] and controls axon guidance via a mechanism called selective fasciculation.[2] Another member in the Ig superfamily, L1, is a homophilic adhesion molecule that stimulates axon growth.[3-5] Integrins have been identified as a neuronal cell-surface receptor that promotes axon elongation on extracellular matrix (ECM) molecules such as laminin and fibronectin.[6,7] Furthermore, cadherins mediate cell-cell adhesion and regulate axon elongation via a homophilic binding mechanism.[8,9] In this way, CAMs on the neuronal surface mediate the interaction between the axon and its environment (neighboring cells or ECM molecules) and direct axon growth along the correct path.[10] However, more recent work revealed that CAMs control neuronal functions in a very complex manner.[11] 1) CAMs have multiple binding partners and form homophilic as well as heterophilic complexes, (2) CAM interactions can occur both in trans (between neighboring cells) and in cis (in the plane of the plasma membrane of one cell), (3) CAMs generate intracellular signals and interact dynamically with the cytoskeleton, (4) CAMs move on the cell surface and inside the cell via vesicular transport, (5) CAM functions can be modified by lipid microenvironment in the cell membrane, and (6) CAMs undergo proteolytic cleavage. Therefore, CAMs can be viewed as adhesive/signaling/trafficking molecules that transmit chemical and physical information across the plasma membrane in both outside-in and inside-out directions. In this

*Hiroyuki Kamiguchi, Laboratory for Neuronal Growth Mechanisms, RIKEN Brain Science Institute, 2-1 Hirosawa, Wako, Saitama 351-0198, Japan. Email: kamiguchi@brain.riken.jp

Axon Growth and Guidance, edited by Dominique Bagnard. ©2007 Landes Bioscience and Springer Science+Business Media.

chapter, I will describe the molecular mechanisms of axon growth and guidance, with a particular focus on the role of NCAM and the L1 subfamily CAMs.

Molecular Structure

Ig superfamily CAMs are either transmembrane or GPI-anchored proteins, with their extracellular region consisting of Ig-like domains and fibronectin type III repeats (Fig. 1). NCAM has three major isoforms in the nervous system, GPI-anchored NCAM (120 kD) and transmembrane NCAM (140 and 180 kD), which are translated from a single gene via alternative splicing.[12] NCAM140 and NCAM180 are present on neurons, whereas glial cells express mainly NCAM140 and myotubes express NCAM120. The NCAM extracellular domain carries carbohydrate chains of varying length of α2-8 linked sialic acid (polysialic acid) that serve as a negative modulator of cell-cell adhesion.[13] Consequently, polysialic acid attenuates axon fasciculation and thereby allows axon reorganization that is important for the proper formation of neuronal projections.[14]

Other important Ig CAMs are L1/NgCAM and its subfamily members (NrCAM, neurofascin, close homolog of L1). L1 is a single-pass transmembrane protein with its extracellular region consisting of Ig-like domains and fibronectin type III repeats.[15] The L1 cytoplasmic domain contains an alternatively spliced sequence, RSLE, which is expressed in neurons but not in the other L1-expressing cells such as Schwann cells and lymphocytes.[16] Members in the TAG-1/axonin-1 family (TAX-1/TAG-1/axonin-1, F3/F11/contactin, BIG-1, BIG-2, NB-2, NB-3) are GPI-anchored CAMs.[17,18] The L1 family members cooperate with the TAG-1/axonin-1 family members to regulate axon growth and guidance, in which CAMs in both families form a hetero-multimeric complex through trans- and *cis*-binding.[19,20] Mutational analyses of the L1 extracellular domain demonstrated that the regions for L1 homophilic and heterophilic binding span multiple Ig and fibronectin domains.[21,22] Furthermore, structural studies of NCAM[23] and axonin-1[24] showed that multiple Ig domains are involved in trans-binding.

More recently, small Ig superfamily molecules, either secreted or membrane bound via a transmembrane region or a GPI anchor, have been identified. The *Drosophila* Beat proteins regulate axon fasciculation,[25] and the *C. elegans* ZIG proteins are involved in the maintenance of axonal positioning.[26] For example, Beat Ia is a secreted protein with two Ig domains, which appears to function as an anti-adhesive factor leading to selective defasciculation at defined choice points.[27] It is likely that these molecules play a role in axon tract development by regulating, rather than mediating, axon adhesion.

Biophysical and Signaling Mechanisms

Intracellular signaling pathways downstream of CAMs have been extensively studied in vitro (Fig. 2). Doherty and Walsh and their colleagues performed a series of experiments that led to a model for CAM-mediated axon growth.[28,29] In response to homophilic binding, N-cadherin and two Ig superfamily CAMs, L1 and NCAM, activate the fibroblast growth factor receptor (FGFr) that subsequently activates phospholipase Cγ to generate diacylglycerol (DAG). DAG is hydrolyzed by DAG lipase to arachidonic acid that increases localized Ca^{2+} influx through N- and L-type channels followed by activation of the Ca^{2+}/calmodulin-dependent kinase.

Studies by several groups have focused attention on the involvement of mitogen-activated protein kinase (MAPK) pathway in CAM-stimulated neurite growth. Crosslinking NCAM on the cell surface activates the MAPK pathway, which is needed for NCAM-stimulated neurite growth.[30,31] A model of NCAM signaling has been proposed that involves two distinct cascades converging on the MAPK pathway: NCAM-Ras-MAPK and NCAM-FGFr-phospholipase Cγ-protein kinase C-Raf-MAPK (Fig. 2). Similarly, L1-mediated neurite growth requires the MAPK activity.[32] Crosslinking L1 activates the MAPK pathway,[33] which involves Src-dependent clathrin-mediated endocytosis of L1, phosphatidylinositol 3-kinase and the small GTP-binding

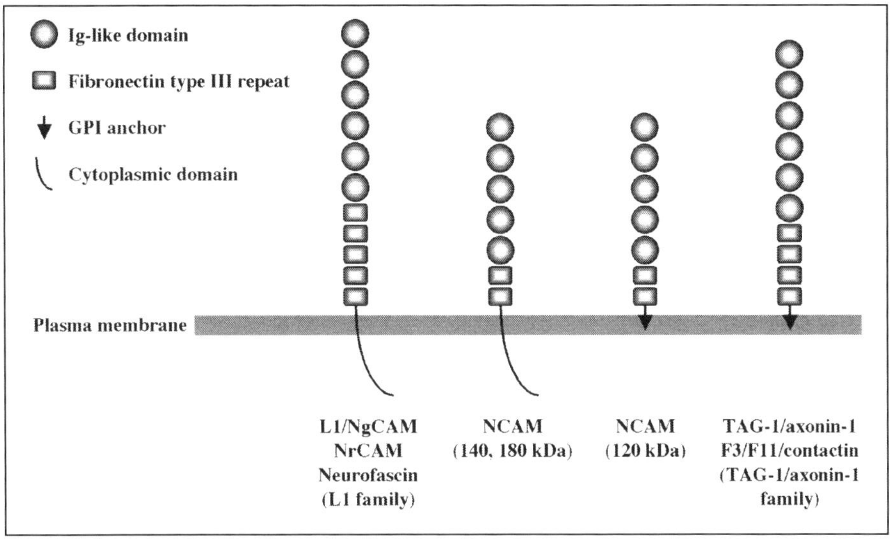

Figure 1. The structure of Ig superfamily CAMs.

protein Rac.[32] Interestingly, the endocytosed L1 colocalizes with the activated MAPK in endosomes,[33] suggesting that membrane trafficking plays a role in localizing CAM-generated signals to specific intracellular compartments. Although signaling pathways for neurite growth on several different substrates converge at MAPK activation,[34] an important question remains to be answered as to what downstream effectors are involved and how the MAPK pathway promotes neurite growth.

CAM signaling has been further complicated by the presence of membrane microdomains. Proteins and lipids are not uniformly distributed in the cell membrane but form spatially differentiated microdomains such as lipid rafts. L1 and N-cadherin are localized to rafts and nonraft membranes whereas β1 integrin is expressed only in nonraft areas. Lipid rafts in growth cones, especially in their peripheral (P) domain, are required for axon growth mediated by L1 and N-cadherin but not by β1 integrin.[35] This observation suggests the importance of CAM-associated signals that are controlled by lipid rafts in the growth cone periphery. Such a mechanism may exist in NCAM signaling. Upon homophilic ligation, NCAM140 triggers distinct signaling cascades through rafts and nonraft membranes.[36] NCAM140 phosphorylates the focal adhesion kinase most likely via the nonreceptor kinase Fyn in rafts, while nonraft NCAM140 facilitates FGFr-activated downstream signals. Both pathways merge in the activation of MAPK (see also Fig. 2) and are necessary for NCAM140 to stimulate neurite growth.

In the growth cone P-domain, filamentous actin (F-actin) continuously moves toward the central (C) domain as a result of spatially localized actin polymerization/depolymerization and actin-myosin interactions. The retrograde movement of F-actin produces a traction force that pulls the growth cone forward.[37,38] The force can be transmitted to extracellular environment by CAMs that link the retrograde F-actin flow with extracellular immobile ligands.[39] The CAM-actin linkage is mediated by a cytoplasmic linker (a molecular clutch) that plays a crucial role in the spatial and temporal regulation of force transmission.[40,41] Ankyrin_B has been identified as a component of the clutch module that mediates L1 coupling with retrograde F-actin flow and promotes the initial formation of neuritis.[42] Understanding growth cone biophysics requires further identification of clutch components and the regulatory mechanisms of clutch engagement/disengagement.

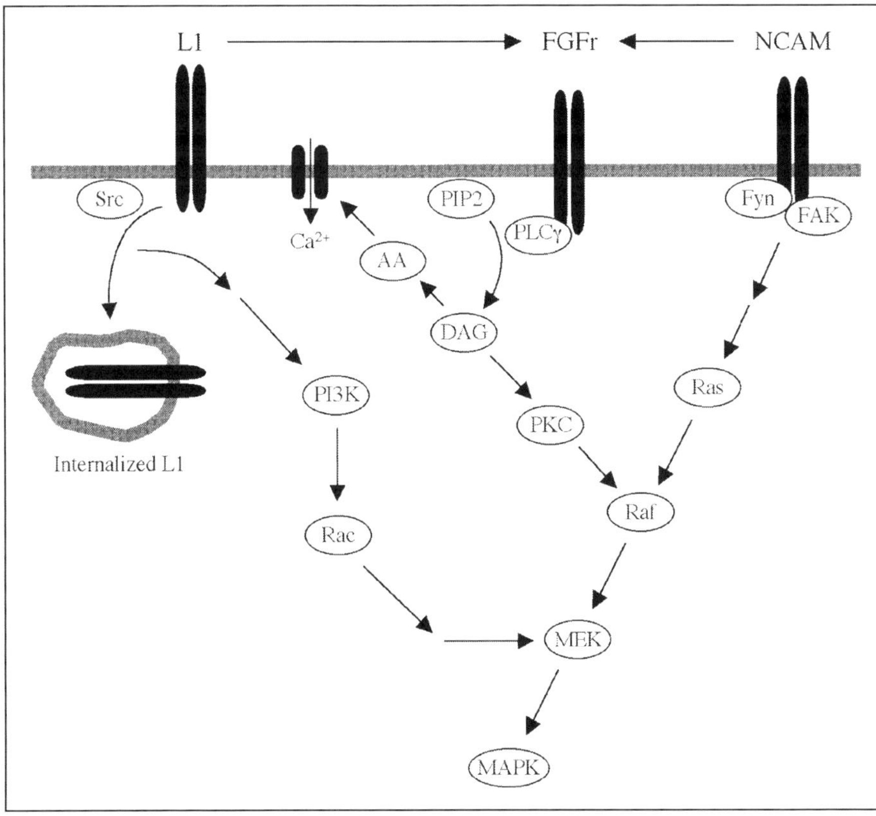

Figure 2. Signaling cascades downstream of L1 and NCAM. This scheme is oversimplified, and a number of intermediate components should be missing. AA: arachidonic acid; DAG: diacylglycerol; FAK: focal adhesion kinase; FGFr: fibroblast growth factor receptor; MAPK: mitogen-activated protein kinase; MEK: MAPK kinase; PI3K: phosphatidylinositol 3-kinase; PIP2: phosphatidylinositol 4,5-bisphosphate; PKC: protein kinase C; PLCγ: phospholipase Cγ.

Forward translocation of the growth cone requires not only the CAM-actin linkage but also a gradient of adhesive interactions with its environment (strong adhesion at the growth cone's leading edge and weak adhesion at the rear).[43] In this way, the cytoskeletal machinery is able to move the growth cone forward as attachments at its rear are released. To create such polarized adhesion, CAMs that have been translocated into the C-domain by coupling to retrograde F-actin flow should be recycled to the leading edge. Indeed, it has been shown that CAMs such as NCAM and β1 integrin undergo bidirectional movement on the growth cone surface, suggesting the centrifugal transport for CAM recycling.[44-46] In addition to this cell-surface pathway, an intracellular pathway for CAM recycling has been demonstrated.[47,48] L1 is endocytosed preferentially at the C-domain followed by centrifugal transport into the P-domain and reinsertion into the plasma membrane of the leading edge (Fig. 3). This recycling pathway is required for the maintenance of polarized growth cone adhesion and axon growth.[49] Therefore, the growth cone is able to regulate its adhesivity by controlling CAM trafficking both on the cell surface and via intracellular vesicular transport. Based on this model (Fig. 3), axon growth should be influenced by three major factors in the growth cone: dynamics of the cytoskeleton, clutch engagement, and CAM recycling. It is very important to understand how these factors are controlled by CAM-associated signals in a spatially defined and coordinated manner.

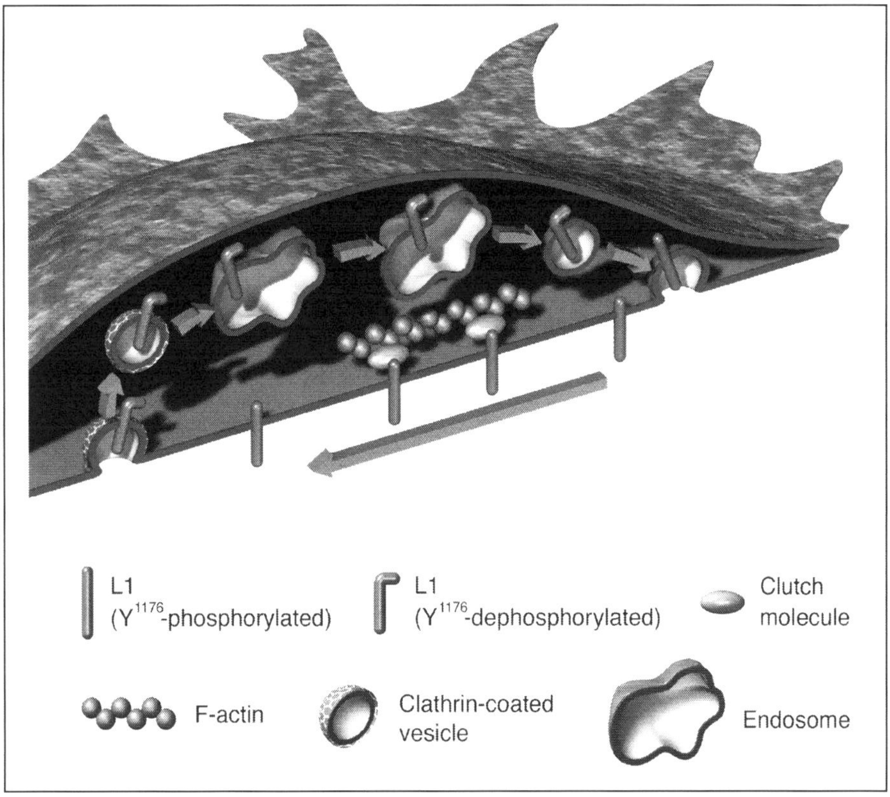

Figure 3. The molecular mechanism of L1-mediated growth cone migration. Reprinted with permission of the publisher.[48]

What is the role of CAMs in axon guidance? CAMs regulate selective fasciculation/defasciculation of axons at defined choice points. CAMs also act as contact-dependent attractive/repulsive cues and their receptors, which delineate a path for elongating axons.[50] Although intracellular molecular mechanisms have not yet been elucidated, CAM-CAM interactions regulate axon guidance in a complex way. Spinal commissural axons express axonin-1 and the ventral midline of the spinal cord (the floor plate) expresses NrCAM. A trans-interaction of axonin-1 with NrCAM is required for commissural axons to enter the floor plate and cross the midline.[51] Involvement of axonin-1 in commissural axon guidance is not dependent on its activity to promote axon elongation.[52] Ig superfamily CAMs are also implicated in guidance of dorsal root ganglion (DRG) axons in the spinal cord. Proprioceptive DRG fibers, which establish connections with motoneurons in the ventral horn, require NrCAM and F11 for their pathfinding, whereas nociceptive DRG fibers, which target to the dorsal horn, depends on NgCAM and axonin-1.[53]

More recent work demonstrated the importance of cross-talk between CAMs and axon guidance receptors/signaling. For example, L1 interacts with neuropilin-1 in cis to form a receptor complex that, in response to semaphorin3A binding, induces repulsive turning of the growth cone.[54] A simultaneous trans-interaction of L1 with neuropilin-1 switches semaphorin3A-triggered repulsion into attraction.[55] CAMs could also influence the turning response of axons without directly interacting with a guidance receptor, as laminin converts netrin-1-induced attraction into repulsion most likely by decreasing the amount of cAMP in the growth cone.[56] Another example comes from studies on N-cadherin and the axon guidance receptor Robo. Engagement of Robo by its ligand, the repulsive guidance cue Slit, results in decoupling of N-cadherin from the cytoskeleton and loss of N-cadherin-mediated adhesion.[57] Thus, a local repulsive signal in the growth cone is converted into decreased adhesion and traction allowing growth cone migration in the contralateral direction.

Functional Implication

The implication of CAMs in human nervous system development has been best illustrated in a congenital disorder, called X-linked hydrocephalus (XLH), which is caused by mutations in the L1 gene.[58,59] Generation and analyses of L1 knockout mice confirmed the similar roles of L1 in neural development.[60-63] XLH patients have ventricular enlargement accompanied by hypoplasia of several major axon tracts such as the corticospinal tract and the corpus callosum, indicating that L1 plays an important role in axon growth and guidance in vivo.[64] In addition, L1 knockout mice show various axon tract phenotypes such as abnormal topographic mapping of retinal axons to their target in the superior colliculus[65] and pathfinding errors of a subset of thalamocortical axons.[66] Also, a substantial proportion of corticospinal axons fail to cross the midline at the pyramidal decussation, which results in abnormal ipsilateral projection and probably in hypoplasia of the corticospinal tract.[61] This guidance error can be explained by in vitro observation: L1-deficient cortical axons do not respond to semaphorin3A secreted from the ventral spinal cord because neuropilin-1 associates with L1 in cis to form the functional receptor.[54] This idea has been supported by recent studies using a new mouse line in which the sixth Ig domain of L1 has been deleted.[67] This L1 mutant loses homophilic binding activity but retains its ability to interact with neuropilin-1, and the mutant animals show ventricular dilatation but no anomalies in axon tract development. This result suggests that L1-L1 homophilic adhesion is important for normal development of the ventricular system but not of axon tracts in mice. However, the real mechanism in humans might not be so simple, because even single amino acid substitutions throughout L1 (including the sixth Ig domain) almost always disrupt axon tract development in XLH.[68]

Analyses of gene-targeted mice revealed the roles of other Ig superfamily CAMs in neural development. The first two reports on NCAM knockout mice demonstrated a size reduction of the olfactory bulb due to impaired migration of olfactory granule cell precursors.[69,70] Later, abnormal fasciculation and pathfinding of intrahippocampal connections were observed in the NCAM knockout.[71] A GPI-anchored CAM, contactin, has been implicated in cerebellar granule cell axon guidance and in dendritic projections from granule and Golgi cells, thus contributing to the formation of cerebellar micro-organization.[72] Recently, the close homolog of L1 has been shown to play significant roles in hippocampal mossy fiber organization, olfactory axon projections, and neuronal positioning and dendritic growth of pyramidal neurons in the posterior region of the cerebral cortex.[73,74] Although the cerebellum of NrCAM knockout mice did not exhibit obvious defects, mice lacking both L1 and NrCAM showed severe cerebellar folial defects and a reduction in the thickness of the inner granule cell layer, suggesting that the L1 family CAMs have overlapping functions.[75] In contrast, neurofascin has been shown to play a unique role in axon guidance in the cerebellum: a subcellular gradient of neurofascin on a Purkinje cell directs basket axons towards the axon initial segment of the Purkinje cell leading to the formation of domain-specific synapses.[76]

Conclusion

Originally CAMs have been identified as cell-surface molecules that recognize specific binding partners and mediate cell-cell and cell-ECM adhesion. However, CAMs have turned out to be surprisingly multifunctional: they transduce signals in both outside-in and inside-out directions, dynamically associate with the cytoskeletons, move inside a cell and along the cell surface, provide anti-adhesive functions, and interact with many functional molecules including axon guidance receptors. These revelations produced significant progress in understanding how CAMs regulate dynamic cell behaviors such as axon growth and guidance. Future work should be devoted to examining how these multiple events are spatially and temporally coordinated in axons and dendrites to produce correct wiring of neuronal networks.

References

1. Rutishauser U, Gall WE, Edelman GM. Adhesion among neural cells of the chick embryo. IV. Role of the cell surface molecule CAM in the formation of neurite bundles in cultures of spinal ganglia. J Cell Biol 1978; 79(2 Pt 1):382-393.
2. Harrelson AL, Goodman CS. Growth cone guidance in insects: Fasciclin II is a member of the immunoglobulin superfamily. Science 1988; 242(4879):700-708.
3. Fischer G, Kunemund V, Schachner M. Neurite outgrowth patterns in cerebellar microexplant cultures are affected by antibodies to the cell surface glycoprotein L1. J Neurosci 1986; 6(2):605-612.
4. Grumet M, Edelman GM. Neuron-glia cell adhesion molecule interacts with neurons and astroglia via different binding mechanisms. J Cell Biol 1988; 106(2):487-503.
5. Lemmon V, Farr KL, Lagenaur C. L1-mediated axon outgrowth occurs via a homophilic binding mechanism. Neuron 1989; 2(6):1597-1603.
6. Bozyczko D, Horwitz AF. The participation of a putative cell surface receptor for laminin and fibronectin in peripheral neurite extension. J Neurosci 1986; 6(5):1241-1251.
7. Tomaselli KJ, Reichardt LF, Bixby JL. Distinct molecular interactions mediate neuronal process outgrowth on nonneuronal cell surfaces and extracellular matrices. J Cell Biol 1986; 103(6 Pt 2):2659-2672.
8. Matsunaga M, Hatta K, Nagafuchi A et al. Guidance of optic nerve fibres by N-cadherin adhesion molecules. Nature 1988; 334(6177):62-64.
9. Bixby JL, Zhang R. Purified N-cadherin is a potent substrate for the rapid induction of neurite outgrowth. J Cell Biol 1990; 110(4):1253-1260.
10. Dodd J, Jessell TM. Axon guidance and the patterning of neuronal projections in vertebrates. Science 1988; 242(4879):692-699.
11. Kamiguchi H, Lemmon V. IgCAMs: Bidirectional signals underlying neurite growth. Curr Opin Cell Biol 2000; 12(5):598-605.
12. Cunningham BA, Hemperly JJ, Murray BA et al. Neural cell adhesion molecule: Structure, immunoglobulin-like domains, cell surface modulation, and alternative RNA splicing. Science 1987; 236(4803):799-806.
13. Rutishauser U. Polysialic acid and the regulation of cell interactions. Curr Opin Cell Biol 1996; 8(5):679-684.
14. Tang J, Rutishauser U, Landmesser L. Polysialic acid regulates growth cone behavior during sorting of motor axons in the plexus region. Neuron 1994; 13(2):405-414.
15. Moos M, Tacke R, Scherer H et al. Neural adhesion molecule L1 as a member of the immunoglobulin superfamily with binding domains similar to fibronectin. Nature 1988; 334(6184):701-703.
16. Takeda Y, Asou H, Murakami Y et al. A nonneuronal isoform of cell adhesion molecule L1: Tissue-specific expression and functional analysis. J Neurochem 1996; 66(6):2338-2349.
17. Brummendorf T, Wolff JM, Frank R et al. Neural cell recognition molecule F11: Homology with fibronectin type III and immunoglobulin type C domains. Neuron 1989; 2(4):1351-1361.
18. Furley AJ, Morton SB, Manalo D et al. The axonal glycoprotein TAG-1 is an immunoglobulin superfamily member with neurite outgrowth-promoting activity. Cell 1990; 61(1):157-170.
19. Volkmer H, Leuschner R, Zacharias U et al. Neurofascin induces neurites by heterophilic interactions with axonal NrCAM while NrCAM requires F11 on the axonal surface to extend neurites. J Cell Biol 1996; 135(4):1059-1069.
20. Kunz S, Spirig M, Ginsburg C et al. Neurite fasciculation mediated by complexes of axonin-1 and Ng cell adhesion molecule. J Cell Biol 1998; 143(6):1673-1690.
21. De Angelis E, MacFarlane J, Du JS et al. Pathological missense mutations of neural cell adhesion molecule L1 affect homophilic and heterophilic binding activities. EMBO J 1999; 18(17):4744-4753.
22. De Angelis E, Watkins A, Schafer M et al. Disease-associated mutations in L1 CAM interfere with ligand interactions and cell-surface expression. Hum Mol Genet 2002; 11(1):1-12.
23. Kasper C, Rasmussen H, Kastrup JS et al. Structural basis of cell-cell adhesion by NCAM. Nat Struct Biol 2000; 7(5):389-393.
24. Freigang J, Proba K, Leder L et al. The crystal structure of the ligand binding module of axonin-1/TAG-1 suggests a zipper mechanism for neural cell adhesion. Cell 2000; 101(4):425-433.
25. Pipes GC, Lin Q, Riley SE et al. The Beat generation: A multigene family encoding IgSF proteins related to the Beat axon guidance molecule in Drosophila. Development 2001; 128(22):4545-4552.
26. Aurelio O, Hall DH, Hobert O. Immunoglobulin-domain proteins required for maintenance of ventral nerve cord organization. Science 2002; 295(5555):686-690.

27. Fambrough D, Goodman CS. The Drosophila beaten path gene encodes a novel secreted protein that regulates defasciculation at motor axon choice points. Cell 1996; 87(6):1049-1058.
28. Doherty P, Walsh FS. CAM-FGF receptor interactions: A model for axonal growth. Mol Cell Neurosci 1996; 8(2-3):99-111.
29. Doherty P, Williams G, Williams EJ. CAMs and axonal growth: A critical evaluation of the role of calcium and the MAPK cascade. Mol Cell Neurosci 2000; 16(4):283-295.
30. Schmid RS, Graff RD, Schaller MD et al. NCAM stimulates the Ras-MAPK pathway and CREB phosphorylation in neuronal cells. J Neurobiol 1999; 38(4):542-558.
31. Kolkova K, Novitskaya V, Pedersen N et al. Neural cell adhesion molecule-stimulated neurite outgrowth depends on activation of protein kinase C and the ras-mitogen-activated protein kinase pathway. J Neurosci 2000; 20(6):2238-2246.
32. Schmid RS, Pruitt WM, Maness PF. A MAP kinase-signaling pathway mediates neurite outgrowth on L1 and requires Src-dependent endocytosis. J Neurosci 2000; 20(11):4177-4188.
33. Schaefer AW, Kamiguchi H, Wong EV et al. Activation of the MAPK signal cascade by the neural cell adhesion molecule L1 requires L1 internalization. J Biol Chem 1999; 274(53):37965-37973.
34. Perron JC, Bixby JL. Distinct neurite outgrowth signaling pathways converge on ERK activation. Mol Cell Neurosci 1999; 13(5):362-378.
35. Nakai Y, Kamiguchi H. Migration of nerve growth cones requires detergent-resistant membranes in a spatially defined and substrate-dependent manner. J Cell Biol 2002; 159(6):1097-1108.
36. Niethammer P, Delling M, Sytnyk V et al. Cosignaling of NCAM via lipid rafts and the FGF receptor is required for neuritogenesis. J Cell Biol 2002; 157(3):521-532.
37. Lin CH, Forscher P. Growth cone advance is inversely proportional to retrograde F-actin flow. Neuron 1995; 14(4):763-771.
38. Dent EW, Gertler FB. Cytoskeletal dynamics and transport in growth cone motility and axon guidance. Neuron 2003; 40(2):209-227.
39. Suter DM, Forscher P. Substrate-cytoskeletal coupling as a mechanism for the regulation of growth cone motility and guidance. J Neurobiol 2000; 44(2):97-113.
40. Mitchison T, Kirschner M. Cytoskeletal dynamics and nerve growth. Neuron 1988; 1(9):761-772.
41. Jay DG. The clutch hypothesis revisited: Ascribing the roles of actin-associated proteins in filopodial protrusion in the nerve growth cone. J Neurobiol 2000; 44(2):114-125.
42. Nishimura K, Yoshihara F, Tojima T et al. L1-dependent neuritogenesis involves ankyrinB that mediates L1-CAM coupling with retrograde actin flow. J Cell Biol 2003; 163(5):1077-1088.
43. Lauffenburger DA, Horwitz AF. Cell migration: A physically integrated molecular process. Cell 1996; 84(3):359-369.
44. Sheetz MP, Baumrind NL, Wayne DB et al. Concentration of membrane antigens by forward transport and trapping in neuronal growth cones. Cell 1990; 61(2):231-241.
45. Schmidt CE, Dai J, Lauffenburger DA et al. Integrin-cytoskeletal interactions in neuronal growth cones. J Neurosci 1995; 15(5 Pt 1):3400-3407.
46. Grabham PW, Foley M, Umeojiako A et al. Nerve growth factor stimulates coupling of beta1 integrin to distinct transport mechanisms in the filopodia of growth cones. J Cell Sci 2000; 113(Pt 17):3003-3012.
47. Kamiguchi H, Long KE, Pendergast M et al. The neural cell adhesion molecule L1 interacts with the AP-2 adaptor and is endocytosed via the clathrin-mediated pathway. J Neurosci 1998; 18(14):5311-5321.
48. Kamiguchi H, Lemmon V. Recycling of the cell adhesion molecule L1 in axonal growth cones. J Neurosci 2000; 20(10):3676-3686.
49. Kamiguchi H, Yoshihara F. The role of endocytic L1 trafficking in polarized adhesion and migration of nerve growth cones. J Neurosci 2001; 21(23):9194-9203.
50. Tessier-Lavigne M, Goodman CS. The molecular biology of axon guidance. Science 1996; 274(5290):1123-1133.
51. Stoeckli ET, Sonderegger P, Pollerberg GE et al. Interference with axonin-1 and NrCAM interactions unmasks a floor-plate activity inhibitory for commissural axons. Neuron 1997; 18(2):209-221.
52. Fitzli D, Stoeckli ET, Kunz S et al. A direct interaction of axonin-1 with NgCAM-related cell adhesion molecule (NrCAM) results in guidance, but not growth of commissural axons. J Cell Biol 2000; 149(4):951-968.
53. Perrin FE, Rathjen FG, Stoeckli ET. Distinct subpopulations of sensory afferents require F11 or axonin-1 for growth to their target layers within the spinal cord of the chick. Neuron 2001; 30(3):707-723.
54. Castellani V, Chedotal A, Schachner M et al. Analysis of the L1-deficient mouse phenotype reveals cross-talk between Sema3A and L1 signaling pathways in axonal guidance. Neuron 2000; 27(2):237-249.

55. Castellani V, De Angelis E, Kenwrick S et al. Cis and trans interactions of L1 with neuropilin-1 control axonal responses to semaphorin 3A. EMBO J 2002; 21(23):6348-6357.
56. Hopker VH, Shewan D, Tessier-Lavigne M et al. Growth-cone attraction to netrin-1 is converted to repulsion by laminin-1. Nature 1999; 401(6748):69-73.
57. Rhee J, Mahfooz NS, Arregui C et al. Activation of the repulsive receptor Roundabout inhibits N-cadherin-mediated cell adhesion. Nat Cell Biol 2002; 4(10):798-805.
58. Rosenthal A, Jouet M, Kenwrick S. Aberrant splicing of neural cell adhesion molecule L1 mRNA in a family with X-linked hydrocephalus [published erratum appears in Nat Genet 1993 Mar;3(3):273]. Nat Genet 1992; 2(2):107-112.
59. Van Camp G, Vits L, Coucke P et al. A duplication in the L1CAM gene associated with X-linked hydrocephalus. Nat Genet 1993; 4(4):421-425.
60. Dahme M, Bartsch U, Martini R et al. Disruption of the mouse L1 gene leads to malformations of the nervous system. Nat Genet 1997; 17(3):346-349.
61. Cohen NR, Taylor JS, Scott LB et al. Errors in corticospinal axon guidance in mice lacking the neural cell adhesion molecule L1. Curr Biol 1998; 8(1):26-33.
62. Fransen E, D'Hooge R, Van Camp G et al. L1 knockout mice show dilated ventricles, vermis hypoplasia and impaired exploration patterns. Hum Mol Genet 1998; 7(6):999-1009.
63. Demyanenko GP, Tsai AY, Maness PF. Abnormalities in neuronal process extension, hippocampal development, and the ventricular system of L1 knockout mice. J Neurosci 1999; 19(12):4907-4920.
64. Kamiguchi H, Hlavin ML, Yamasaki M et al. Adhesion molecules and inherited diseases of the human nervous system. Annu Rev Neurosci 1998; 21:97-125.
65. Demyanenko GP, Maness PF. The L1 cell adhesion molecule is essential for topographic mapping of retinal axons. J Neurosci 2003; 23(2):530-538.
66. Wiencken-Barger AE, Mavity-Hudson J, Bartsch U et al Casagrande VA. The role of L1 in axon pathfinding and fasciculation. Cereb Cortex 2004;14(2):121-131.
67. Itoh K, Cheng L, Kamei Y et al. Brain development in mice lacking L1-L1 homophilic adhesion. J Cell Biol 2004; 165(1):145-154.
68. Yamasaki M, Thompson P, Lemmon V. CRASH syndrome: Mutations in L1CAM correlate with severity of the disease. Neuropediatrics 1997; 28(3):175-178.
69. Tomasiewicz H, Ono K, Yee D et al. Genetic deletion of a neural cell adhesion molecule variant (N-CAM-180) produces distinct defects in the central nervous system. Neuron 1993; 11(6):1163-1174.
70. Cremer H, Lange R, Christoph A et al. Inactivation of the N-CAM gene in mice results in size reduction of the olfactory bulb and deficits in spatial learning. Nature 1994; 367(6462):455-459.
71. Cremer H, Chazal G, Goridis C et al. NCAM is essential for axonal growth and fasciculation in the hippocampus. Mol Cell Neurosci 1997; 8(5):323-335.
72. Berglund EO, Murai KK, Fredette B et al. Ataxia and abnormal cerebellar microorganization in mice with ablated contactin gene expression. Neuron 1999; 24(3):739-750.
73. Montag-Sallaz M, Schachner M, Montag D. Misguided axonal projections, neural cell adhesion molecule 180 mRNA upregulation, and altered behavior in mice deficient for the close homolog of L1. Mol Cell Biol 2002; 22(22):7967-7981.
74. Demyanenko GP, Schachner M, Anton E et al. Close homolog of L1 modulates area-specific neuronal positioning and dendrite orientation in the cerebral cortex. Neuron 2004; 44(3):423-437.
75. Sakurai T, Lustig M, Babiarz J et al. Overlapping functions of the cell adhesion molecules Nr-CAM and L1 in cerebellar granule cell development. J Cell Biol 2001; 154(6):1259-1273.
76. Ango F, di Cristo G, Higashiyama H et al. Ankyrin-based subcellular gradient of neurofascin, an immunoglobulin family protein, directs GABAergic innervation at purkinje axon initial segment. Cell 2004; 119(2):257-272.

Chapter 8

Neurotransmitters and the Development of Neuronal Circuits

Tina Ruediger and Juergen Bolz*

Abstract

In the mature brain, neurotransmitters are used for synaptic communication between neurons. But during nervous system development, neurons often express and release transmitters before their axons establish contacts with their target cells. While much is known about the synaptic effects of neurotransmitters, their extrasynaptic effects are less understood. There is increasing evidence that neurotransmitters in the immature nervous system can act as trophic factors that influence different developmental events such as cell proliferation and differentiation. However, more recent work demonstrates that neurotransmitters can also influence the targeting of migrating neurons and growing axons during the formation of neuronal circuits. This chapter will focus on such guidance effects of neurotransmitters during the development of the nervous system. Elucidating extrasynaptic functions during the nervous system development might also provide insights in their potential roles for plasticity and regeneration in the adult nervous system.

Introduction

In the adult nervous system, neurotransmitters released from presynaptic terminals bind to receptors on postsynaptic cells, which lead either to an excitation or inhibition of these neurons. However, during the development of the nervous system, neurotransmitters, their synthesizing enzymes and their receptors are often expressed before synapses are being formed. Even at the earliest stages of development, progenitor cells of the nervous system can express receptors for almost all neurotransmitter classes.[1] Because neurotransmitters are secreted molecules, it has been suggested that they might be important developmental signals that influence proliferation, differentiation, synapse maturation and survival of neurons.[2,3] In addition to such trophic effects, neurotransmitters have also been implicated to act as tropic signals that guide migrating neurons and growing axons to their target cells. Neurotransmitter-induced tropic effects can be chemoattractive, resulting in cell migration or axonal growth toward the transmitter source or chemorepulsive, causing migration or growth away from the secreted transmitter.

In this review we summarize some of the more recent work on neuronal guidance functions of neurotransmitters (for earlier reviews see refs. 3,4). We describe studies which provide evidence that all major transmitters, such as glutamate, acetylcholine (ACh), γ-aminobutyric acid (GABA), dopamine and serotonin can act as regulators for growing neurites and migrating neurons in many different species (see Table 1). These effects are mediated by different ionotropic

*Corresponding Author: Juergen Bolz—Universitaet Jena, Institut für Allgemeine Zoologie und Tierphysiologie, Erbertstrasse 1, 07743 Jena, Germany. Email: bolz@pan.zoo.uni-jena.de

Axon Growth and Guidance, edited by Dominique Bagnard.

Table 1. Neurotransmitters and their effects on various neuronal tissues.

NT	Neuronal Preparation	Effect	Reference
Acetylcholine	Leech embryo	retraction of neurites	49
	Xenopus spinal cord neurons	attractive turning response induction of growth cone filopodia	53
	Neuroblastoma cells and lamellipodia	induction of growth cone filopodia	51, 52
	Chick sympathic neurons	inhibition of neurite outgrowth	45
	Chick ciliary ganglion neurons	retraction of neurites	48
	Rat retinal ganglion cell neurons	growth inhibition	46
	Mouse spinal cord neurons	inhibition of neurite outgrowth and growth cone motility	47
	Mouse thalamic neurons	induction of growth cone pause behavior	Rüdiger and Bolz (unpubl.)
Dopamine	Helisoma neurons	inhibition of growth cone motility and neurite elongation	10
	Lymnaea stagnalis presynaptic neurons	induction of growth cone attraction and growth cone collapse	9
	Chick retina cells	retraction of neurites inhibition of growth cone motility	11
GABA	Rat superior ganglia	promotion of dendritic outgrowth and synapse formation	64
	Rat spinal cord	neurons induction of cell migration	75
	Rat cortical neurons	induction and stop of cell migration	77, 78
	Rat cerebellar neurons	repulsive turning response	80
	Mouse spinal cord neurons	inhibition of neurite outgrowth and growth cone motility	84
Glutamate	Xenopus spinal cord neurons	attractive turning response induction of growth cone filopodia	21, 22
	Rat cortical neurons	induction of cell migration	79
	Rat cerebellar granule cells	induction of cell migration	15
	Rat hippocampal neurons	inhibition of dendritic outgrowth	32
	Rat spinal motoneurons	inhibition of dendritic outgrowth	33
	Rat hippocampal neurons	inhibition of growth cone motility	35
	Serotonin Helisoma neurons	inhibition of growth cone motility and neurite elongation	7
	Rat thalamic neurons	promotion of neurite outgrowth	85

List of the effects of neurotransmitters on growing neurites and migrating neurons in different species with the references discussed this review.

or metabotropic receptors, and the activation of different receptors by the same neurotransmitter can have very different effects on the guidance of neuronal growth cones. A summary of neurotransmitter mediated guidance effects are illustrated in Figure 1. We also discuss possible mechanisms of how neurotransmitter-receptor interactions exert their guidance effects and we provide specific examples for the roles of neurotransmitters during the assembly of neuronal circuits in different systems.

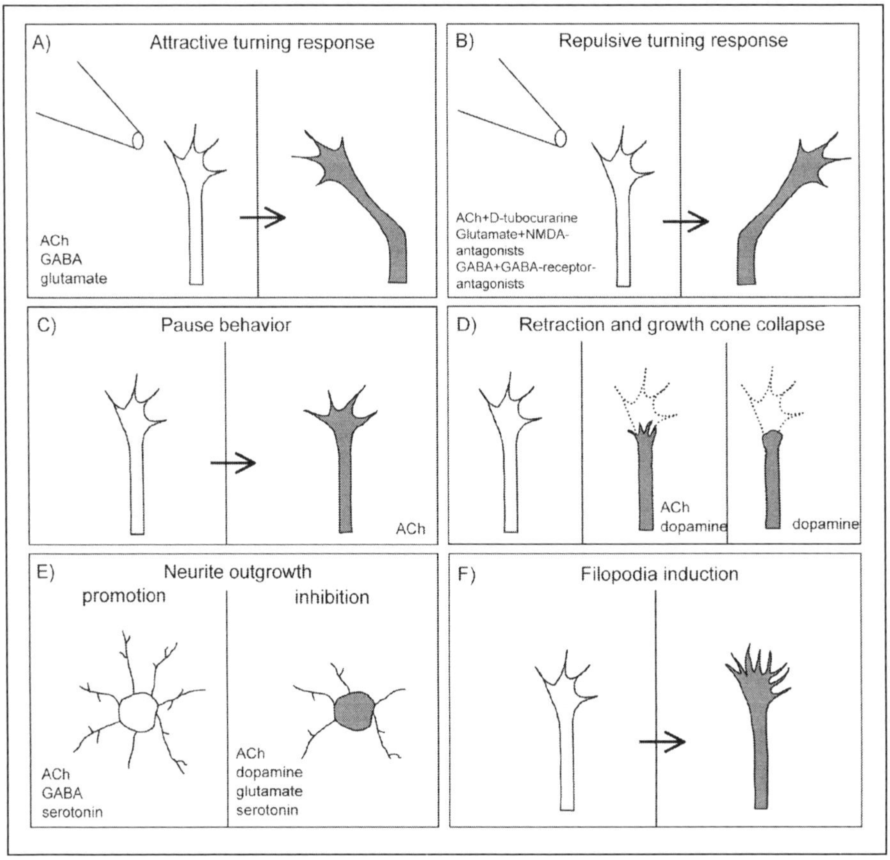

Figure 1. Schematic illustration of the effects of different neurotransmitters on growth cone behavior, neurite outgrowth and axon guidance.

Effects of Neurotransmitters on Growth Cone Steering

Growth cones are extensions of the tip of developing dendrites and axons which have finger-like protrusions that actively extend and retract during neurite outgrowth. Growth cones also possess receptors for different extracellular signals, and the activation of these receptors can influence growth cone motility and lead to changes in the rate or direction of neurite extension. During development, these guidance signals allow growth cones to navigate through a complex environment to innervate distant target neurons. Over the last decade, several families of axonal guidance molecules, including semaphorins, netrins and ephrins have been discovered.[5,6] However, one of the first identified molecular cues that affect growth cone steering were neurotransmitters. Early studies in the snail *Helisoma* demonstrated that serotonin and dopamine inhibit growth cone motility and arrest neurite outgrowth of specific sets of neurons.[7,8]

More recent findings in the snail *Lymnea stagnalis* demonstrated that dopamine spontaneously released from an identified neuron in culture differentially regulates the growth cone behavior of its in vivo target and nontarget neurons. While dopamine enhanced the rate of growth cone advance from target cells it also induced growth cone retraction and turning away from nontarget cells.[9] The results from these invertebrate studies then suggest that a neurotransmitter released from a postsynaptic neuron might act as chemoattractant for target growth cones, but as chemorepellent for nontarget growth cones. Effects of serotonin and dopamine on growth cones and extending axons have also been described in the central nervous system of vertebrates.[7,10,11]

Glutamate: An Excitatory Transmitter and a Versatile Developmental Signal

Glutamate is the most prevalent excitatory neurotransmitter in the nervous system. Glutamate can activate several classes of specific receptors, including ionotropic NMDA (N-methyl-D-aspartate), AMPA (amino-3-hydroxy-5-methylisoxazoleproprionic acid) receptors and G-protein coupled metabotropic receptors which excert their effects via second messengers like cyclic adenosine monophosphate (cAMP) and inositol triphosphate (IP_3).[12] In the adult nervous system, glutamatergic synapses have been most intensively studied in pyramidal neurons of the hippocampus and cerebral cortex, because specific activity patterns of their afferent inputs can lead either to long term potentiation (LTP) or long term depression (LTD) of these synapses. These functional changes at glutamatergic synapses also result in morphological alterations in dendritic spines, the specialized protrusions on dendritic processes that constitute the postsynaptic element of these synapses. Glutamate has also been implicated to mediate activity-dependent processes during development and during the so-called critical periods. Such processes influence the elaboration, pruning and stabilization of axonal and dendritic arbors.[13,14]

Glutamate released from active cells or constitutively released glutamate can also act as a guidance cue for migrating cells and growing axons. Several lines of evidence have indicated that activation of the NMDA type of glutamate receptors modulates that rate of granule cell migration in the developing cerebellum. Patch clamp analysis revealed a large increase in the frequency of spontaneous NMDA receptor activity in migrating neurons compared to premigratory cells.[15] Furthermore, migrating granule cells express different NMDA subunits than postmigratory neurons.[16] Blocking NMDA receptor activity significantly decreased the rate of granule cell migration in cerebellar slices in a dose-dependent manner.[17] In contrast, the rate of migration was not altered by blocking nonNMDA receptors or GABA-A and GABA-B receptors.

How does glutamate control the rate of migration? Since activation of NMDA receptors induces substantial Ca^{2+} influxes into cells, intracellular Ca^{2+} was monitored with Ca^{2+} indicator dyes simultaneously with cell migration.[18] These experiments revealed periodic fluctuations of Ca^{2+} levels in cell somata that were closely related to cell movement. Granule cells moved forward during the phase of transient Ca^{2+} increase and remained stationary during Ca^{2+} decrease. Blocking NMDA receptors reduced the Ca^{2+} fluctuations and the rate of cell movement in a dose-dependent manner. Ca^{2+} fluctuations and cell movements were also reduced when N-type Ca^{2+} channels were blocked by specific antagonist,[19] whereas transient elevation of Ca^{2+} increased the forward movements of migrating granule cells. Taken together, these results demonstrate that neurotransmitter receptors and ion channels control the motility of migrating granule cells by modifying Ca^{2+} fluctuations.[20]

In growth cone turning assays, microscopic gradients of glutamate were produced by ejecting these substances from a pipette placed near the growth cone, but at an angle of 45° with respect to the initial direction of the axon.[21,22] Glutamate gradients induced a marked attraction of axons from *Xenopus* spinal neurons. The turning response of growth cones required the activation of NMDA and nonNMDA receptors, because either NMDA or nonNMDA antagonists abolished the turning induced by glutamate. It is known that extracellular Mg^{2+} blocks NMDA receptor activity in a voltage dependent manner.[23,24] In the absence of extracelullar Mg^{2+}, the glutamate-induced turning response was blocked by NMDA antagonists but not by nonNMDA antagonists. This then suggests that NMDA receptors are primarily responsible for growth cone turning. In the presence of Mg^{2+}, depolarization via nonNMDA receptors is required to remove the Mg^{2+} block of NMDA receptors. Ca^{2+} influx through NMDA receptors then induces growth cone turning, because the chemoattractive effect of glutamate is abolished in Ca^{2+} free medium. An additional requirement for the chemotropic effect of glutamate is that

Ca^{2+} influx into the growth cone occurs asymmetrically. At high glutamate concentrations, when all receptors have been activated and Ca^{2+} elevations are uniformly high across the growth cone, turning responses were no longer elicited by glutamate gradients.[21,25]

Metabotropic glutamate receptors have also been implicated to contribute to growth cone guidance and axonal pathfinding. Mossy fibers, the axons of granule cells in the dentate gyrus, project to pyramidal cells in the CA3 region of the hippocampus. This projection is restricted to a narrow zone on the apical dendrites of pyramidal cells, the striatum lucidum. This layer-specific projection is recapitulated in organotypic slice cultures of the hippocampus.[26] However, in the presence of antagonists of metabotropic glutamate receptors, but not with ionotropic antagonists, mossy fibers in slice cultures were defasciculated and their axon terminals were no longer restricted to their target layer in the CA3 region.[27] How glutamate regulates the guidance of mossy fibers is not clear. Since mossy fiber guidance was unaffected by chronic application of agonists for metabotropic glutamate receptors, it is unlikely that glutamate gradients operate in mossy fiber pathfinding. Granule cells in the dentate gyrus express metabotropic glutamate receptors,[28,29] and their activation might be necessary for axon fasciculation and pathfinding. Alternatively, since glia cells throughout the CA1-CA2 region of the hippocampus also possess metabotropic glutamate receptors, when stimulated with glutamate, these cells might secrete guidance cues for mossy fiber pathfinding. Still another possibility is that activation of metabotropic glutamate receptors on mossy fibers interferes with the responsiveness to other guidance cues for axon targeting.

Evidence for such indirect effects of glutamate on axon guidance was provided by a recent study.[30] When axons in vitro are exposed to repellent guidance signals of the semaphorin and slit family, their growth cones respond to these repellents by withdrawing filopodia and contracting lamelipodia, and thereby displaying a collapsed morphology.[31] However, the growth cone collapse induced by semaphorin 3A, semaphorin 3C and slit-2 was strongly reduced when glutamate was added to the medium. This modulatory effect of glutamate on the responsiveness to these repellents was blocked by antagonists to metabotropic glutamate receptors, whereas agonists for these receptors mimicked the effect of glutamate. These indirect effects of glutamate might also be relevant to activity-dependent changes later in brain development and in brain plasticity, when neuronal connections are refined or become modified by the advance and retraction of axonal and dendritic processes. The release of glutamate from active neurons could reduce their own response to target derived repellents. In addition to such an autocrine mechanism, the release of glutamate from presynaptic elements could also attract postsynaptic elements by making them resistant to repulsive cues in their environment. The modulatory effects of glutamate on repellent signals might therefore provide a mechanism how activity-dependent changes lead to the remodelling of neuronal circuits.

In addition to such indirect effects of glutamate on cell morphology, numerous studies have shown that this transmitter can also have direct effects by stimulating ionotropic glutamate receptors. For example, in cultured hippocampal pyramidal cells and spinal motor neurons dendritic extension is inhibited through glutamate-induced stimulation of nonNMDA receptors.[32,33] In contrast, NMDA receptor activation in cerebellar granule neurons can reduce axon extension and act as a "stop signal" for their afferent fiber from the pons.[34] Glutamate has also profound effects during synaptogenesis by regulating the shape and motility of both axonal filipodia, which emerge from the shaft of axonal branches and contain synaptic vesicles, and dendritic filipodia, which later transform into spines.[14,35-37]

The effects of glutamate-induced stimulation of NMDA receptors have been most extensively studied in the context of activity-dependent axonal targeting and experience-dependent remodelling of sensory circuits. Although, as indicated above, metabotropic glutamate receptors might also participate in these processes, the NMDA receptor is considered as the prime candidate to mediate the effects of patterned neuronal activity in the refinement and modifications of neuronal connections. Because the Mg^{2+} block of NMDA receptor transmission is relieved when the neuron is depolarized by excitatory input at its other synapses, NMDA receptors can therefore detect correlations in the firing between pre- and postsynaptic cells as well as coactivity between multiple converging inputs to the neuron. According to a hypotheses proposed by Hebb (1949), the fine tuning of axonal and dendritic arbors is based on correlated firing patterns.[38] There are several recent reviews which describe the requirement of NMDA receptor stimulation for the activity-dependent development and plasticity of sensory systems.[39]

Acetylcholine Acts as a Neuronal Guidance Signal during Brain Development

Acetylcholine was the first neurotransmitter to be discovered. It is synthesized from choline and acetyl-CoA through the action of choline acetyltransferase (ChAT). According to their main agonists, muscarine and nicotine, acetylcholine receptors are divided into two classes: muscarinic and nicotinic acetylcholine receptors (mAChR and nAChR). Both receptor classes are abundant in the brain. The nAChR is a member of the ligand-gated superfamily of ion channels and is composed of five polypeptide subunits that surround a central channel pore, which is selective for most cations. While nicotinic receptors are ionotropic receptors, all muscarinic receptors are metabotropic and coupled to G proteins which activate second messenger pathways. Five muscarinic subtypes M1-M5 do exist.[1,40-42] Both, the acetylcholine synthesizing enzyme (ChAT) and the AChR are present in the early embryonic nervous system, suggesting extrasynaptic functions of ACh during brain development.[43,44]

Several in vitro studies demonstrated that ACh can influence neurite extension of invertebrates as well as in vertebrates neurons. For example, the work of Small et al (1995) showed that the presence of ACh in cultures of isolated embryonic chick sympathetic neurons inhibited neurite outgrowth and decreased the percentage of neurons bearing neurites. Application of muscarinic and nicotinic receptor agonists caused also an inhibition of neurite outgrowth, indicating that in this case the effect of acetylcholine was mediated through muscarinic and nicotinic receptors.[45] Other studies report growth inhibiting effects of ACh only after stimulation of nicotinic receptors.[46-49] In contrast, treatment with carbachol, a mixed cholinergic agonist, increased neurite length of cultured rat olfactory bulb neurons, suggesting that ACh can also promote neurite outgrowth.[50] Studies with neuroblastoma cell lines demonstrate that ACh can induce the formation of filopodia and lamellipodia when applied locally by a micropipette.[51,52] Interestingly, no effects were observed by adding acetylcholine to the medium, suggesting that acetylcholine needs to be present in a concentration gradient to be effective.[52]

A direct guidance effect of ACh gradients has been described for neurites from *Xenopus* spinal neurons.[53] Application of an acetylcholine gradient by a micropipette increased the number of growth cone filopodia on the side facing the pipette and the neurites extended towards the ACh gradient. This attractive turning response of growth cones towards ACh required the presence of extracellular Ca^{2+} and was blocked by nicotinic, but not by muscarinic receptor antagonists. Application of Rp-cAMPS, an analogue competitor of cAMP, the attractive turning response of the growth cone towards ACh was converted into a repulsive turning response. Thus, ACh, like other axonal guidance cues, can elicit either attraction or repulsion depending on the level of cyclic nucleotides (see chapter of this book by Piper et al).

Since there is a prominent cholinergic innervation of the cerebral cortex from the basal forebrain,[44,54] several studies examined the effects of ACh on the development of cortical neurons. Coculture studies indicated that cholinergic basal forebrain neurons provide trophic support for cortical.[55] In vivo studies after neonatal lesions of the nucleus basalis Meynert in the forebrain demonstrated that ACh has a profound effect on the development of cortical neurons. For example, layer 4 and 5 neurons of the somatosensory cortex were smaller and more densely packed during first postnatal week. In addition, neuronal maturation and dendritic development after basal forebrain lesions is delayed.[56-59]

Recently, our laboratory investigated the possible role of ACh on the growth of thalamic axons, which innervate the cortex at a time when ACh is already expressed in cortical neurons.

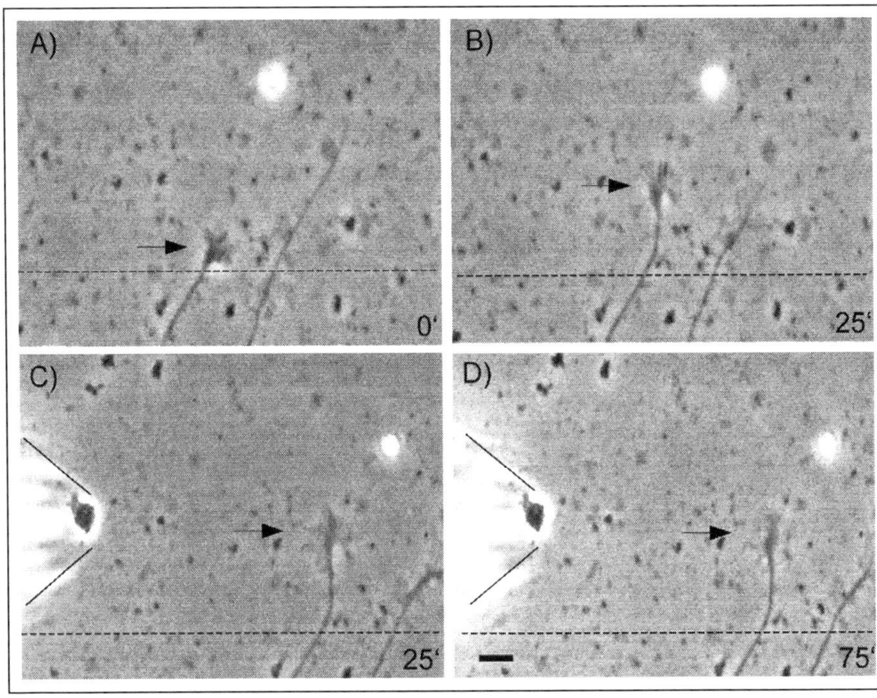

Figure 2. Effects of ACh on growing thalamic axons. This sequence of video-micrographs shows the advance of a thalamic growth cone during 25 minutes in vitro (A,B). During the next 50 minutes, when ACh was applied from a micropipette, the growth cone showed the typical pausing behavior: although lamellipodia and filopodia were continuously formed and retracted, there was no forward movement of the growth cone (C,D). Scale bar: 5 μm.

After application of ACh gradients from a pipette, thalamic axons almost immediately stopped their growth. This growth arrest was not accompanied by a collapse or a backward movement of the growth cone; rather filopodia and lamellipodia were continuously formed and retracted. But the growth cones did not move forward, instead they exhibited a specific pausing behavior (Fig. 2). The presence of atropine or d-tubocurarine abolished the growth arrest of thalamic axons induced by ACh, suggesting that muscarinic and nicotinic ACh receptors are involved. Since thalamic axons stop their growth in vivo for several hours when they first reach the cortical intermediate zone and innervate the cortical plate only after a "waiting period",[60,61] we propose that acetylcholine is one of the signals which regulates the timing of thalamocortical projections.

There Is More to GABA Than Synaptic Inhibition

GABA was the first clear example for a transmitter substance that mediates direct synaptic inhibition in the mature nervous system. It is synthesized from glutamate by glutamic acid decarboxylase (GAD) and there are two GAD enzymes, GAD65 and GAD67, encoded by different genes. GABA activates either ionotropic GABA-A and GABA-C receptors or metabotropic GABA-B receptors. Ionotropic GABA receptors are members of the ligand-gated ion channel superfamily; they consist of five subunits derived from several related gene families that form chloride channels. GABA-B receptors are members of the seven-transmembrane domain protein superfamiliy with close homology to the metabotropic glutamate receptors.[62] They can be localized both pre- and postsynaptically and they are coupled to calcium or potassium channels or adenylate cyclase via G-protein activation.[63] Several studies showed that in the developing nervous system GABA can act as a trophic substance and influence different aspects of neuronal differentiation. For example, early work of Wolff et al (1978) showed that continuous application of GABA in rat superior ganglia in vivo promotes dendritic outgrowth and synapse formation.[64] Involvement of GABA receptor activation on neurite outgrowth has now been observed in several other systems.[65-68] In addition there is evidence that GABA application promotes the maturation of GABAergic interneurons and regulates DNA synthesis and proliferation of neuronal progenitor cells.[69-71]

GABA also plays a role as a tropic signals that provides directional cues for migrating neurons and growing axons. In the developing neocortex, cells in the proliferative ventricular zone (VZ) start to migrate after their last cell division through the intermediate zone (IZ) into the cortical plate (CP), where they organize into layers. Cells that migrate out of the VZ move along radial glia cells which extend from the VZ to the pial surface and most of these cells differentiate into projections neurons.[72,73] Cortical interneurons are generated in the ganglionic eminences, from where they migrate along distinct routes towards the cortex and, once they arrive in the cortex, they then migrate perpendicular to the axis of the glia cells into the CP.[74] Because GABAergic cells are present in the CP at the time when radial migration occurs and because GABA receptors are expressed by migrating neurons, it has been suggested that GABA could exert chemotropic effects, inducing migration out of the VZ. To test this hypothesis, in a series of in vitro studies it has been shown that GABA receptor activation influences migration of immature cortical neurons. First, using a microchemotaxis chamber it was demonstrated that dissociated cortical neurons move from an upper chamber through a membrane filter pores to a lower chamber containing femtomolar GABA concentrations. When the GABA solution in the lower chamber was replaced by dissociated cells from the CP, it also induced cell migration. HPLC measurements indicated that CP cells release GABA, and the migration of VZ cells in the chemotaxis chamber towards the GABA secreting CP cells was blocked by saclofen, a GABA-B receptor antagonist.[75,76] Thus migrating cortical neurons are attracted by exogenously applied GABA as wells as by endogenous GABA released from CP cells.

In a second set of experiments, migrating cells were examined in cortical slice cultures in the presence of different GABA antagonists to examine the possible role of endogenous GABA on cell movement.[77,78] For this, slices from rat cortex at embryonic day 18 (E18) were incubated with bromodeoxyuridine (BrdU) to label VZ cells as they underwent their final mitosis. Under control conditions, after 2 days in vitro, most BrdU labelled cells migrated out of theCP and reached the CP. However, in the presence of picrotoxin, an antagonist for GABA-A and GABA-C receptors, the majority of the labelled cells were still found in the VZ. In cultures treated with the GABA-B receptor antagonist saclofen, BrdU cells migrated out of the CP, but then accumulated in the IZ without reaching the CP. Finally, the GABA-A receptor antagonist bicuculline did not block, but rather enhanced cortical migration into the CP. These experiments lead to the proposal that changing GABA receptor expression may regulate the radial migration of cortical neurons in response to GABA released from their target destination. This attractive hypothesis for the radial migration, however, has several limitations. First, blocking of GABA-B or GABA-C receptors for 6 days in vitro had very little effects on cell migration, suggesting that blocking GABA receptors delays, but does not arrest cell migration.[78] Therefore, GABA-mediated signalling for neuronal migration might be compensated by other mechanisms or might not be essential for this process. Moreover, in similar studies with slice cultures from mouse cortex it was found that GABA had very little effects on cell movement, whereas glutamate acting via NMDA receptors was a much more potent attractant for migrating neurons.[79] The reasons for these discrepancies are not clear and it would be important to examine the combined effects of GABA and glutamate on cortical neuron migration.

GABA has also been implicated as a guidance signal for growth cones of elongating axons. When *Xenopus* spinal neurons were tested in the growth cone turning assay, their axons turned toward the GABA gradient.[80] However, in the presence of GABA-A receptor antagonist, growth cones turned away from GABA gradients. Similar repulsive turning responses were observed with baclofen gradients which activate GABA-B receptors. The repulsive effects were abolished by the GABA-B receptor antagonist saclofen and by pertussis toxin, a blocker of G protein signalling. These results indicate that GABA can elicit opposing effects on growth cone turning of *Xenopus* neurons, an attractive response after GABA-A and a repulsive response after GABA-B receptor activation. Under normal culture conditions chemoattraction appears to dominate over chemorepulsion, but the situation is even more complex. Several previous studies have shown that cytosolic levels of cyclic nucleotides are critical in determine the growth cone behavior in gradients of many guidance molecules.[81] When intracellular levels of cyclic guanosine monophosphate (cGMP), but not cAMP, were raised, the growth cone repulsion by GABA-B receptor activation was converted into attraction. Thus GABA-mediated guidance effects can depend on receptor activation and on cAMP levels, which in turn often depend on other incoming signals to the neurons.[82]

Studies in the olfactory system suggest that GABA-B receptor function might contribute to target recognition of olfactory receptor neurons (ORNs) during development and regeneration.[83] Immunostaining of ORNs in the neuroepithelium of the mouse demonstrated that they express GABA-B receptors from E14 to adulthood. GABA-B receptor immunoreactivity was also observed on ORN growth cones in vitro. In the olfactory bulb, neurons surrounding the glomeruli, the projection targets of ORNs, express GABA and its synthesizing enzyme GAD from E16 to adulthood. In vitro assays demonstrated that in ORN neurite outgrowth was inhibited by GABA-B receptor agonists. Thus, ORN growth cones are responsive to GABA prior synapse formation. Taken together, these results suggest that GABAergic neurons that surround glomeruli could release GABA as a stop signal for developing and regenerating ORN axons and restrict these fibers to the glomerula target layer. A "stop growth" response of neurites after GABA-B receptor activation has also been observed in other systems.[66,84]

Conclusions

There is now convincing evidence that there is more to neurotransmitters than regulating synaptic transmission. They can also act as extrasynaptic signalling molecules in the developing nervous system, contributing to the assembly of neuronal circuits. This dual function of neurotransmitters occurred probably very early during the evolution of the nervous system. As neuronal assemblies evolved, many additional signalling molecules were required for the precise wiring of highly complex brains. The first evidence for attractive or repulsive guidance effects of neurotransmitters for axon navigation came from studies in invertebrates. However, it is now clear that neurotransmitters can also guide growing axons and migrating neurons in highly evolved brain structures of vertebrates, including the mammalian neocortex. A challenge for future studies will be to clarify how the extrasynaptic functions of neurotransmitters during normal brain development are related to functional recovery after injury or diseases in the adult brain.

References

1. Nguyen L, Rigo JM, Rocher V et al. Neurotransmitters as early signals for central nervous system development. Cell Tissue Res 2001; 305(2):187-202.
2. Mattson MP. Neurotransmitters in the regulation of neuronal cytoarchitecture. Brain Res 1988; 472(2):179-212.
3. Lauder JM. Neurotransmitters as growth regulatory signals: Role of receptors and second messengers. Trends Neurosci 1993; 16(6):233-240.
4. Lipton SA, Kater SB. Neurotransmitter regulation of neuronal outgrowth, plasticity and survival. Trends Neurosci 1989; 12(7):265-270.
5. Tessier-Lavigne M, Goodman CS. The molecular biology of axon guidance. Science 1996; 274(5290):1123-1133.
6. Dickson BJ. Molecular mechanisms of axon guidance. Science 2002; 298(5600):1959-1964.
7. McCobb DP, Cohan CS, Connor JA et al. Interactive effects of serotonin and acetylcholine on neurite elongation. Neuron 1988; 1(5):377-385.
8. Haydon PG, McCobb DP, Kater SB. Serotonin selectively inhibits growth cone motility and synaptogenesis of specific identified neurons. Science 1984; 226(4674):561-564.

9. Spencer GE, Klumperman J, Syed NI. Neurotransmitters and neurodevelopment. Role of dopamine in neurite outgrowth, target selection and specific synapse formation. Perspect Dev Neurobiol 1998; 5(4):451-467.
10. McCobb DP, Haydon PG, Kater SB. Dopamine and serotonin inhibition of neurite elongation of different identified neurons. J Neurosci Res 1988; 19(1):19-26.
11. Lankford KL, DeMello FG, Klein WL. D1-type dopamine receptors inhibit growth cone motility in cultured retina neurons: Evidence that neurotransmitters act as morphogenic growth regulators in the developing central nervous system. Proc Natl Acad Sci USA 1988; 85(8):2839-2843.
12. Simeone TA, Sanchez RM, Rho JM. Molecular biology and ontogeny of glutamate receptors in the mammalian central nervous system. J Child Neurol 2004; 19(5):343-360, (discussion 361).
13. Ruthazer ES, Akerman CJ, Cline HT. Control of axon branch dynamics by correlated activity in vivo. Science 2003; 301(5629):66-70.
14. Fischer M, Kaech S, Wagner U et al. Glutamate receptors regulate actin-based plasticity in dendritic spines. Nat Neurosci 2000; 3(9):887-894.
15. Rossi DJ, Slater NT. The developmental onset of NMDA receptor-channel activity during neuronal migration. Neuropharmacology 1993; 32(11):1239-1248.
16. Farrant M, Feldmeyer D, Takahashi T et al. NMDA-receptor channel diversity in the developing cerebellum. Nature 1994; 368(6469):335-339.
17. Komuro H, Rakic P. Modulation of neuronal migration by NMDA receptors. Science 1993; 260(5104):95-97.
18. Komuro H, Rakic P. Intracellular Ca2+ fluctuations modulate the rate of neuronal migration. Neuron 1996; 17(2):275-285.
19. Komuro H, Rakic P. Selective role of N-type calcium channels in neuronal migration. Science 1992; 257(5071):806-809.
20. Komuro H, Rakic P. Orchestration of neuronal migration by activity of ion channels, neurotransmitter receptors, and intracellular Ca2+ fluctuations. J Neurobiol 1998; 37(1):110-130.
21. Zheng JQ, Wan JJ, Poo MM. Essential role of filopodia in chemotropic turning of nerve growth cone induced by a glutamate gradient. J Neurosci 1996; 16(3):1140-1149.
22. Guirland C, Suzuki S, Kojima M et al. Lipid rafts mediate chemotropic guidance of nerve growth cones. Neuron 2004; 42(1):51-62.
23. Nowak L, Bregestovski P, Ascher P et al. Magnesium gates glutamate-activated channels in mouse central neurones. Nature 1984; 307(5950):462-465.
24. Jahr CE, Stevens CF. Glutamate activates multiple single channel conductances in hippocampal neurons. Nature 1987; 325(6104):522-525.
25. Henley J, Poo MM. Guiding neuronal growth cones using Ca2+ signals. Trends Cell Biol 2004; 14(6):320-330.
26. Deller T, Haas CA, Frotscher M. Sprouting in the hippocampus after entorhinal cortex lesion is layer- specific but not translaminar: Which molecules may be involved? Restor Neurol Neurosci 2001; 19(3-4):159-167.
27. Koyama R, Yamada MK, Nishiyama N et al. Group II metabotropic glutamate receptor activation is required for normal hippocampal mossy fibre development in the rat. J Physiol 2002; 539(Pt 1):157-162.
28. Ohishi H, Shigemoto R, Nakanishi S et al. Distribution of the mRNA for a metabotropic glutamate receptor (mGluR3) in the rat brain: An in situ hybridization study. J Comp Neurol 1993; 335(2):252-266.
29. Shigemoto R, Kinoshita A, Wada E et al. Differential presynaptic localization of metabotropic glutamate receptor subtypes in the rat hippocampus. J Neurosci 1997; 17(19):7503-7522.
30. Kreibich TA, Chalasani SH, Raper JA. The neurotransmitter glutamate reduces axonal responsiveness to multiple repellents through the activation of metabotropic glutamate receptor 1. J Neurosci 2004; 24(32):7085-7095.
31. Kapfhammer JP, Raper JA. Collapse of growth cone structure on contact with specific neurites in culture. J Neurosci 1987; 7(1):201-212.
32. Mattson MP, Dou P, Kater SB. Outgrowth-regulating actions of glutamate in isolated hippocampal pyramidal neurons. J Neurosci 1988; 8(6):2087-2100.
33. Metzger F, Wiese S, Sendtner M. Effect of glutamate on dendritic growth in embryonic rat motoneurons. J Neurosci 1998; 18(5):1735-1742.
34. Baird DH, Trenkner E, Mason CA. Arrest of afferent axon extension by target neurons in vitro is regulated by the NMDA receptor. J Neurosci 1996; 16(8):2642-2648.
35. Chang S, De Camilli P. Glutamate regulates actin-based motility in axonal filopodia. Nat Neurosci 2001; 4(8):787-793.

36. Vanderklish PW, Edelman GM. Dendritic spines elongate after stimulation of group 1 metabotropic glutamate receptors in cultured hippocampal neurons. Proc Natl Acad Sci USA 2002; 99(3):1639-1644.
37. McKinney RA, Capogna M, Durr R et al. Miniature synaptic events maintain dendritic spines via AMPA receptor activation. Nat Neurosci 1999; 2(1):44-49.
38. Hebb. The organization of behavior: A neuropsychological theory. 1949.
39. Ruthazer ES, Cline HT. Insights into activity-dependent map formation from the retinotectal system: A middle-of-the-brain perspective. J Neurobiol 2004; 59(1):134-146.
40. Caulfield MP, Birdsall NJ. International Union of Pharmacology. XVII. Classification of muscarinic acetylcholine receptors. Pharmacol Rev 1998; 50(2):279-290.
41. Gu Q. Neuromodulatory transmitter systems in the cortex and their role in cortical plasticity. Neuroscience 2002; 111(4):815-835.
42. Lukas RJ, Changeux JP, Le Novere N et al. International Union of Pharmacology. XX. Current status of the nomenclature for nicotinic acetylcholine receptors and their subunits. Pharmacol Rev 1999; 51(2):397-401.
43. Zoli M, Le Novere N, Hill Jr JA et al. Developmental regulation of nicotinic ACh receptor subunit mRNAs in the rat central and peripheral nervous systems. J Neurosci 1995; 15(3 Pt 1):1912-1939.
44. Dori I, Parnavelas JG. The cholinergic innervation of the rat cerebral cortex shows two distinct phases in development. Exp Brain Res 1989; 76(2):417-423.
45. Small DH, Reed G, Whitefield B et al. Cholinergic regulation of neurite outgrowth from isolated chick sympathetic neurons in culture. J Neurosci 1995; 15(1 Pt 1):144-151.
46. Lipton SA, Frosch MP, Phillips MD et al. Nicotinic antagonists enhance process outgrowth by rat retinal ganglion cells in culture. Science 1988; 239(4845):1293-1296.
47. Owen A, Bird M. Acetylcholine as a regulator of neurite outgrowth and motility in cultured embryonic mouse spinal cord. Neuroreport 1995; 6(17):2269-2272.
48. Pugh PC, Berg DK. Neuronal acetylcholine receptors that bind alpha-bungarotoxin mediate neurite retraction in a calcium-dependent manner. J Neurosci 1994; 14(2):889-896.
49. Elsas SM, Kwak EM, Stent GS. Acetylcholine-induced retraction of an identified axon in the developing leech embryo. J Neurosci 1995; 15(2):1419-1436.
50. Coronas V, Durand M, Chabot JG et al. Acetylcholine induces neuritic outgrowth in rat primary olfactory bulb cultures. Neuroscience 2000; 98(2):213-219.
51. Rosner H, Fischer H. In growth cones of rat cerebral neurons and human neuroblastoma cells, activation of protein kinase C causes a shift from filopodial to lamellipodial actin dynamics. Neurosci Lett 1996; 219(3):175-178.
52. Kozma R, Sarner S, Ahmed S et al. Rho family GTPases and neuronal growth cone remodelling: Relationship between increased complexity induced by Cdc42Hs, Rac1, and acetylcholine and collapse induced by RhoA and lysophosphatidic acid. Mol Cell Biol 1997; 17(3):1201-1211.
53. Zheng JQ, Felder M, Connor JA et al. Turning of nerve growth cones induced by neurotransmitters. Nature 1994; 368(6467):140-144.
54. Kolb B, Tees R, Ables M. The cerebral cortex of the rat. 1990.
55. Ha DH, Robertson RT, Weiss JH. Distinctive morphological features of a subset of cortical neurons grown in the presence of basal forebrain neurons in vitro. J Neurosci 1998; 18(11):4201-4215.
56. Lauder JM, Schambra UB. Morphogenetic roles of acetylcholine. Environ Health Perspect 1999; 107(Suppl 1):65-69.
57. Hohmann CF. A morphogenetic role for acetylcholine in mouse cerebral neocortex. Neurosci Biobehav Rev 2003; 27(4):351-363.
58. Hohmann CF, Brooks AR, Coyle JT. Neonatal lesions of the basal forebrain cholinergic neurons result in abnormal cortical development. Brain Res 1988; 470(2):253-264.
59. Hohmann CF, Berger-Sweeney J. Cholinergic regulation of cortical development and plasticity. New twists to an old story. Perspect Dev Neurobiol 1998; 5(4):401-425.
60. Allendoerfer KL, Shatz CJ. The subplate, a transient neocortical structure: Its role in the development of connections between thalamus and cortex. Annu Rev Neurosci 1994; 17:185-218.
61. Skaliora I, Adams R, Blakemore C. Morphology and growth patterns of developing thalamocortical axons. J Neurosci 2000; 20(10):3650-3662.
62. Franek M. History and the present of metabotropic GABAB receptor. Cesk Fysiol 2004; 53(3):117-124.
63. Marshall FH, Jones KA, Kaupmann K et al. GABAB receptors - The first 7TM heterodimers. Trends Pharmacol Sci 1999; 20(10):396-399.
64. Wolff JR, Joo F, Dames W. Plasticity in dendrites shown by continuous GABA administration in superior cervical ganglion of adult rat. Nature 1978; 274(5666):72-74.

65. Matsutani S, Yamamoto N. GABAergic neuron-to-astrocyte signaling regulates dendritic branching in coculture. J Neurobiol 1998; 37(2):251-264.
66. Michler A. Involvement of GABA receptors in the regulation of neurite growth in cultured embryonic chick tectum. Int J Dev Neurosci 1990; 8(4):463-472.
67. Obrietan K, van den Pol AN. Growth cone calcium elevation by GABA. J Comp Neurol 1996; 372(2):167-175.
68. Barbin G, Pollard H, Gaiarsa JL et al. Involvement of GABAA receptors in the outgrowth of cultured hippocampal neurons. Neurosci Lett 1993; 152(1-2):150-154.
69. LoTurco JJ, Owens DF, Heath MJ et al. GABA and glutamate depolarize cortical progenitor cells and inhibit DNA synthesis. Neuron 1995; 15(6):1287-1298.
70. Marty S, Berninger B, Carroll P et al. GABAergic stimulation regulates the phenotype of hippocampal interneurons through the regulation of brain-derived neurotrophic factor. Neuron 1996; 16(3):565-570.
71. Haydar TF, Wang F, Schwartz ML et al. Differential modulation of proliferation in the neocortical ventricular and subventricular zones. J Neurosci 2000; 20(15):5764-5774.
72. Nadarajah B, Parnavelas JG. Modes of neuronal migration in the developing cerebral cortex. Nat Rev Neurosci 2002; 3(6):423-432.
73. Rakic P. Mode of cell migration to the superficial layers of fetal monkey neocortex. J Comp Neurol 1972; 145(1):61-83.
74. Anderson SA, Marin O, Horn C et al. Distinct cortical migrations from the medial and lateral ganglionic eminences. Development 2001; 128(3):353-363.
75. Behar TN, Schaffner AE, Tran HT et al. GABA-induced motility of spinal neuroblasts develops along a ventrodorsal gradient and can be mimicked by agonists of GABAA and GABAB receptors. J Neurosci Res 1995; 42(1):97-108.
76. Behar TN, Schaffner AE, Scott CA et al. Differential response of cortical plate and ventricular zone cells to GABA as a migration stimulus. J Neurosci 1998; 18(16):6378-6387.
77. Behar TN, Schaffner AE, Scott CA et al. GABA receptor antagonists modulate postmitotic cell migration in slice cultures of embryonic rat cortex. Cereb Cortex 2000; 10(9):899-909.
78. Behar TN, Smith SV, Kennedy RT et al. GABA(B) receptors mediate motility signals for migrating embryonic cortical cells. Cereb Cortex 2001; 11(8):744-753.
79. Behar TN, Scott CA, Greene CL et al. Glutamate acting at NMDA receptors stimulates embryonic cortical neuronal migration. J Neurosci 1999; 19(11):4449-4461.
80. Xiang Y, Li Y, Zhang Z et al. Nerve growth cone guidance mediated by G protein-coupled receptors. Nat Neurosci 2002; 5(9):843-848.
81. Song H, Ming G, He Z et al. Conversion of neuronal growth cone responses from repulsion to attraction by cyclic nucleotides. Science 1998; 281(5382):1515-1518.
82. Höpker VH, Shewan D, Tessier-Lavigne M et al. Growth-cone attraction to netrin-1 is converted to repulsion by laminin- 1. Nature 1999; 401(6748):69-73.
83. Priest CA, Puche AC. GABAB receptor expression and function in olfactory receptor neuron axon growth. J Neurobiol 2004; 60(2):154-165.
84. Bird M, Owen A. Neurite outgrowth-regulating properties of GABA and the effect of serum on mouse spinal cord neurons in culture. J Anat 1998; 193 (Pt 4):503-508.
85. Lieske V, Bennett-Clarke CA, Rhoades RW. Effects of serotonin on neurite outgrowth from thalamic neurons in vitro. Neuroscience 1999; 90(3):967-974.

Chapter 9

The Hedgehog, TGF-β/BMP and Wnt Families of Morphogens in Axon Guidance

Frédéric Charron* and Marc Tessier-Lavigne

Summary

During embryonic development, morphogens act as graded positional cues to dictate cell fate specification and tissue patterning. Recent findings indicate that morphogen gradients also serve to guide axonal pathfinding during development of the nervous system. These findings challenge our previous notions about morphogens and axon guidance molecules and suggest that these proteins, rather than having sharply divergent functions, act more globally to provide graded positional information that can be interpreted by responding cells either to specify cell fate or to direct axonal pathfinding. This chapter presents the roles identified for members of three prominent morphogen families—the Hedgehog, Wnt and TGF-β/BMP families—in axon guidance, and discusses potential implications for the molecular mechanisms underlying their guidance functions.

Introduction

In the 1990s, genetic, biochemical and molecular approaches let to the identification of four major conserved families of guidance cues with prominent developmental effects: the Netrins, Slits, Semaphorins and Ephrins.[1,2] Although the identification of these major families of guidance cues has increased our understanding of how the nervous system is wired, many guidance events observed during development do not appear to be accounted for by these molecules. Moreover, the number of guidance cues and receptors identified seemed small relative to the immense complexity of nervous system wiring; thus, it seems likely that additional guidance cues and receptors remain to be discovered. Remarkably, over the last few years, members from three other families of secreted signaling molecules were shown to act as guidance cues: the Wingless/Wnt, Hedgehog (Hh) and Decapentaplegic/Bone Morphogenic Protein/Transforming Growth Factor-β (Dpp/BMP/TGF-β) families. In addition to their axon guidance properties, these molecules share a common characteristic of having been previously identified as morphogens controlling cell fate and tissue patterning. This discovery has opened the door to the study of an entirely new set of axon guidance cues. In addition, these findings change our current notions about morphogenic and axon guidance molecules and suggest that these proteins can be thought of more generally as providing graded positional information that can be interpreted by responding cells for either the specification of cell fate or for axonal pathfinding.

*Corresponding Author: Frédéric Charron—Molecular Biology of Neural Development, Institut de Recherches Cliniques de Montréal (IRCM), 110 Pine Ave West, Montreal, Quebec, H2W 1R7, Canada. Email: frederic.charron@ircm.qc.ca

After briefly introducing the roles of the three morphogen families in cell fate specification, we will focus on the emerging evidence that they are reused later in development to guide axons. We conclude by comparing similarities and differences in positional information interpretation for axon guidance versus cell fate specification.

Morphogens, Cell Fate Specification and Tissue Patterning

Morphogens are signaling molecules produced in a restricted region of a tissue that provide positional information by diffusing from their source to form a long-range concentration gradient. A cell's program of differentiation in response to a morphogen is dictated by its position within the gradient and thus on its distance from the morphogen source. Two criteria have gained acceptance as the evidence needed to qualify a secreted signaling protein as a morphogen: it must have a concentration-dependent effect on its target cells and it must exert a direct action at a distance. To date, only three protein families have members that fulfill these criteria: the Wingless/Wnt, Hedgehog (Hh) and Dpp/BMP/TGF-β families.[3] Although there is abundant evidence for concentration-dependent activity of signaling proteins during development (reviewed in ref. 4), evidence for direct action at a distance has only been provided recently in some vertebrate systems.[5,6] In the following section, we summarize briefly some of the biological processes involving members from each of the morphogen families, with a special emphasis on vertebrate neural tube development, which provides a convenient system to compare and contrast roles of classic guidance molecules and morphogens in axon guidance.

In vertebrate embryos, one of the first steps in the development of the nervous system is the specification of the diverse neural cell fates. Members of each of the three morphogen families are expressed in the developing neural tube and are implicated in its patterning, as we now summarize.

The Hedgehog Family

Hedgehog proteins are found in insects and vertebrates, but not nematodes. There is a single Hedgehog gene in flies, and three in mammals: Sonic hedgehog (Shh), Indian hedgehog (Ihh) and Desert hedgehog (Dhh). Shh is secreted by the notochord and by floor plate cells at the ventral midline of the neural tube, and functions as a graded signal for the generation of distinct classes of ventral neurons along the dorsoventral axis of the neural tube (Fig. 1A; reviewed in refs. 7-9). In agreement with its role as a morphogen, Shh is able to induce a range of ventral spinal cord cell fates in a concentration-dependent manner[10] and has been shown to exert a direct action at a distance to specify neural tube cell fate.[6]

Much evidence indicates that these cell fate specification and tissue patterning activities of Hhs are mediated by members of the Ci/Gli transcription factor family, but the signaling mechanisms that lead to activation of these transcription factors are not fully elucidated.[8] Genetic and biochemical experiments have shown that Hhs activate signaling by binding to their receptor Patched (Ptc), which leads to the relief of Ptc-mediated inhibition of Smoothened (Smo), also a transmembrane protein, which can then activate downstream signaling (Fig. 2A). Smo associates directly with a Ci-containing complex, which contains the atypical kinesin Costal-2 (Cos2) and the protein kinase Fused (Fu).[11] This complex constitutively suppresses pathway activity. Activation of Hh signaling reverses this regulatory effect and allows Ci to activate transcription of Hh target genes, thus specifying cell fate.

The Decapentaplegic/Bone Morphogenic Protein/Transforming Growth Factor-β Family

The roof plate at the dorsal midline of the neural tube is the major source of inductive signals controlling the generation of dorsal cell types.[12] Around the time when dorsal neurons are generated, the roof plate expresses many members of the Dpp/BMP/TGF-β family and some of them are required for the normal specification of dorsal neurons[13] (Fig. 1A). Whether they function specifically as morphogens in this system remains to be determined.[7]

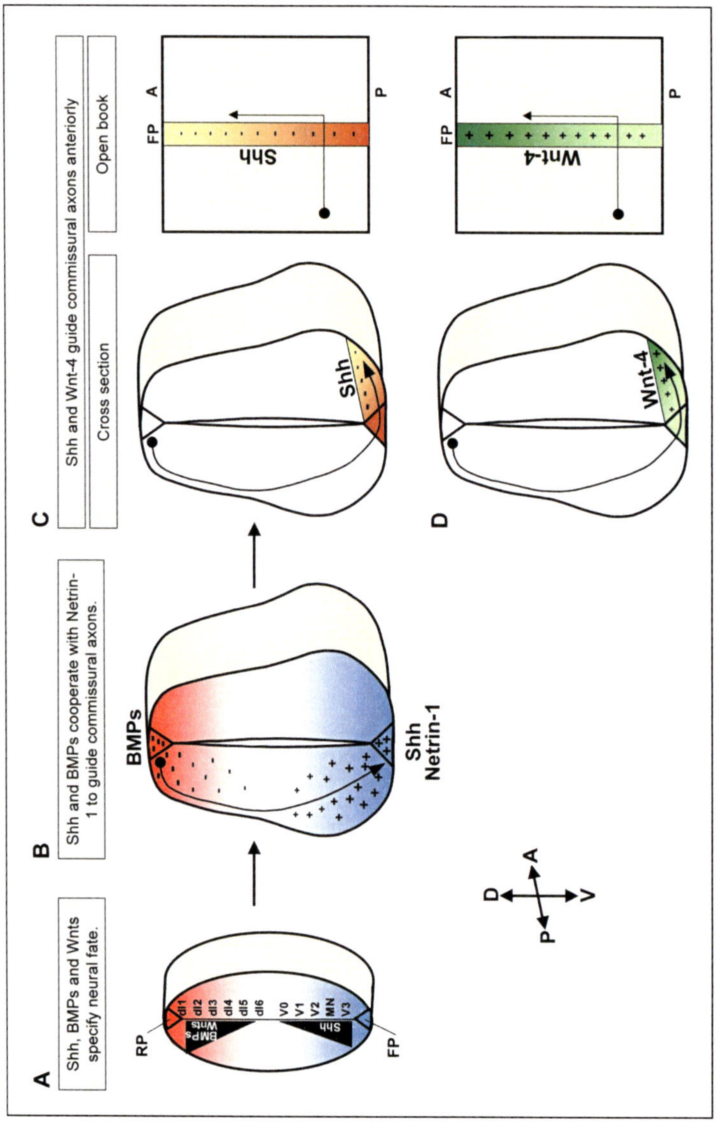

Figure 1. Neuronal cell fate specification and guidance of commissural axons by morphogens and Netrin-1. Three sets of morphogens, Shh, BMPs and Wnts, are first used to pattern neural progenitors in the spinal cord, and then appear to be reused as guidance cues for commissural axons. A) In the early neural tube, Shh, BMP and Wnt protein concentration gradients act to specify neural cell fate in the ventral and dorsal spinal cord. B) Later, the axons of differentiated commissural neurons are repelled from the dorsal midline by BMPs (red) and attracted to the ventral midline by the combined chemoattractant effects of Netrin-1 and Shh (blue). C) and D) After crossing the floor plate, commissural axons are repelled from the posterior pole by a Shh gradient (C; orange) and attracted anteriorly by a Wnt-4 gradient (D; green). A, B and left panels in C and D are cross section representations of the developing spinal cord and right panels in C and D are open book representations. V0-3, ventral interneuron sub-populations; dI1-6, dorsal interneuron sub-populations; MN, motoneurons; RP, roof plate; FP, floor plate; D, dorsal; V, ventral; P, posterior; A, anterior. Reproduced with permission of the Company of Biologists.

Members of the Dpp/BMP/TGF-β family regulate cell fate by inducing the dimerization of type I and type II TGF-β receptors, resulting in phosphorylation and activation of the intracellular kinase domain of the type I receptor (Fig. 2B). Targets of the type I receptor are the receptor-regulated Smads (R-Smads) which, upon phosphorylation, associate with co-Smads and translocate to the nucleus where they activate transcription.

The Wingless/Wnt Family

Roof plate cells also express several members of the Wnt family (reviewed in ref. 12). Although Wnt-1 and Wnt-3a are required for normal specification of dorsal neurons,[14] it also remains an open question whether they function specifically as morphogens in this system.

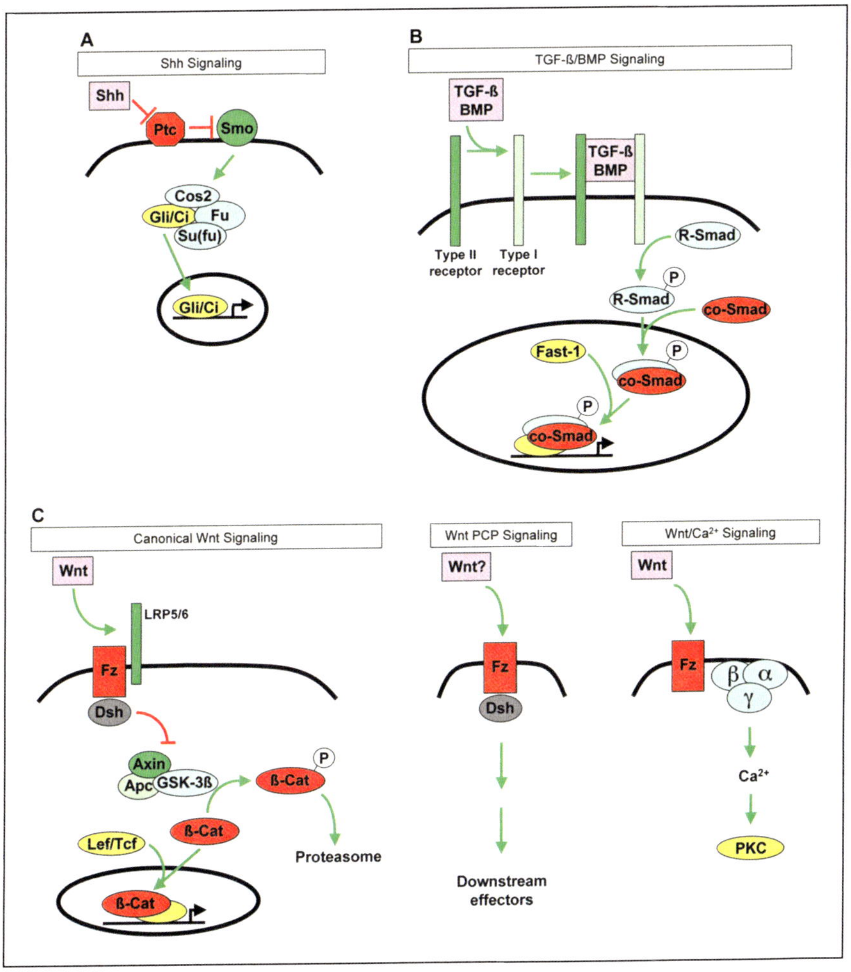

Figure 2. Overview of the Shh, TGF-β/BMP and Wnt morphogenic signaling pathways. A) Shh signaling pathway. Genetic and biochemical experiments have shown that Hhs activate signaling by binding to their receptor Patched (Ptc; a 12-pass transmembrane protein), which leads to the relief of Ptc-mediated inhibition of Smoothened (Smo), a 7-pass transmembrane protein, which can then activate downstream signaling. Smo associates directly with a Ci-containing complex which contains the atypical kinesin Costal-2 (Cos2), the protein kinase Fused (Fu) and the Suppressor of Fused [Su(fu)]. Continued on next page.

Figure 2, continued. This complex constitutively suppresses pathway activity by leading to the proteolytic cleavage of Ci, which acts as a transcriptional repressor. Activation of Hh signaling reverses this regulatory effect and leads to the production of full length Ci, which activates transcription of Hh target genes. B) TGF-β/BMP signaling pathway. Members of the Dpp/BMP/TGF-β family regulate cell fate by inducing the dimerization of type I and type II TGF-β receptors, resulting in phosphorylation and activation of the intracellular kinase domain of the type I receptor. Targets of the type I receptor are the receptor-regulated Smads (R-Smads) which, upon phosphorylation, associate with co-Smads and translocate to the nucleus where, together with DNA-binding partners such as Fast-1, they activate transcription. C) Wnt signaling pathway. Wnt ligands can activate several different signal transduction pathways. The canonical Wnt pathway controls gene expression by stabilizing β-Catenin (β-Cat). Frizzled (Fz) proteins are seven transmembrane domains molecules which, together with the members of the low density lipoprotein (LDL) receptor-related protein 5 and 6 (LRP5/6; Arrow in Drosophila) family of co-receptors, function as Wnt receptors. When Wnts are absent, β-Catenin is phosphorylated by GSK-3β leading to its degradation by the proteasome. This process requires the formation of a complex scaffolded by Axin and adenomatous polyposis coli (Apc). Binding of Wnts to their receptors results in Dishevelled (Dsh) activation and suppression of GSK-3β activity, thus protecting β-Catenin from targeting for degradation. Accumulated β-Catenin converts the lymphoid enhancer factor (Lef)/Tcf from a transcriptional repressor to an activator. Two non-canonical Wnt pathways have been described: the Wnt/Ca^{2+} pathway and the planar cell polarity (PCP) pathway. The PCP pathway involves a non-canonical β-Catenin-independent Wnt/Fz pathway that requires Dsh. A Wnt ligand for this pathway has yet to be identified in *Drosophila*, but Wnt ligands have been found to activate an analogous pathway in vertebrates. The Wnt/Ca^{2+} pathway probably signals via heterotrimeric G-proteins (α, β and γ subunits) to mobilize intracellular Ca^{2+} and, in some contexts, to stimulate protein kinase C (PKC). In vertebrates, Wnt/Ca^{2+} signaling is activated by the same ligands as the PCP pathway, suggesting that these pathways may overlap to some extent. Reproduced with permission of the Company of Biologists.

Wnt ligands can activate several different signal transduction pathways. The most extensively studied is the canonical Wnt pathway, which controls gene expression by stabilizing β-Catenin (Fig. 2C). This pathway involves evolutionarily-conserved cellular components (reviewed in refs. 15,16). Frizzled (Fz) proteins are seven transmembrane domains molecules which function as Wnt receptors. When Wnts are absent, β-Catenin is phosphorylated by GSK-3β leading to its degradation. Binding of Wnts to their receptors results in Dishevelled (Dsh) activation and suppression of GSK-3β activity, thus stabilizing β-Catenin. Accumulated β-Catenin converts the lymphoid enhancer factor (Lef)/Tcf from a transcriptional repressor to an activator.

Recently, much attention has been given to the non-canonical Wnt pathways, i.e., those that are β-Catenin-independent. Two have been described: the Wnt/Ca^{2+} pathway and the planar cell polarity (PCP) pathway (Fig. 2C). The PCP pathway involves a non-canonical β-Catenin-independent Wnt/Fz pathway that requires Dsh. The Wnt/Ca^{2+} pathway is thought to signal via heterotrimeric G-proteins to mobilize intracellular Ca^{2+} and, in some contexts, to stimulate protein kinase C (PKC).

Below, we discuss the functions of these morphogen families in axon guidance.

Morphogens in Axon Guidance

The Hedgehog Family

Shh is a Chemoattractant for Commissural Axons

During spinal cord development, commissural neurons, which differentiate in the dorsal neural tube, send axons that project toward and subsequently across the floor plate, forming axon commissures (Fig. 1B and see also ref 17). As discussed in this book (see chapter by T. Kennedy and colleagues), these axons project toward the midline in part because they are attracted by Netrin-1, a long-range chemoattractant secreted by the floor plate refs. 18-22). In mice mutant for *Netrin-1* or its receptor *DCC* many commissural axon trajectories are foreshortened, fail to invade the ventral spinal cord, and are misguided refs. 20,23). However, some of them do reach the midline, indicating that other guidance cues cooperate with Netrin-1 to guide these axons. Further analysis of *Netrin-1* knock-out mice suggested that the floor plate might actually express an additional diffusible attractant(s) for commissural axons.[20,24]

Given its expression by the floor plate and its long-range effects in the spinal cord, Shh was a candidate for a midline-derived axonal guidance cue. Shh was indeed shown to function as an axonal chemoattractant that can mimic the Netrin-1-independent chemoattractant activity of the floor plate in in vitro assays.[24] The chemoattractant activity of Shh, like the chemoattractant activity of floor plate derived from *Netrin-1* mutants, can be blocked by cyclopamine, which blocks the actions of Shh in cell fate determination by inhibiting the Shh signaling mediator Smo. This shows that Smo is required for Shh-mediated axon attraction and, importantly, that the Netrin-1-independent chemoattractant activity of the floor plate also requires Hh signaling. Since Shh is the only Hh family member expressed in the spinal cord at this stage, these results suggest that Shh is functioning as a floor plate-derived chemoattractant for commissural axons.

While the reorienting effect of Shh could be due to a direct chemoattractant effect, an alternative explanation is suggested by the fact that Shh is a potent morphogen. Since in these assays commissural axon turning occurs within the spinal cord tissue explant, it seemed possible that Shh was not acting directly on the axons but rather repatterning and altering the expression of guidance cues by cells within the explant, which then secondarily guided the axons to the Shh source. Arguing against this possibility is the finding that the spinal cord explants used to assess chemoattractant activity are at a developmental stage at which they have apparently lost the competence to be repatterned by Shh, as assessed using a battery of markers of dorsoventral patterning.[24]

A direct action of Shh in attracting the axons was supported further by two sets of experiments.[24] First, Shh was shown to attract the growth cones of isolated *Xenopus* spinal axons in dispersed cell culture in a cyclopamine-dependent manner, proving that Shh, acting via Smo, can function as a chemoattractant at least for these *Xenopus* axons. A second way of providing evidence that Shh can act directly on commissural axons to guide them relied on blocking Shh signaling selectively in commissural neurons without blocking it in the terrain through which their axons course. This was achieved by conditional inactivation of a floxed allele of *Smo* using the Cre recombinase expressed under the control of the Wnt1 promoter, which drives expression in the dorsal spinal cord (as well as in neural crest progenitors). When Cre, driven by this promoter, was used to delete a floxed *Smo* allele in the dorsal spinal cord, commissural axon trajectories were defective in the ventral spinal cord, where Cre is not expressed (Table 1). This result strongly implies that the axonal misrouting is not due to repatterning of the ventral spinal cord, and must instead reflect a guidance defect arising from loss of Smo function in commissural neurons. Taken together, these results suggest that Shh functions to guide commissural axons both in vitro and in vivo by acting directly as a chemoattractant on these axons through a Smo-dependent signaling mechanism.

Shh Guides Commissural Axons Along the Longitudinal Axis of the Spinal Cord

After commissural axons have reached and crossed the floor plate, they make a sharp anterior turn toward the brain (Fig. 1). The molecules involved in the dorso-ventral projection of commissural axons to and at the floor plate have been well described, but it is only recently that cues controlling anteroposterior guidance have been identified. Remarkably, the guidance of commissural axons to the floor plate is not, apparently, the only effect of Shh on commissural axons: recent evidence suggests that Shh also guides post-crossing commissural axons in the rostral direction along the longitudinal axis of the spinal cord.[25] Using a subtractive hybridization approach to identify guidance cues responsible for the rostral turn of post-crossing commissural axons in chick embryos, Bourikas and colleagues identified differentially-expressed candidates whose function they investigated by RNA interference (RNAi)-mediated in ovo

Table 1. In vivo experiments supporting a role for morphogens in axon guidance

Morphogen Pathways	Genes	Species	Experiments	Phenotypes	References
Hedgehog	*Smo*	Mouse	Conditional inactivation of *Smo* in commissural neurons	Commissural axons invade the motor columns	24
	Shh	Chick	Ectopic expression of Shhin the optic chiasm	Retinal axons are prevented from crossing the chiasm	27
	Shh	Chick	Silencing of Shh by RNA interference	Commissural axons stall at the contralateral floor plate border, with some axons randomly turning caudally or rostrally	25
TGFβ/BMP	*Bmp7 and Gdf7*	Mouse	*Bmp7* and *Gdf7* gene inactivation	Initial trajectory of commissural axons aberrant, axons invade the roof plate	32, 33
	unc-129	*C. elegans*	*unc-129* mutation	Defects in the dorsally oriented trajectories of motor axons	38, 39
Wnt	*drl*	*Drosophila*	*drl* mutation	Axons that normally project into the anterior commissure now project into the posterior commissure	44
	drl	*Drosophila*	Ectopic expression of Drl in posterior commissure neurons	Forces posterior commissureneurons to cross in the anterior commissure	44
	wnt5	*Drosophila*	*wnt5* mutation	Axons that normally project into the anterior commissure project into the posterior commissure	45
	wnt5	*Drosophila*	Ectopic expression of Wnt5 throughout the CNS midline	Prevents the anterior commissure from forming	45
	Fz3	Mouse	*Fz3* gene inactivation	Defects in anteroposterior guidance of commissural axons after midline crossing	47
	Ryk	Mouse	Injection of anti-Ryk blocking antibodies in the spinal cord	Blocks the posterior growth of CST axons in the spinal cord	50
	Wnt3	chick	Wnt3 overexpression in the tectum	RGC axons avoid the site of ectopic Wnt3	55
	Ryk	chick	Expression of a dominant-negative Ryk in dorsal RGC axons	Medial shift of RGC axon termination zone in the tectum	55

gene silencing. Unexpectedly, one of their candidates turned out to be Shh. In agreement with these results, silencing of the *Shh* gene by a different RNAi construct or injection of a hybridoma producing a function-blocking Shh antibody led to axon stalling at the contralateral floor plate border, with some axons turning caudally or rostrally, apparently in a random manner. Importantly, marker analysis revealed that the patterning of the spinal cord was not apparently affected by these manipulations, suggesting that these experiments were done after neural cell fate specification by Shh has occurred. Finally, post-crossing commissural axons were shown to avoid ectopic Shh in vivo. Together, these results provide strong evidence that Shh is essential for the normal guidance of commissural axons along the longitudinal axis of the spinal cord.

A Shh gradient could be guiding commissural axons directly, or could alternatively be acting only indirectly by controlling a graded distribution of a distinct guidance cue. Two lines of evidence, however, were provided for a direct role of Shh.[25] The first came from an investigation of the receptor mechanism for this guidance. Interestingly, neither cyclopamine nor Smo RNAi interfered with the rostral turn of commissural axons along the longitudinal axis, suggesting that Smo might not be involved in this process. Instead, RNAi-mediated silencing of *Hip1*, a gene encoding a Shh-binding membrane protein transiently expressed in commissural neurons at the time when they cross the floor plate as well as in the peri-ventricular region, resulted in the same post-crossing phenotype as Shh RNAi. These results, which contrast with the essential role of Smo in Shh-mediated attraction of commissural axons to the floor plate,[24] suggest that Hip1 might be involved in transducing a Shh guidance signal in post-crossing commissural neurons. The relatively restricted expression of Hip1 mRNA to commissural neurons would be consistent with a direct action of Shh on these axons. A second line of evidence that supports a direct role for Shh was obtained by in vitro experiments, which showed that post-crossing commissural axons from spinal cord explants can be repelled by Shh beads in vitro. Taken together, these results suggest a model in which Shh could be functioning directly through Hip1 as a chemorepellent for post-crossing commissural axons (Fig. 1C).

Shh is a Negative Regulator of Retinal Ganglion Cell Axon Growth

Retinal ganglion cell (RGC) axons growing towards the diencephalic ventral midline are faced with the decision to project either contralaterally or ipsilaterally in response to guidance cues at the optic chiasm (Fig. 3). Homozygous inactivation of the mouse *Pax2* gene alters the development of the optic chiasm and RGC axons never cross the midline in these mice. Interestingly, whereas in wild-type mice Shh expression is downregulated in the chiasm as RGC axons are migrating towards this region, Shh expression is ectopically maintained along the ventral midline in *Pax2*$^{-/-}$ mice.[26] These obervations raised the possibility that the continuous expression of Shh at the ventral midline might contribute to preventing RGC axon crossing. In agreement with this idea, Trousse et al (2001) found that ectopic expression of Shh in the midline region interferes with RGC axon growth and prevents them from crossing the midline[27] (Fig. 3). Consistent with the idea that Shh might be acting directly on RGC axons, it was shown that these manipulations do not affect patterning and neural differentiation in the eye. Further experiments will be required to determine whether the chiasm region is repatterned in these experiments, but in vitro experiments supported the idea that Shh acts directly to control RGC axon migration: addition of exogenous recombinant Shh to retinal explants decreases the number and length of growing axons, without interfering with the rate of proliferation and differentiation of cells in the explant, and time-lapse analysis showed that addition of Shh to retinal explants rapidly causes growth cone arrest and subsequent retraction of RGC axons.[27] Since the response of the growth cone to many extracellular guidance cues appears to be modulated and in some cases perhaps even mediated by intracellular cyclic nucleotide levels (cAMP and cGMP),[28,29] the possibility was explored that the effect of Shh on retinal axons in vitro might be due to a change in cAMP levels. In agreement with this, addition of Shh to retinal growth cones was shown to decrease intracellular levels of cAMP, a finding consistent with the

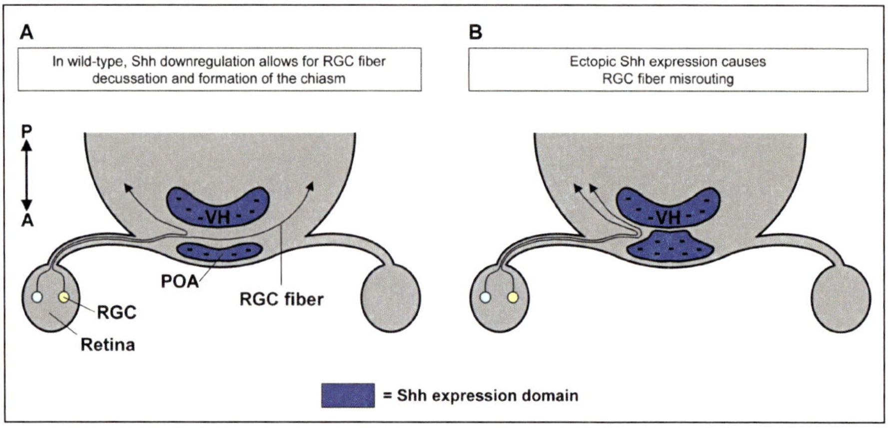

Figure 3. Shh expression at the chiasm border defines a barrier within the ventral midline implicated in guiding RGC axons. Retinal ganglion cell (RGC) axons growing towards the diencephalic ventral midline are faced with the decision to project either contralaterally or ipsilaterally in response to guidance cues at the optic chiasm. A) In wild-type animals, downregulation of Shh expression allows RGC axon decussation and formation of the chiasm. B) Ectopic expression of Shh in the ventral midline region interferes with RGC axon growth and prevents them from crossing the midline. The Shh expression domain is shown in blue. Shh can inhibit retinal axons in vitro, suggesting that in vivo it may be acting on the axons directly rather than by altering the expression of distinct guidance cues in the chiasm, although conclusive evidence for this guidance function in vivo remains to be obtained. A, anterior; P, posterior; POA, pre-optic area; VH, ventral hypothalamus. Reproduced with permission of the Company of Biologists.

observation that lowering cAMP levels favors growth inhibition[30] (see also the chapter by Holt and colleagues in this book).

Taken together, these results provide evidence that Shh expression at the chiasm border helps define a barrier within the ventral midline that serves to guide RGC axons, and suggest that Shh may be acting on the axons directly, rather than indirectly by repatterning the chiasm. Proving that the effect in vivo is direct will, however, require additional studies, such as identifying the mechanism that mediates retinal growth cone responses to Shh and showing that cell-autonomous inhibition of this signaling pathway in the neurons results in guidance defects in vivo.

The opposite effects of Shh on pre- and post-crossing commissural and retinal axons (attraction and repulsion) might be due to an intrinsic or extrinsic factor that modulates cyclic nucleotide levels, much as extrinsic factors can convert Netrin-attraction to repulsion by modulating cyclic nucleotide levels.[31] Alternatively, as the molecular mechanisms underlying the effects of Shh on commissural and retinal axons are poorly understood, it is also possible that these two effects are mediated by distinct signaling pathways that result in opposite guidance effects—a possibility that also has a precedent in the case of Netrins, which can attract axons by activating DCC family receptors and repel them by activating UNC5 family receptors.[1,2]

The Dpp/BMP/TGF-β family

BMPs Are Chemorepellents for Commissural Axons

In *Netrin-1* and *DCC* mutants, commissural axons initially migrate ventrally for approximately the first third of their normal trajectory before becoming misrouted,[20,23] suggesting that an additional guidance cue might be acting to control their dorsal migration.

The proximity of commissural neurons to the roof plate and their initial growth away from the dorsal midline suggested that the roof plate might repel commissural axons away. A direct test of this possibility showed that the roof plate expresses a diffusible activity that repels commissural axons in vitro[32] (Fig. 1B). Testing a battery of candidate diffusible molecules that might act as repellent signals, it was found that BMP7 and BMP6, two BMP family members expressed by the roof plate, can each mimic the chemorepellent activity of the roof plate in vitro without causing changes in spinal cord cell fate at the doses used for chemorepulsion. Inhibition of BMP7 activity using soluble inhibitors of BMP activity, BMP7 blocking antibodies and genetic inactivation of *Bmp7* showed that BMP7 contributes to the chemorepellent activity of the roof plate for commissural axons. Moreover, BMP7 was shown to induce the collapse of commissural axon growth cones, providing evidence that it can act directly on growth cones to elicit a rapid change in cytoskeletal organization.[32] Further evidence indicated that the roof plate chemorepellent BMP complex is likely a heterodimer composed of BMP7 and GDF7, as genetic inactivation studies showed that expression of both *Bmp7* and *Gdf7* by roof plate cells is required for the fidelity of commissural axon growth in vivo.[33] Together these results support a model in which a GDF7:BMP7 heterodimer mediates the roof plate chemorepellent activity that guides the initial trajectory of commissural axons in the developing spinal cord.

The molecular mechanisms underlying the effect of BMPs on growth cones are not known. Although BMPs typically activate signaling through type I and II receptors, it is not known whether these receptors also transduce the BMP chemorepellent activity. Activation of BMP receptors normally leads to transcriptional activation by Smads, but the time course of commissural growth cone collapse is difficult to reconcile with the idea that a transcriptional effect mediates the chemorepellent effects of BMPs. Thus, the guidance effect of BMPs could be mediated independently of Smads or may involve non-transcriptional effects. Indeed, some of the cytoplasmic components implicated in guidance responses to other pathfinding cues are activated in response to TGFβ/BMP family members. For example, BMPs activate PKA and LIM kinase[34,35] and other TGFβ family members stimulate phospholipase C, PKC and Rho GTPase.[36,37]

It is interesting to note that gradients of BMPs and Shh appear to cooperate at least twice during neural tube development: first to specify cell fate, and later to guide commissural axons to the ventral midline (see Fig. 1). In the case of Shh, the same molecule plays both roles. In the case of BMPs, it remains to be determined whether the same family member can play both roles or whether different BMPs independently perform each role.

The TGFβ Family Member Unc-129 Is Required for Motor Axon Guidance

Intriguing evidence for a role of *TGFβ* family members in axon guidance has come from studies on the C. elegans gene *unc-129*, a TGFβ family member that is required for proper guidance of pioneer motor axons along the dorsoventral axis.[38,39] Mutations in *unc-129* cause defects in the dorsally oriented trajectories of motor axons that resemble those present in *unc-5*, *unc-6/Netrin* and *unc-40/DCC* mutants, without causing other overt patterning or morphological defects.[38-40] The dorsal expression of *unc-129* suggests that it might be acting as a chemoattractant for motor axons; however, whether UNC-129 acts directly on growth cones remains to be established. Interestingly, UNC-129 function does not require DAF-4, the only known type II TGFβ receptor in *C. elegans*, suggesting that UNC-129 may act through a novel receptor mechanism. Identification of the UNC-129 receptor will help determine whether UNC-129 functions directly to guide motor axons, and will help identify the mechanisms of UNC-129 signaling.

Interestingly, *unc-129* appears to act in parallel to the *unc-6/Netrin* pathway.[39,41] Thus, if UNC-129 acts directly as a guidance cue, it would provide another example of a BMP collaborating with a Netrin to guide axons; in this case, however, the sign of the guidance would be inverted compared to the spinal cord, with the BMP attracting and the Netrin repelling. As in the spinal cord, the presence of the two cooperative pathways providing a push from behind and a pull from afar would then help ensure the necessary fidelity in axon guidance required for invariant and robust development.

The Wg/Wnt Family

Wnt5 Repels Commissural Axons from the Posterior Commissure

The ability of Wnt proteins to stimulate a reorganization of the cytoskeleton during axonal growth and growth cone extension[42] suggested that Wnt proteins might also be involved in guiding axons to their targets. The first direct demonstration of a guidance role was obtained in studies of commissural neurons in the fly central nervous system (CNS). During Drosophila development, the array of axons composing the CNS has a ladder-like structure: each body segment comprises an anterior and a posterior commissural tract that cross the midline and join one of the two lateral longitudinal tracts that extend the length of the embryo (Fig. 4). The attractive and repulsive signals regulating the decision of commissural axons to cross or not have been well characterized; however, how axons choose between the two major subdivisions of the crossing pathways—the anterior or posterior commissure—was only recently elucidated.

Based on the initial observation that the expression of the Derailed (Drl) receptor tyrosine kinase, a relative of the vertebrate RYK family, is restricted to the growth cone and axons of neurons that project in the anterior commissure,[43] the role of Drl in guiding these neurons through the anterior commissure was explored.[44] In the absence of Drl, the neurons that normally project into the anterior commissure were found often to project into the posterior commissure. Conversely, misexpression of Drl in posterior commissure neurons forced them to cross in the anterior commissure. Remarkably, Drl appears to be able to redirect any crossing axon into the anterior commissure, as misexpression of Drl in neurons that do not normally cross the midline but that have been genetically engineered to do so directs them to the anterior commissure. Thus, Drl appears to be both necessary and sufficient for guiding axons in the anterior commissure.

To explore the mechanism underlying Drl function, a soluble, labeled version of the extracellular domain of Drl was used to detect potential cell-surface ligands for Drl in the fly ventral nerve cord.[44,45] Drl binding sites were observed specifically in the posterior commissure, sug-

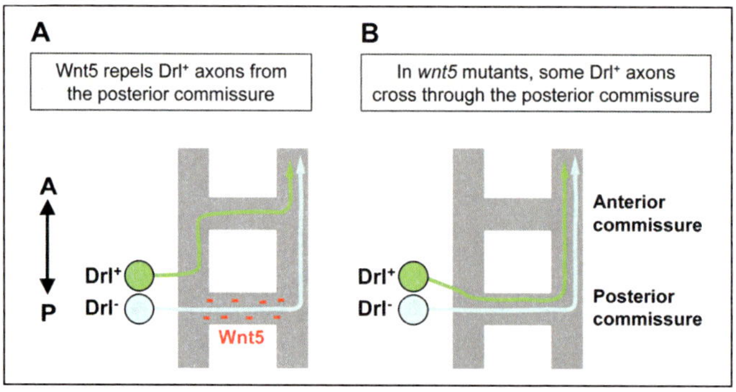

Figure 4. Wnt5 repels commissural axons from the posterior commissure. During Drosophila development, the axons composing the CNS have a ladder-like structure: each segment comprises an anterior and a posterior commissural tract that cross the midline and join the longitudinal tracts. A) In wild-type animals, the Wnt5 ligand, which is restricted to the posterior commissure (red), repels axons expressing Derailed (Drl^+; light green) from the posterior commissure into the anterior commissure. B) In *wnt5* mutants (as well as in *drl* mutants, not shown), the Drl^+ neurons that normally project into the anterior commissure project into the posterior commissure. A, anterior; P, posterior. Reproduced with permission of the Company of Biologists.

gesting that Drl functions to guide axons into the anterior commissure by repelling them away from the region of ligand expression in the posterior commissure. The fact that Drl, like other Ryk family members, possesses a so-called Wnt inhibitory factor (WIF) domain, which in other proteins functions to bind Wnt proteins,[46] suggested that a Wnt might be the repellent in the posterior commissure that repels the axons by binding Drl. Indeed loss of *wnt5* function resulted in commissural axon defects similar to those in *drl* mutants, and decreased the ability of misexpressed Drl to force axons into the anterior commissure[45] (Fig. 4). Moreover, overexpressing Wnt5 throughout the midline prevented the anterior commissure from forming, while the overexpression of Wnt5 in *drl* mutants did not. Taken together, these results imply that Wnt5 repels Drl-expressing axons and suggest that Drl might function as a receptor for Wnt5. In a direct test of this possibility, the soluble Drl extracellular domain was shown to bind to the endogenous Wnt5 protein from fly extracts, and its binding to the fly ventral nerve cord was found to disappear in *wnt5* mutants.[45] Thus, biochemical and genetic data indicate that Wnt5 is a Drl ligand responsible for repelling axons from the posterior commissure.

Importantly, this work is the first to identify a ligand for the Drl family of receptors and suggests that the other member of the family, Drl2, might also act as a Wnt receptor. This receptor-ligand interaction appears to be specific for Wnt5 since Drl does not interact with the two other Wnt family members tested, Wingless and Wnt4, a finding consistent with the lack of genetic interaction between *drl* and either *wg* or *wnt4*. It remains to be determined whether Drl acts to transduce the repulsive Wnt signal directly, or functions to prevent or reverse attraction through an alternative receptor, possibly of the Fz family.

Wnt4 Controls the Antero-Posterior Guidance of Ascending Commissural Axons

Prior the finding that Shh controls the antero-posterior guidance of commissural axons, Wnt4 was also reported to play a role in this process. Using a novel in vitro assay, evidence was obtained that the activity responsible for the anterior guidance of post-crossing commissural axons is an increasing posterior to anterior gradient of a diffusible attractant.[47] Several members of the Wnt family were then shown to be able to affect the growth of post-crossing commissural axons. Among them, Wnt4 was found to be expressed in an increasing posterior to anterior gradient, at least at the RNA level. Importantly, an ectopic posterior source of Wnt4 was found to redirect post-crossing axons posteriorly in vitro, whereas the Wnt inhibitors sFRP1, sFRP2, and sFRP3 (secreted frizzled-related proteins; soluble proteins that block the interaction of Wnts with their receptor) made post-crossing commissural axons stall and turn randomly along the antero-posterior axis. In the presence of Wnt4, the growth cones of post-crossing commissural axons were enlarged and more complex; addition of sFRP2 reduced this effect within one hour, suggesting that Wnt4 might be acting directly on the growth cone. These results indicate that Wnt activity is essential for the normal guidance of post-crossing commissural axons and that Wnt4 can act as an instructive post-crossing commissural axon attractant (Fig. 1D).

In agreement with a role for Wnt factors in the control of post-crossing guidance of commissural axons, it was found that mice lacking the Wnt receptor Frizzled3 (Fz3) have normal pre-crossing commissural axon behavior but display defects in antero-posterior guidance of commissural axons after midline crossing.[47] It will, of course, be important to determine whether Fz3 is required specifically in commissural neurons for this effect; however, a lack of apparent patterning defects in the neural tube of *fz3* mutants,[47] combined with the in vitro experiments described above, already provide strong evidence that Wnt-Frizzled signaling directly guides commissural axons along the antero-posterior axis of the spinal cord.

Of note, the inactivation of *fz3* in mice also results in other axonal abnormalities.[48] These animals display severe defects in many major axon tracts within the forebrain, including complete loss of the thalamocortical, corticothalamic and nigrostriatal tracts and of the anterior commissure and a variable loss of the corpus callosum. These results suggest that, in addition to guiding commissural axons at the spinal cord midline, Wnt-Frizzled signaling might also play a much broader role in axonal development, as discussed in the next section of this chapter.

LRP6 is a Fz co-receptor required for the canonical Wnt/β-catenin signaling pathway.[49] The axon guidance signaling pathway downstream of Fz3 has not been investigated but, interestingly, the pathfinding of commissural axons is reported to be normal in *LRP6* mutant embryos, suggesting that the canonical Wnt signaling pathway is not required for Wnt-mediated commissural axon guidance.[47]

Although it is not yet known whether Wnt4 guides post-crossing commissural axons in chicks and whether Shh guides post-crossing commissural axons in rodents, it is nonetheless interesting to note that the complementary Wnt4 and Shh gradients might act cooperatively in the rostral guidance of commissural axons.

Wnts Repel Corticospinal Tract Axons Down the Spinal Cord

Remarkably, while Wnt4 attracts ascending commissural interneurons toward the brain, other Wnts repels corticospinal tract (CST) axons down the spinal cord.[50] CST neurons are located in the cortex and project axons that migrate through the mid- and hindbrain, cross the midline and then grow down the spinal cord in the dorsal funiculus (Fig. 5). By looking at the expression of all *wnt* genes, Liu et al found that *wnt1* and *wnt5a* are expressed in an anterior to posterior decreasing gradient in a region neighboring the dorsal midline and dorsal funiculus at stages when CST axons migrate down the spinal cord. Consistent with a repulsive role for Wnts on CST axons, Wnt1 and Wnt5a were found to repel motor cortical axons in collagen gel assays. While the same group found that the attractive effect of Wnt4 on ascending com-

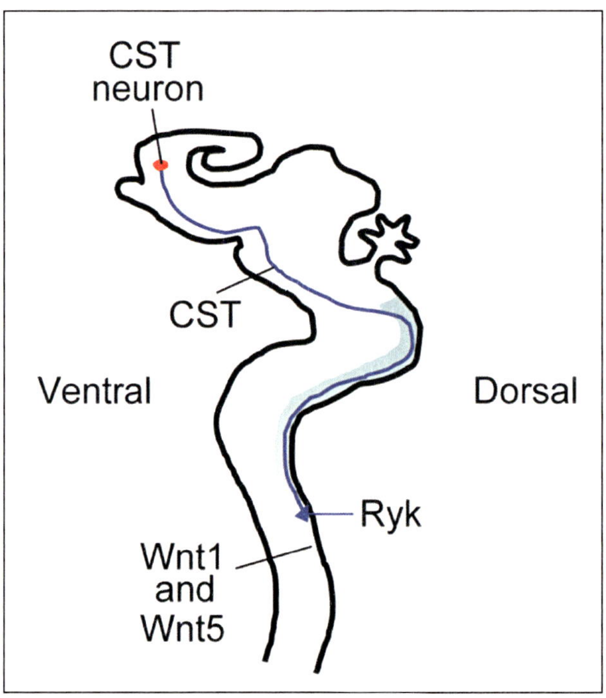

Figure 5. Wnts repel corticospinal tract axons down the spinal cord. After migrating through the midbrain and hindbrain, CST axons cross the midline and grow down the spinal cord in the dorsal funiculus. Wnt1 and Wnt5a are expressed in an anterior-posterior decreasing gradient in the dorsal spinal cord and repel CST axons through the Ryk receptor. CST, corticospinal tract.

missural axons was mediated by Fz3,[47] they also found that Ryk, the vertebrate homologue of the Drosophila *drl* gene product, is expressed by CST axons and mediates the repulsive effect of Wnt1 and Wnt5a in vitro.[50] Moreover, injection of anti-Ryk blocking antibodies directly into the spinal cord blocks the posterior growth of CST axons. Together, these results suggest that Wnt1 and Wnt5a repel CST axons down the spinal cord through their Ryk receptor.

Wnt3 Mediates Medial-Lateral Retinotectal Topographic Mapping

After having crossed the chiasm, RGC axons follow their journey to the brain where they will make contact with their targets in the optic tectum (or the superior colliculus in the mouse). The projections of RGC axons form a topographic map on the tectum such that the image projected on the retina is recapitulated in the tectum. Along the anterior-posterior axis, gradients of repulsive EphrinA in the tectum specify where RGC axons will target through their EphA receptors.[51,52] Along the medial-lateral axis, gradients of attractive EphrinB play this role through their EphB receptors.[53,54] Experimental and modeling studies suggest that an additional activity is necessary in addition to EphrinB to account for RGC axons guidance along

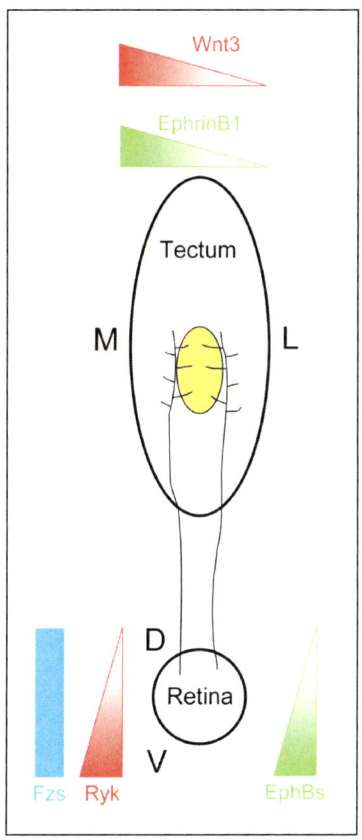

Figure 6. Wnt3 mediates medial-lateral retinotectal topographic mapping. Wnt3 is expressed in a medial-lateral decreasing gradient in the chick tectum. RGCs express two Wnt receptors: Fz5 and Ryk. While Fz5 appears to be expressed evenly throughout RGCs, Ryk is expressed in a ventral-dorsal decreasing gradient, similarly to EphB receptors. Wnt3 repels RGC axons laterally through the Ryk receptor. Thus, the Wnt3 gradient provides a repulsive force that counterbalances the medially directing attractive force of EphrinB. M, medial; L, lateral; D, dorsal; V, ventral.

the medial-lateral axis.[53] A recent study suggested that a gradient of Wnt3 might account for this activity.[55]

Similarly to EphrinB molecules, Schmitt et al found that *wnt3* is expressed in a medial-lateral decreasing gradient in the chick tectum[55] (Fig. 6). RGCs express two Wnt receptors: *fz5* and Ryk. While *fz5* appears to be expressed evenly throughout RGCs, Ryk is expressed in a ventral-dorsal decreasing gradient, similarly to EphB receptors. Interestingly, in vitro assays suggest that Wnt3 inhibits RGC axon outgrowth via Ryk and, conversely, stimulates RGC axon outgrowth via Fz. Two sets of experiments were performed to provide evidence that Ryk-mediated repulsion in response to Wnt3 plays a role in RGC axon targeting along the medial-lateral axis in vivo. First, when Wnt3 was overexpressed in the tectum, RGC axons avoided the site of ectopic Wnt3. Second, when a dominant-negative Ryk was expressed in dorsal RGC axons, a medial shift of the termination zone was observed. This phenotype is opposite to RGC axons mutant for *ephB*. Together, these results suggest that Wnt3 repels RGC axons laterally through the Ryk receptor. Thus, the Wnt3 gradient provides a repulsive force that counterbalances the medially directing attractive force of EphrinB.

Interpreting Positional Information: Signaling Components in Axon Guidance and Cell Fate Specification

In the studies summarized above, members of all three morphogen families were shown to act rapidly (in an hour or less) to affect growth cone morphology. Although these results appear inconsistent with the model that their axon guidance effect function through the canonical, transcriptional signaling pathways to the nucleus, this needs to be formally proven, as none of the studies described above have addressed this issue directly. Nonetheless, even if a transcriptional response is found to be required for their guidance effects, additional local signaling would still be required to be elicited in the growth cone in order to generate a polarized response leading to growth cone turning in a specific direction. Evidently, a purely transcriptional response consisting of a retrograde signal to the nucleus followed by an anterograde signal back to the growth cone cannot account for the polarized turning effect of a guidance cue. Studies aimed at understanding the molecular mechanisms underlying growth cone turning by morphogens will be necessary to identify the molecules linking morphogen signaling to localized growth cone effects.

In this regard, at least three possible models may account for the effects of morphogens in axon guidance. The first is based on the fact that, despite many efforts, the Wnt, BMP and Hh signaling pathways are only beginning to be understood and many intermediate signaling molecules remain to be identified and characterized. Thus, it is possible that the signaling proteins eliciting the growth cone effects are simply components of the signaling pathways required for cell fate specification.

A second model is that the same cue might be acting through entirely different signaling pathways, including the use of a different receptor. In the case of BMP/TGF-β family members, no known receptor has so far been implicated in the commissural axon guidance activity in vertebrates. In worms the evidence indicates that UNC-129 does not require the classical TGF-β receptors, suggesting that it may be functioning through an alternative family of receptors, and it will be exciting to determine whether the classical BMP receptors are required for the guidance activity of the BMPs on vertebrate commissural axons or whether they signal through non-classical BMP receptors. In the case of Shh, although Smo is required for Shh-mediated commissural axon guidance to the floor plate, it is not known whether Ptc, the Shh-binding component of the Shh receptor, is involved. This finding contrasts with chick post-crossing commissural axon guidance where Smo does not appear to be required for the rostral turn away from the Shh gradient. Additional experiments on commissural and retinal axons are required to determine the receptor components mediating the effects of Shh on axon guidance. Finally, for Wnt-mediated axon guidance, where the identity of the receptors involved was more thoroughly investigated, an unexpected situation was uncovered. The non-classical Wnt receptor Drl is required for repulsion from the posterior commissure in

Drosophila and, similarly, the Drl homolog Ryk is required for the repulsive effect of Wnt1 and Wnt5a in the spinal cord and for the repulsive effect of Wnt3 in the optic tectum. Interestingly, the classical receptor Fz3 is required for attraction towards the anterior pole of the spinal cord in mouse and Fz5 (or another Fz receptor) mediates attraction of RGC axons by Wnt3. Taken together, these results suggest that Drl/Ryk receptors might be generally involved in mediating repulsion while the classical Fz receptors might be generally involved in mediating attraction.

The third model for how morphogens guide axons is a combination of the first two models described above. In this case, a morphogen might use the upstream part of the classical cell fate signaling cascade but then diverge and use a non-classical pathway to elicit its effects on the growth cone. Antero-posterior commissural axon guidance by Wnt4 through Fz3 might use such a mechanism: although this effect requires a Fz receptor, the fact that the Fz co-receptor LRP6 is not required suggests that Wnt4 may regulate commissural axon guidance through a non-canonical pathway. In this case, the non-canonical—and potentially overlapping—Wnt/Ca^{2+} and PCP pathways are potential candidate signaling cascades to mediate the guidance effects of Wnt4. Indeed, in *Xenopus*, Wnt4 and its related family members Wnt5a and Wnt11 were found to regulate various morphological events by activating signaling by heterotrimeric G-protein, Ca^{2+} and PKC pathways,[16] and Fz3 was shown to activate PKC.[56] Together, these results raise the possibility that axon guidance mediated by Wnt4 might be operating through a non-canonical, PKC-dependent Wnt/Fz signaling pathway and, more importantly, that axon guidance and PCP might overlap not only conceptually—by controlling the polarity of a growth cone or an entire cell, respectively—but also mechanistically. Indirect evidence supporting mechanistic links between PCP and axon guidance is provided by the finding that the receptor tyrosine kinase-like protein PTK7 regulates planar cell polarity in vertebrates,[57] whereas its Drosophila homologue Off-track (Otk) participates in Semaphorin signaling in axon guidance.[58]

Conclusions and Perspectives

The discovery that morphogens can be reused to guide axons has generated considerable excitement in the field. It remains an open question how widespread these guidance effects are. At one extreme, the examples of guidance by morphogens may be isolated instances. At the other extreme, morphogens may prove to be as important as the classic axon guidance molecules (Netrins, Slits, Semaphorins, Ephrins and Growth Factors) in guiding axons. Elucidating the precise contribution of morphogens will, however, continue to be difficult for some time, given the significant difficulty in determining in any particular situation whether a morphogen is functioning directly or indirectly to regulate axonal guidance. In any gain- or loss-of-function experiment in vivo, the morphogen may be altering the expression of guidance cues in the environment where the guidance effects are observed, or the fate of cells (and hence their responses to guidance cues) that are showing guidance responses. Thus, several tests are required to prove that altered guidance in such experiments reflect direct guidance effects of the morphogen: (1) evidence against a change in the fate of cells or expression of other guidance cues in the environment, (2) evidence that cell autonomous manipulation of the signal transduction pathway for the morphogen in the responsive neuron results in similar guidance deficits, and (3) evidence that growth cones of responsive axons can respond directly to the morphogen—which may most frequently be obtained in vitro in growth cone collapse or turning assays. It can be expected that the level of proof that is obtained will increase over time as the signaling pathways linking the morphogens to the cytoskeleton for growth cone turning are elucidated, which should provide entry points to interfere selectively with guidance effects of the morphogens in the responsive neurons without altering their transcriptional effects either in those neurons or in the environment. Nonetheless, the collective weight of the experiments summarized above, many of which attempted and succeeded at least partly in distinguishing between direct and indirect effects of the morphogens, already provide strong evidence that morphogens have widespread roles in axon guidance, no doubt with more to come.

References

1. Dickson BJ. Molecular mechanisms of axon guidance. Science 2002; 298(5600):1959-1964.
2. Tessier-Lavigne M, Goodman CS. The molecular biology of axon guidance. Science 1996; 274(5290):1123-1133.
3. Teleman AA, Strigini M, Cohen SM. Shaping morphogen gradients. Cell 2001; 105(5):559-562.
4. Gurdon JB, Dyson S, St Johnston D. Cells' perception of position in a concentration gradient. Cell 1998; 95(2):159-162.
5. Chen Y, Schier AF. The zebrafish Nodal signal Squint functions as a morphogen. Nature 2001; 411(6837):607-610.
6. Briscoe J, Chen Y, Jessell TM et al. A hedgehog-insensitive form of patched provides evidence for direct long-range morphogen activity of sonic hedgehog in the neural tube. Mol Cell 2001; 7(6):1279-1291.
7. Jessell TM. Neuronal specification in the spinal cord: Inductive signals and transcriptional codes. Nat Rev Genet 2000; 1(1):20-29.
8. Ingham PW, McMahon AP. Hedgehog signaling in animal development: Paradigms and principles. Genes Dev 2001; 15(23):3059-3087.
9. Marti E, Bovolenta P. Sonic hedgehog in CNS development: One signal, multiple outputs. Trends Neurosci 2002; 25(2):89-96.
10. Roelink H, Porter JA, Chiang C et al. Floor plate and motor neuron induction by different concentrations of the amino-terminal cleavage product of sonic hedgehog autoproteolysis. Cell 1995; 81(3):445-455.
11. Lum L, Beachy PA. The Hedgehog response network: Sensors, switches, and routers. Science 2004; 304(5678):1755-1759.
12. Lee KJ, Jessell TM. The specification of dorsal cell fates in the vertebrate central nervous system. Annu Rev Neurosci 1999; 22:261-294.
13. Lee KJ, Mendelsohn M, Jessell TM. Neuronal patterning by BMPs: A requirement for GDF7 in the generation of a discrete class of commissural interneurons in the mouse spinal cord. Genes Dev 1998; 12(21):3394-3407.
14. Muroyama Y, Fujihara M, Ikeya M et al. Wnt signaling plays an essential role in neuronal specification of the dorsal spinal cord. Genes Dev 2002; 16(5):548-553.
15. Nelson WJ, Nusse R. Convergence of Wnt, beta-catenin, and cadherin pathways. Science 2004; 303(5663):1483-1487.
16. Strutt D. Frizzled signalling and cell polarisation in Drosophila and vertebrates. Development 2003; 130(19):4501-4513.
17. Colamarino SA, Tessier-Lavigne M. The role of the floor plate in axon guidance. Annu Rev Neurosci 1995; 18:497-529.
18. Kennedy TE, Serafini T, de la Torre JR et al. Netrins are diffusible chemotropic factors for commissural axons in the embryonic spinal cord. Cell 1994; 78(3):425-435.
19. Placzek M, Tessier-Lavigne M, Jessell T et al. Orientation of commissural axons in vitro in response to a floor plate-derived chemoattractant. Development 1990; 110(1):19-30.
20. Serafini T, Colamarino SA, Leonardo ED et al. Netrin-1 is required for commissural axon guidance in the developing vertebrate nervous system. Cell 1996; 87(6):1001-1014.
21. Serafini T, Kennedy TE, Galko MJ et al. The netrins define a family of axon outgrowth-promoting proteins homologous to C. elegans UNC-6. Cell 1994; 78(3):409-424.
22. Tessier-Lavigne M, Placzek M, Lumsden AG et al. Chemotropic guidance of developing axons in the mammalian central nervous system. Nature 1988; 336(6201):775-778.
23. Fazeli A, Dickinson SL, Hermiston ML et al. Phenotype of mice lacking functional Deleted in colorectal cancer (Dcc) gene. Nature 1997; 386(6627):796-804.
24. Charron F, Stein E, Jeong J et al. The morphogen sonic hedgehog is an axonal chemoattractant that collaborates with netrin-1 in midline axon guidance. Cell 2003; 113(1):11-23.
25. Bourikas D, Pekarik V, Baeriswyl T et al. Sonic hedgehog guides commissural axons along the longitudinal axis of the spinal cord. Nat Neurosci 2005.
26. Torres M, Gomez-Pardo E, Gruss P. Pax2 contributes to inner ear patterning and optic nerve trajectory. Development 1996; 122(11):3381-3391.
27. Trousse F, Marti E, Gruss P et al. Control of retinal ganglion cell axon growth: A new role for Sonic hedgehog. Development 2001; 128(20):3927-3936.
28. Song HJ, Ming GL, Poo MM. cAMP-induced switching in turning direction of nerve growth cones. Nature 1997; 388(6639):275-279.
29. Song H, Ming G, He Z et al. Conversion of neuronal growth cone responses from repulsion to attraction by cyclic nucleotides. Science 1998; 281(5382):1515-1518.
30. Song HJ, Poo MM. Signal transduction underlying growth cone guidance by diffusible factors. Curr Opin Neurobiol 1999; 9(3):355-363.

31. Hopker VH, Shewan D, Tessier-Lavigne M et al. Growth-cone attraction to netrin-1 is converted to repulsion by laminin-1. Nature 1999; 401(6748):69-73.
32. Augsburger A, Schuchardt A, Hoskins S et al. BMPs as mediators of roof plate repulsion of commissural neurons. Neuron 1999; 24(1):127-141.
33. Butler SJ, Dodd J. A role for BMP heterodimers in roof plate-mediated repulsion of commissural axons. Neuron 2003; 38(3):389-401.
34. Lee YS, Chuong CM. Activation of protein kinase A is a pivotal step involved in both BMP-2- and cyclic AMP-induced chondrogenesis. J Cell Physiol 1997; 170(2):153-165.
35. Foletta VC, Lim MA, Soosairajah J et al. Direct signaling by the BMP type II receptor via the cytoskeletal regulator LIMK1. J Cell Biol 2003; 162(6):1089-1098.
36. Halstead J, Kemp K, Ignotz RA. Evidence for involvement of phosphatidylcholine-phospholipase C and protein kinase C in transforming growth factor-beta signaling. J Biol Chem 1995; 270(23):13600-13603.
37. Choi SE, Choi EY, Kim PH et al. Involvement of protein kinase C and rho GTPase in the nuclear signalling pathway by transforming growth factor-beta1 in rat-2 fibroblast cells. Cell Signal 1999; 11(1):71-76.
38. Colavita A, Culotti JG. Suppressors of ectopic UNC-5 growth cone steering identify eight genes involved in axon guidance in Caenorhabditis elegans. Dev Biol 1998; 194(1):72-85.
39. Colavita A, Krishna S, Zheng H et al. Pioneer axon guidance by UNC-129, a C. elegans TGF-beta. Science 1998; 281(5377):706-709.
40. Hedgecock EM, Culotti JG, Hall DH. The unc-5, unc-6, and unc-40 genes guide circumferential migrations of pioneer axons and mesodermal cells on the epidermis in C. elegans. Neuron 1990; 4(1):61-85.
41. Nash B, Colavita A, Zheng H et al. The forkhead transcription factor UNC-130 is required for the graded spatial expression of the UNC-129 TGF-beta guidance factor in C. elegans. Genes Dev 2000; 14(19):2486-2500.
42. Hall AC, Lucas FR, Salinas PC. Axonal remodeling and synaptic differentiation in the cerebellum is regulated by WNT-7a signaling. Cell 2000; 100(5):525-535.
43. Callahan CA, Muralidhar MG, Lundgren SE et al. Control of neuronal pathway selection by a Drosophila receptor protein-tyrosine kinase family member. Nature 1995; 376(6536):171-174.
44. Bonkowsky JL, Yoshikawa S, O'Keefe DD et al. Axon routing across the midline controlled by the Drosophila Derailed receptor. Nature 1999; 402(6761):540-544.
45. Yoshikawa S, McKinnon RD, Kokel M et al. Wnt-mediated axon guidance via the Drosophila Derailed receptor. Nature 2003; 422(6932):583-588.
46. Patthy L. The WIF module. Trends Biochem Sci 2000; 25(1):12-13.
47. Lyuksyutova AI, Lu CC, Milanesio N et al. Anterior-posterior guidance of commissural axons by Wnt-frizzled signaling. Science 2003; 302(5652):1984-1988.
48. Wang Y, Thekdi N, Smallwood PM et al. Frizzled-3 is required for the development of major fiber tracts in the rostral CNS. J Neurosci 2002; 22(19):8563-8573.
49. He X, Semenov M, Tamai K et al. LDL receptor-related proteins 5 and 6 in Wnt/beta-catenin signaling: Arrows point the way. Development 2004; 131(8):1663-1677.
50. Liu Y, Shi J, Lu CC et al. Ryk-mediated Wnt repulsion regulates posterior-directed growth of corticospinal tract. Nat Neurosci 2005; 8(9):1151-1159.
51. Cheng HJ, Nakamoto M, Bergemann AD et al. Complementary gradients in expression and binding of ELF-1 and Mek4 in development of the topographic retinotectal projection map. Cell 1995; 82(3):371-381.
52. Drescher U, Kremoser C, Handwerker C et al. In vitro guidance of retinal ganglion cell axons by RAGS, a 25 kDa tectal protein related to ligands for Eph receptor tyrosine kinases. Cell 1995; 82(3):359-370.
53. Hindges R, McLaughlin T, Genoud N et al. EphB forward signaling controls directional branch extension and arborization required for dorsal-ventral retinotopic mapping. Neuron 2002; 35(3):475-487.
54. Mann F, Ray S, Harris W et al. Topographic mapping in dorsoventral axis of the Xenopus retinotectal system depends on signaling through ephrin-B ligands. Neuron 2002; 35(3):461-473.
55. Schmitt AM, Shi J, Wolf AM et al. Wnt-Ryk signalling mediates medial-lateral retinotectal topographic mapping. Nature 2006; 439(7072):31-37.
56. Kuhl M, Sheldahl LC, Park M et al. The Wnt/Ca2+ pathway: A new vertebrate Wnt signaling pathway takes shape. Trends Genet 2000; 16(7):279-283.
57. Lu X, Borchers AG, Jolicoeur C et al. PTK7/CCK-4 is a novel regulator of planar cell polarity in vertebrates. Nature 2004; 430(6995):93-98.
58. Winberg ML, Tamagnone L, Bai J et al. The transmembrane protein Off-track associates with Plexins and functions downstream of Semaphorin signaling during axon guidance. Neuron 2001; 32(1):53-62.

CHAPTER 10

The Role of Cyclic Nucleotides in Axon Guidance

Michael Piper,[1] Francis van Horck[1] and Christine Holt*

Abstract

During the formation of the nervous system, axonal growth cones navigate through the complex environment of the developing embryo to innervate their targets. Growth cones achieve this formidable feat by responding to attractive or repulsive guidance cues expressed at specific points along the trajectory of their growth, which impart the directional information required for accurate pathfinding. While much is known about guidance molecules and their receptors, many questions remain unanswered. Which signal transduction pathways are activated within the growth cone after encountering a guidance cue? How is this related to rearrangement of the growth cone cytoskeleton? Do different cues use different signal transduction pathways? This chapter will review some of the work that has addressed these fundamental questions, with a specific focus on the role of the cyclic nucleotides, cyclic adenosine monophosphate (cAMP) and cyclic guanosine monophosphate (cGMP), in axon guidance.

Introduction

Throughout embryonic development, elongating axons are in constant communication with the extracellular environment, which provides signals necessary for axonal growth and survival. The local environment also provides specific positional information, facilitating directional pathfinding. With respect to this, significant advances have been made towards understanding how locally expressed molecules can act as axon guidance cues, mediating attraction or repulsion. This has culminated in the discovery of many families of axon guidance molecules whose roles have been conserved to a remarkable extent during evolution. These include the netrins, Slits, Semaphorins and ephrins.[1] The tip of the developing axon, known as the growth cone, is pivotal in the process of recognising guidance cues expressed in the environmental milieu and integrating this information into a coordinated response.

To do this, the growth cone must be able to control its cytoskeletal assembly and disassembly, membrane dynamics and adhesion to the extracellular matrix. A number of signal transduction pathways have been shown to underlie this, including the mitogen activated protein kinases (MAPK) and the Rho GTPase family.[1,2] Another signal transduction system involved in regulating growth cone cytoskeletal dynamics in response to axon guidance cues centres on the cyclic nucleotides, cyclic adenosine monophosphate (cAMP) and cyclic guanosine monophosphate (cGMP). The first indication that cyclic nucleotides may be involved in axon guidance

*Corresponding Author: Christine Holt—Department of Anatomy, University of Cambridge, Downing Street, Cambridge, CB2 3DY, U.K. Email: ceh@mole.bio.cam.ac.uk

[1]These authors contributed equally to this work.

Axon Growth and Guidance, edited by Dominique Bagnard. ©2007 Landes Bioscience and Springer Science+Business Media.

came from studies of embryonic chick dorsal root ganglion (DRG) neurites exposed to gradients of cGMP or dB cAMP (a lipid soluble cAMP analogue) in vitro.[3] These neurites turned towards the source of these molecules, suggesting that local asymmetries in cAMP and cGMP within the growth cone may control axonal turning. Subsequent studies have demonstrated that cyclic nucleotides play an important role in guidance in response to many common cues. This review will provide a brief overview of the structure of cyclic nucleotides and their mechanism of signal transduction, before focussing on the current understanding of the role played by cyclic nucleotides in response to guidance cues and the functional implications for this in nervous system development.

Molecular Structure of Cyclic Nucleotides

cAMP

cAMP is a small cytoplasmic molecule whose function as an intracellular messenger has been conserved during evolution. Indeed, cAMP is found ubiquitously in both prokaryotes and eukaryotes. Membrane-bound enzymes called adenylyl cyclases catalyse the formation of cAMP from ATP, whilst cAMP phosphodiesterases prevent the accumulation of cAMP by converting it to AMP (Fig. 1A).

cGMP

cGMP is another common intracellular messenger, whose cycle of synthesis and degradation is similar to that of cAMP. Guanylyl cyclases, which may be soluble or membrane-bound, convert GTP to cGMP, and cGMP phosphodiesterases catalyse the conversion of cGMP to GMP (Fig. 1B).

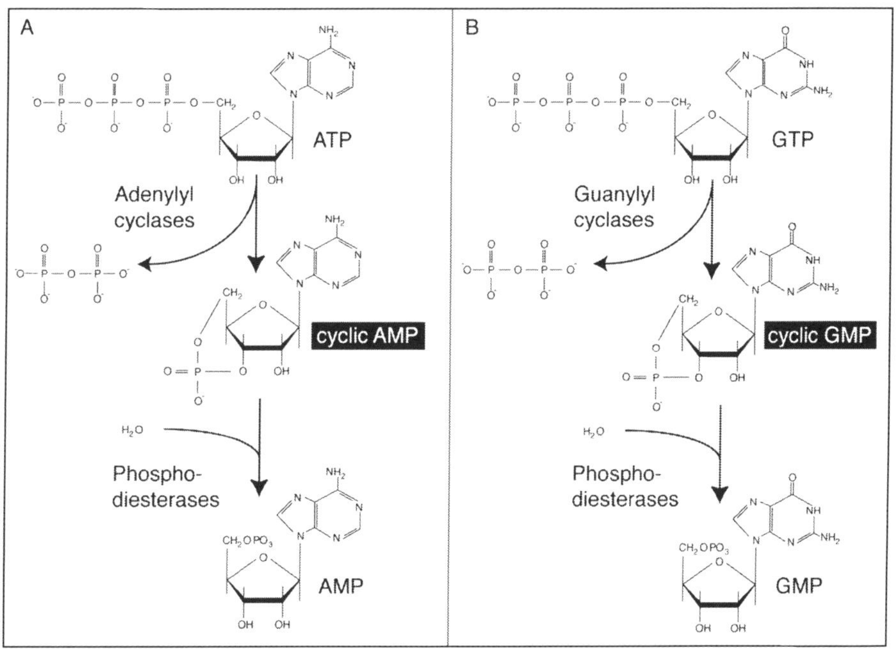

Figure 1. Molecular structure of cyclic nucleotides. A) Cyclic AMP (cAMP) is generated from ATP through the action of the enzyme adenylyl cyclase and is degraded by cAMP phosphodiesterases, which catalyse its conversion to AMP. B) Similarly, guanylyl cyclases convert GTP to cyclic GMP (cGMP), and cGMP phosphodiesterases convert this to GMP.

Mechanisms of Signal Transduction

cAMP Pathway

cAMP plays a central role in the mediation of many cellular events, and much of our understanding of how cAMP signalling occurs has been generated in a nonneuronal context. Binding of an extracellular ligand to a G protein-coupled receptor (GPCR) leads to disassembly of the heterotrimeric G protein complex from the receptor, allowing the stimulatory Gs_{α} subunit to activate adenylyl cyclase. This leads to a local elevation of intracellular cAMP (Fig. 2A).[4] The primary physiological target of cAMP is cAMP-dependent protein kinase (PKA). This tetrameric complex consists of two catalytic and two regulatory subunits. Binding of two cAMP molecules to each regulatory domain abolishes the inhibition of the catalytic subunits, allowing them to phosphorylate downstream targets. These include a very broad range of substrates, as PKA has targets in the cytoskeleton, nucleus, cytoplasm, mitochondria and cell membrane. It should also be noted that cAMP also exhibits some PKA-independent effects, including activating cyclic nucleotide gated ion channels and binding to the cAMP-interacting proteins EPAC1 and EPAC2. These guanine nucleotide exchange factors regulate the small GTP binding protein Rap1, which is involved in regulating cell adhesion.[5] The relevance of these nonPKA mediated functions in axon guidance, however, remains undefined.

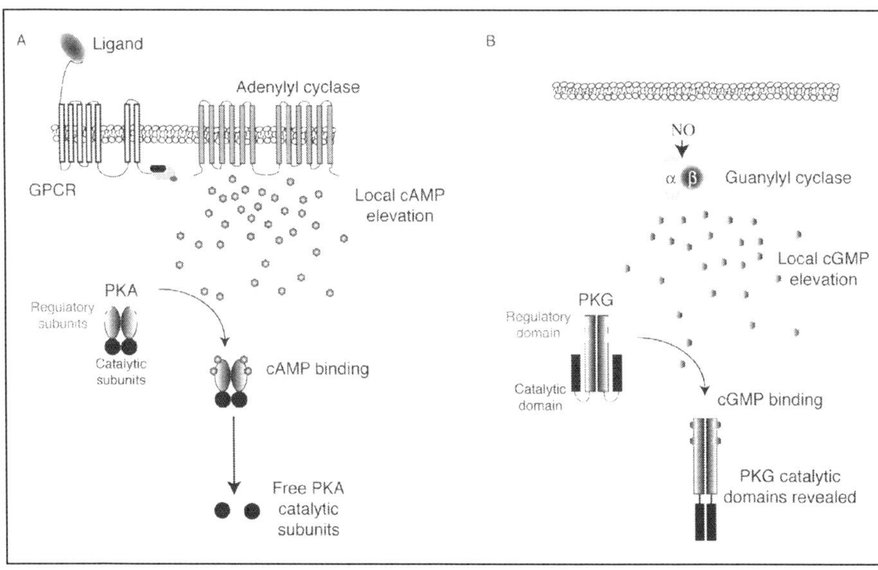

Figure 2. Mechanisms of signal transduction. A) Adenylyl cyclases are dimeric transmembrane proteins that interact with G protein-coupled receptors (GPCRs). Binding of ligand to receptor results in the release of the $G\alpha_s$ subunit of the G protein, which activates adenylyl cyclase. The local elevation of cAMP results in increased binding of cAMP to the regulatory subunits of cAMP-dependent protein kinase (PKA), and subsequent release of the active catalytic kinase domains of PKA. PKA can modulate cytoskeletal dymanics via multiple pathways. B) Soluble guanylyl cyclase is activated by nitric oxide (NO), resulting in a local increase in cGMP concentration. cGMP-dependent protein kinase (PKG) is a homodimer, with each subunit possessing an autoinhibitory regulatory domain, 2 cGMP binding sites and a catalytic domain. Binding of cGMP releases autoinhibition and reveals the catalytic domain. PKG is able to modulate cytoskeletal dynamics.

cGMP Pathway

The cGMP signalling cascade is similar to that of cAMP. cGMP is produced by guanylyl cyclases, which may be soluble or membrane-bound. The best-known activators of (soluble) guanylyl cyclases are small gaseous molecules like nitric oxide. The main target of cGMP is cGMP-dependent protein kinase (PKG). This homodimeric protein is comprised of multiple functional domains. In the absence of cGMP, the autoinhibitory domain suppresses the activity of the catalytic domain. Binding of two cGMP molecules to each inhibitory domain causes a conformational shift that reveals the catalytic site (Fig. 2B).

Functional Implications during Nervous System Development

The Role of Cyclic Nucleotides in Chemotropic Responses

The finding that local asymmetries in cyclic nucleotides within the growth cone may mediate turning decisions[3] has been further explored using in vitro turning assays first elaborated in the Poo laboratory.[6] To conduct a growth cone turning assay, a defined microscopic gradient of a chemical is produced near a growth cone by repetitive pulsatile application of a concentrated source of the cue from a nearby pipette. In turning assays conducted with embryonic *Xenopus* spinal neurons, a gradient of dB-cAMP induces turning towards the pipette (Fig. 3).[6] Furthermore, local elevation of intracellular cAMP by photolytic release of caged cAMP induces turning of *Xenopus* spinal neurons towards the side of the growth cone in which cAMP is elevated.[7] This illustrates that a cytoplasmic gradient of cAMP across the growth cone is sufficient to initiate a turning response.

Poo and colleagues have gone on to translate these findings into a developmental context by demonstrating that cyclic nucleotides are capable of modulating the response of growth cones to guidance cues. For example, embryonic *Xenopus* spinal neurons are attracted to a gradient of brain-derived neurotrophic factor (BDNF), but inhibiting cAMP production within the growth cone converts this attraction into repulsion.[8] Similarly, the attractive response of these neurons to netrin-1[9] and achetylcholine (ACh)[8] can be switched to repulsion by inhibiting PKA, while the repulsive response from myelin associated glycoprotein (MAG)[10] can be reversed by raising the intracellular levels of cAMP. On the other hand, attraction of *Xenopus* spinal neurons to neurotrophin-3 (NT-3) and repulsion from Sema III requires cGMP signalling. Thus, increasing levels of growth cone cGMP converts repulsive turning

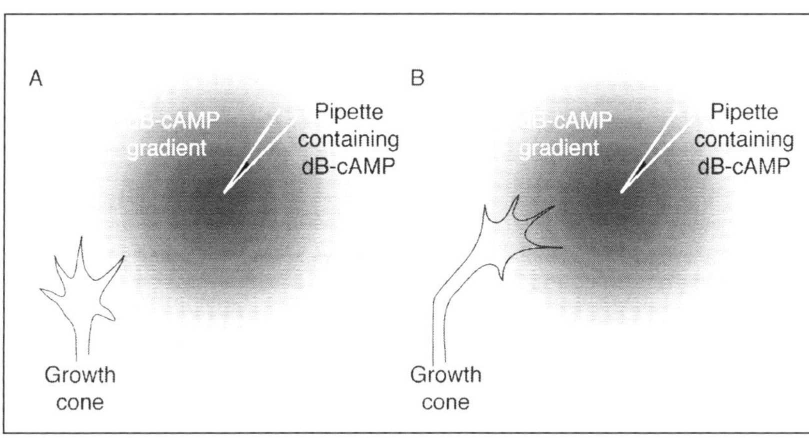

Figure 3. Growth cone turning assay. In the growth cone turning assay elaborated by Poo and colleagues,[6] a microscopic gradient of dB-cAMP (a lipid-soluble cAMP analogue) was produced through pulsatile ejection from a pipette. Growth cones from embryonic *Xenopus* spinal neurons encountering the gradient (A) turned and grew towards the pipette (B).

induced by Sema III into attraction, whereas lowering cGMP levels converts NT-3 induced attraction into repulsion.[10] The repulsion of embryonic rat DRG axons from Slit2 also appears to function via a cGMP-dependent pathway.[11]

These observations led to the idea that guidance cues can be classified into two groups, in which levels of either cAMP (netrin-1, BDNF, ACh, MAG) or cGMP (Sema III, NT-3, Slit2) are critical determinants for the polarity of the turning response.[2] However, recent results have caused this concept to be revised with respect to the guidance cue netrin-1. By modulating the ratio of cAMP to cGMP in embryonic *Xenopus* spinal axons exposed to a gradient of netrin-1, Nishiyama et al[12] have demonstrated that it is the ratio between these cyclic nucleotides that is the key in determining the polarity of the turning response. A high cAMP/cGMP ratio favours growth cone attraction, while a low ratio favours repulsion. This was shown by comparing axons expressing the DCC receptor (which mediates attraction to netrin-1) to those axons expressing DCC as well as over-expressing UNC5 (the DCC-UNC5 complex mediates repulsion to netrin-1). In those axons expressing DCC, inhibiting cAMP or elevating cGMP, thus lowering the cAMP/cGMP ratio, converted attraction to repulsion. Conversely, in axons over-expressing UNC-5, repulsion was converted to attraction by raising the cAMP/cGMP ratio. Furthermore, when varying ratios of the membrane-permeable analogues Sp-8-Br-cAMPS and 8-Br-cGMP (Table 1) were bath-applied to the cultures, high proportions of the cAMP analogue promoted attraction to netrin-1 gradients, while high proportions of the cGMP analogue favoured repulsion.[12]

Thus, the ratio of the cyclic nucleotides cAMP and cGMP within the growth cone is thought to underlie turning responses to netrin-1 in embryonic *Xenopus* spinal neurons (Fig. 4). However, caution must be used when generalising from these data, as it is clear that the role of cyclic nucleotides during guidance varies greatly depending on the types of signal, axonal population and age of neuron involved. For instance, in cultured embryonic rat spinal commissural neurons, intracellular cAMP levels do not increase after stimulation of the growth cone with netrin-1.[13]

Cyclic Nucleotide Signalling Pathways in Axons

An obvious conclusion to draw from these data is that many axon guidance cues elicit changes in growth cone behaviour by activating cyclic nucleotide signal transduction pathways. Although much remains unknown, the components of these pathways are gradually being identified, and it is apparent that growth cones use the classical PKA/PKG signalling pathways to activate and transduce signals via cyclic nucleotides. For example, mice deficient in adenylyl cyclase I activity display patterning defects in the somatosensory cortex of the brain,[14] and studies in *Drosophila* have shown that both membrane-bound receptor guanylyl cyclases (motor axons)[15] and soluble guanylyl cyclases (retinal axons)[16] have the potential to mediate axon pathfinding. Activation of soluble guanylyl cyclases via the lipid 12-HPETE, a metabolite in the arachidonic acid pathway, has also been implicated in the response of *Xenopus* spinal axons to netrin-1 in vitro.[12]

The targets of cAMP and cGMP, PKA and PKG respectively, are also intimately involved in the control of axon guidance. Activators and inhibitors of PKA and PKG (Table 1) often mimic the responses seen with cAMP and cGMP analogues indicating that the cyclic nucleotides exert their effects on growth cone turning via these enzymes. One potential mechanism to link guidance cues to PKA activation is by the direct recruitment of PKA to the guidance receptor via scaffolding proteins like PKA anchoring proteins (AKAPs). In *Drosophila* motor axons, the protein Nervy has been suggested to act as an AKAP, coupling cAMP-PKA signalling to the plexin-A receptor to regulate Sema-1a mediated axonal repulsion.[17] However, a recent report casts doubt on this suggestion, indicating that the axon guidance defects seen in Nervy mutants may arise from changes in gene expression rather than cytoplasmic control of PKA anchoring.[18]

Perhaps the most important role of PKA in the context of axon guidance is its capacity to regulate components of the cytoskeleton, including actin filaments, intermediate filaments

Table 1. Activators and inhibitors of cyclic nucleotide signalling pathways

		Compound	Mode of Action	References
cAMP	Activators	dB-cAMP	cAMP analogue that activates PKA	3,6,30,31,33,34
		8-Br-cAMP	cAMP analogue that activates PKA	32
		Sp-cAMP	cAMP analogue that specifically activates PKA	8,10-12,22,23, 26-29,31
		forskolin	Activator of adenylyl cyclases	6-8,13,28,32,33
	Inhibitors	Rp-cAMP	cAMP analogue that specifically inhibits PKA	7-10,22,23,26,31
		KT5720	Specific inhibitor of PKA	7-9,12,13,23,27, 30,34
		H89	Specific inhibitor of PKA	31,34
		PKI	Myristoylated inhibitor protein of PKA	22
		IBMX	Non-specific inhibitor of cAMP and cGMP phosphodiesterases	6
cGMP	Activators	dB-cGMP	cGMP analogue that activates PKG	6,31
		8-Br-cGMP	cGMP analogue that activates PKG	10,12,22
		Sp-cGMP	cGMP analogue that specifically activates PKG	10,23,26
		PP-9	Activator of soluble guanylyl cyclases	10
		SNAP	Nitric oxide donor, activates soluble guanylyl cyclases	11
	Inhibitors	Rp-cGMP	cGMP analogue that specifically inhibits PKG	10,22,23,26
		KT5823	Specific inhibitor of PKG	11,12,31,34
		LY83583	Blocks cGMP production, inhibits extracellular Ca^{2+}, blocks effects of nitric oxide	6

Listed are several activators and inhibitors of components of cAMP- and cGMP-dependent signalling pathways. The references indicate the studies in which these compounds have been used to address the role of cyclic nucleotide signalling in axons.

and microtubules.[5,19] For example, the Ena/VASP family of proteins, which act as enhancers of actin filament formation, are substrates for PKA.[20,21] The Rho family of small GTPases, including Rho, Rac and Cdc42, control many aspects of actin function, and are also regulated by PKA, both directly (Rho) and indirectly (Rac, Cdc42).[21] For instance cAMP-mediated regulation of Rho activity has been implicated in the response of embryonic chick DRG axons to the chemokine SDF1,[22] and of *Xenopus* spinal axons to the peptide PACAP.[7] Furthermore, the regenerative capacity of injured rat DRG axons can be enhanced through elevation of cAMP (see below), possibly via modulation of RhoA and Rac1 activity.[23] A third potential cytoskeletal target for PKA is myosin, an actin-associated motor protein, via the action of myosin light chain kinase (MLCK). SDF1 has been suggested to alter MLCK activity in a cAMP-dependent fashion.[22] Thus, by activating PKA, cAMP may be able to regulate actin dynamics in a variety of ways, and so influence axon pathfinding during development.

The role of PKG in mediating aspects of neuronal development has received far less attention than PKA, and as such, how it acts to mediate cytoskeletal dynamics during axon guidance

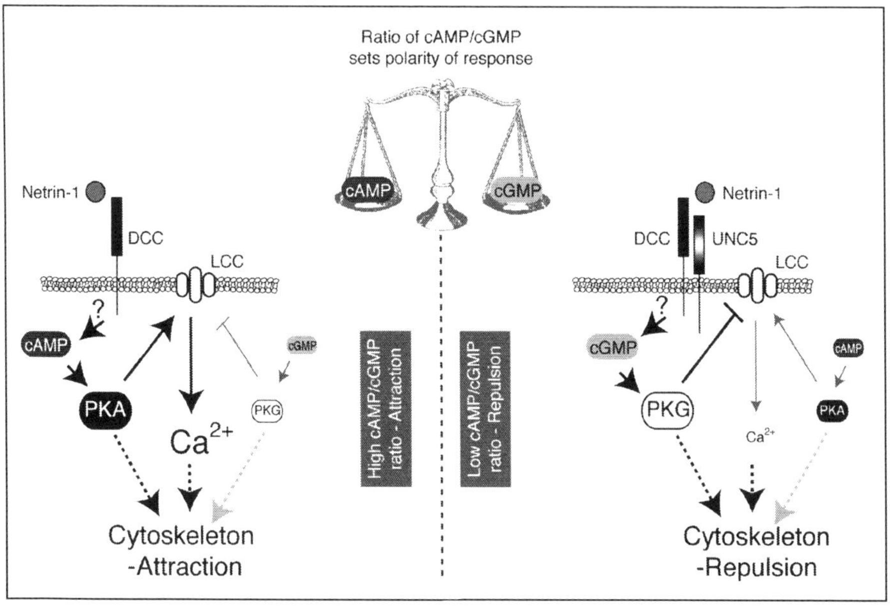

Figure 4. Model for netrin-1-induced turning of embryonic *Xenopus* spinal neurites. In response to netrin-1, embryonic *Xenopus* spinal neurites exhibit either attraction or repulsion. The ratio of cAMP to cGMP within the growth cone is postulated to determine which outcome is seen. For attraction, binding of netrin-1 to DCC elicits a rise in intracellular cAMP. PKA is subsequently activated and may stimulate an increase in calcium (Ca^{2+}) by activating L-type Ca^{2+} channels (LCC) in the membrane. For repulsion, binding of netrin-1 to the DCC-UNC5 complex results in a rise in cGMP levels and PKG activity. PKG may inactivate LCC, lowering growth cone Ca^{2+}. How differential patterns of intracellular calcium induced by cAMP/cGMP induce attractive or repulsive turning responses remains unclear. Adapted from reference 12.

remains unclear. However, the target motif for PKG is similar to that of PKA, and the two are thought to have overlapping substrate specificities. Thus, PKG also has the potential to modulate cytoskeletal behaviour in an analogous fashion. Members of the Ena/VASP family, for instance, have been implicated as substrates of PKG in vitro.[24,25]

Cyclic nucleotide signalling is also unequivocally linked to Ca^{2+} signalling, which itself plays a central role in growth cone dynamics. Indeed, local asymmetries in Ca^{2+} across the growth cone have been shown to elicit turning responses in cultured axons similar to those induced by cyclic nucleotides. For instance, a localised Ca^{2+} signal in the growth cone generated by photolytic release of caged Ca^{2+} or induction of Ca^{2+} release from internal stores is sufficient to induce growth cone turning in embryonic *Xenopus* spinal neurons,[26,27] and preventing cytoplasmic Ca^{2+} elevation abolishes netrin-1 induced attractive and repulsive responses.[27] The Ca^{2+} and cyclic nucleotide signalling cascades have the potential to interact and cross regulate each other at a number of different levels. In some instances, cyclic nucleotides act upstream of Ca^{2+}, directly regulating the level of Ca^{2+} within the growth cone. An example of this is seen in *Xenopus* spinal neurons, where the cAMP/cGMP ratio can directly affect the activity of L-type Ca^{2+} channels (LCC) to alter intracellular Ca^{2+} signals induced by netrin-1. When signalling via DCC, netrin-1 stimulates a high cAMP/cGMP ratio and thus elicits PKA activation. PKA activates LCC in the plasma membrane and Ca^{2+} channels in the endoplasmic reticulum, eventually leading to a steep Ca^{2+} gradient across the growth cone, favouring attraction. When signalling via DCC-UNC5, netrin-1 stimulates the cGMP pathway, which, through PKG, closes LCCs and inhibits Ca^{2+} release from internal stores, therefore creating a gradient of Ca^{2+} across the growth cone which is lowest on the side facing the guidance cue, resulting in repulsion.[12] In this model cyclic nucleotide signalling directly regulates Ca^{2+} signalling to determine bi-directional turning responses (Fig. 4).

Recent studies have shown that Ca^{2+} gradients regulate another switch-like mechanism by activating Ca^{2+}-calmodulin-dependent protein kinase II (CaMKII) and calcineurin (CaN)-phosphatase (PP1) to mediate attraction and repulsion respectively.[26] However, in this study, the cAMP pathway was shown to act downstream of Ca^{2+} signalling by negatively regulating CaN-PP1, emphasising the complexity and potential for multiple interactions between these two signal transduction cascades. Moreover, it should be noted that some axon guidance cues also affect cyclic nucleotide levels without affecting the Ca^{2+} pathway. For example, in *Xenopus* spinal neurons Sema3A has no effect on calcium currents and PACAP induces attractive turning independently of Ca^{2+}, although both of these cues require cyclic nucleotide signalling.[7,12]

Thus, these studies have started to elucidate the role of cyclic nucleotides in the control of growth cone cytoskeletal dynamics in response to guidance cues. While it is apparent that cue-induced activation of cAMP or cGMP can elicit cytoskeletal changes that manifest as attractive or repulsive turning in vitro, a great deal remains to be discovered. The ways in which adenylyl and guanylyl cyclases are activated is, in many cases, completely unknown, and, in vivo, the relevant cytoskeletal targets for PKA and PKG within the growth cone remain poorly defined. Furthermore, although many studies have focussed on the effect of netrin-1 on embryonic *Xenopus* spinal neurons, it is clear that not all cues act in the same fashion, nor even that all developing axonal populations respond to this cue in the same way. More work is needed to clarify the molecular components of the cyclic nucleotide signalling pathways in growth cones and to discover how they are activated and regulated to coordinate axon pathfinding during axonogenesis.

Modulation of Guidance Cue Activity via Cyclic Nucleotides

The capacity to control cue-induced turning by manipulating the levels of cyclic nucleotides within the growth cone also implies that the polarity of turning in response to guidance signals may be modulated by other extrinsic or intrinsic factors that alter cAMP or cGMP levels. This may underlie how some axons change responsiveness to the same guidance cue over time. For example, a switch in responsiveness from attraction to repulsion has been documented for embryonic *Xenopus* retinal axons exposed to netrin-1 in vitro. When cultured on a fibronectin substrate, these neurites turn towards netrin-1, but when cultured on high levels of laminin-1, repulsive turning is seen. Experimental evidence suggests this is due to laminin-1 decreasing cAMP levels within the growth cone.[28] This may be relevant in vivo, as, within the retina, these axons are attracted to the netrin-1 expressing optic nerve head (ONH). At the ONH, laminin-1 is expressed at the retinal surface, suggesting that a laminin-1-mediated reduction in cAMP levels at this point may cause repulsion from the area of netrin-1/laminin-1 coexpression, so driving axons from the retinal surface into the optic nerve.[28] Intrinsic factors might also contribute to this process as the change in netrin-1 responsiveness of old (repulsive) versus young (attractive) neurons, concomitant with a decline in cAMP levels, is also seen in pathway-naïve neurons. Cyclic AMP elevators and adenosine A2b receptor agonists can rejuvenate the behaviour of the old growth cones, causing them to regain attraction to netrin-1, whereas antagonists cause young growth cones to be repelled. Thus, age-related intrinsic changes might also modulate cAMP levels to control the developmental switch in netrin-1 responsiveness.[29] A second example of the modulation of cyclic nucleotide levels by extrinsic factors affecting responses to axon guidance cues is seen in developing chick axons. In vitro, repulsion of these axons induced by Slit2, Sema3A or Sema3C can be reduced by application of SDF1, which through its receptor, CXCR4 (a GPCR), elicits an increase in cAMP in these neurons.[22]

Cyclic Nucleotides and Axon Regeneration

The evidence supporting a role for cyclic nucleotides in axon guidance has mainly come from in vitro experiments, such as the growth cone turning assay. Whether cyclic nucleotides are relevant for axonal guidance responses in vivo, where multiple signals are likely to impinge simultaneously on a growth cone, remains unclear. However, recent data have demonstrated an important role for cAMP in the regulation of axonal growth cone responses in response to axonal lesioning in vivo, implicating cyclic nucleotides as potential targets to enhance spinal cord regeneration.

In the adult mammalian peripheral nervous system, injured axons are able to regenerate, whereas those of the central nervous system (CNS) are not. The damaged myelin surrounding the site of injury accounts for at least some of the failure of CNS axons to regenerate by producing inhibitory molecules like myelin-associated glycoprotein (MAG) and Nogo. In vitro studies have shown that elevating cAMP levels converts MAG induced repulsion into attraction in *Xenopus* spinal neurons,[10] and enhances the ability of neonatal mammalian axons to grow on substrates of MAG and myelin.[30,31] Together with the growth promoting effect of cAMP on cultured embryonic axons,[32] these observations make cAMP an attractive candidate to potentially overcome inhibitory regenerative responses in vivo. This has been addressed by two recent studies investigating the response of rat DRG CNS axons after spinal cord injury. After lesioning of the dorsal column, the centrally projecting axons from the DRG normally fail to regenerate into the lesion. However, injection of the lipid soluble analogue db-cAMP into rat DRG neurons prior to lesioning results in extensive regeneration into the lesion site.[33,34] Removal of DRG neurons treated with db-cAMP in vivo, followed by culturing in vitro, demonstrates that these neurons have an increased intrinsic growth capacity and the ability to overcome the effects of inhibitory factors like MAG.[33,34] Although these studies are somewhat artificial in that db-cAMP was applied prior to injury, they show that regulating cyclic nucleotide levels in vivo may be a potentially useful approach in promoting regeneration after spinal cord injury and stress the importance of understanding cyclic nucleotide signalling in neuronal behaviour.

Conclusion and Perspectives

The importance of cyclic nucleotides in mediating responses of axonal growth cones to guidance cues during development is now well established. It is clear that guidance cues can activate the cAMP and/or the cGMP signalling cascades, potentiating changes in the growth cone cytoskeleton and leading to attractive or repulsive responses. Furthermore, cAMP and cGMP are themselves able to modulate responses of growth cones to environmental signals, so directly controlling the polarity of growth. In this way extrinsic and intrinsic factors may be able to temporally regulate axonal responsiveness. However, we only possess a very general hypothesis of the role of these molecules during development of neuronal axons, as the use of cyclic nucleotide signalling seems to differ widely among axonal populations. It may be that intracellular signal transduction pathways in general depend on developmental context, age and the neuronal type involved. More work is required to clarify the molecular components of the cAMP and cGMP pathways and the targets of PKA and PKG that elicit cytoskeletal changes. From a clinical perspective, these developmental studies have the potential to be of great benefit in understanding why, after axonal injury, neurons of the CNS are unable to regenerate in vivo.

References

1. Dickson BJ. Molecular mechanisms of axon guidance. Science 2002; 298(5600):1959-1964.
2. Song H, Poo M. The cell biology of neuronal navigation. Nat Cell Biol 2001; 3(3):E81-88.
3. Gundersen RW, Barrett JN. Characterization of the turning response of dorsal root neurites toward nerve growth factor. J Cell Biol 1980; 87(3 Pt 1):546-554.
4. Linder JU, Schultz JE. The class III adenylyl cyclases: Multi-purpose signalling modules. Cell Signal 2003; 15(12):1081-1089.
5. Howe AK. Regulation of actin-based cell migration by cAMP/PKA. Biochim Biophys Acta 2004; 1692(2-3):159-174.

6. Lohof AM, Quillan M, Dan Y et al. Asymmetric modulation of cytosolic cAMP activity induces growth cone turning. J Neurosci 1992; 12(4):1253-1261.
7. Guirland C, Buck KB, Gibney JA et al. Direct cAMP signaling through G-protein-coupled receptors mediates growth cone attraction induced by pituitary adenylate cyclase-activating polypeptide. J Neurosci 2003; 23(6):2274-2283.
8. Song HJ, Ming GL, Poo MM. cAMP-induced switching in turning direction of nerve growth cones. Nature 1997; 388(6639):275-279.
9. Ming GL, Song HJ, Berninger B et al. cAMP-dependent growth cone guidance by netrin-1. Neuron 1997; 19(6):1225-1235.
10. Song H, Ming G, He Z et al. Conversion of neuronal growth cone responses from repulsion to attraction by cyclic nucleotides. Science 1998; 281(5382):1515-1518.
11. Nguyen-Ba-Charvet KT, Brose K, Marillat V et al. Sensory axon response to substrate-bound Slit2 is modulated by laminin and cyclic GMP. Mol Cell Neurosci 2001; 17(6):1048-1058.
12. Nishiyama M, Hoshino A, Tsai L et al. Cyclic AMP/GMP-dependent modulation of Ca2+ channels sets the polarity of nerve growth-cone turning. Nature 2003; 423(6943):990-995.
13. Bouchard JF, Moore SW, Tritsch NX et al. Protein kinase A activation promotes plasma membrane insertion of DCC from an intracellular pool: A novel mechanism regulating commissural axon extension. J Neurosci 2004; 24(12):3040-3050.
14. Abdel-Majid RM, Leong WL, Schalkwyk LC et al. Loss of adenylyl cyclase I activity disrupts patterning of mouse somatosensory cortex. Nat Genet 1998; 19(3):289-291.
15. Ayoob JC, Yu HH, Terman JR et al. The Drosophila receptor guanylyl cyclase Gyc76C is required for semaphorin-1a-plexin A-mediated axonal repulsion. J Neurosci 2004; 24(30):6639-6649.
16. Gibbs SM, Becker A, Hardy RW et al. Soluble guanylate cyclase is required during development for visual system function in Drosophila. J Neurosci 2001; 21(19):7705-7714.
17. Terman JR, Kolodkin AL. Nervy links protein kinase a to plexin-mediated semaphorin repulsion. Science 2004; 303(5661):1204-1207.
18. Ice RJ, Wildonger J, Mann RS et al. Comment on "Nervy links protein kinase A to plexin-mediated semaphorin repulsion". Science 2005; 309(5734):558, (author reply 558).
19. Tasken K, Aandahl EM. Localized effects of cAMP mediated by distinct routes of protein kinase A. Physiol Rev 2004; 84(1):137-167.
20. Lebrand C, Dent EW, Strasser GA et al. Critical role of Ena/VASP proteins for filopodia formation in neurons and in function downstream of netrin-1. Neuron 2004; 42(1):37-49.
21. Krause M, Dent EW, Bear JE et al. Ena/VASP proteins: Regulators of the actin cytoskeleton and cell migration. Annu Rev Cell Dev Biol 2003; 19:541-564.
22. Chalasani SH, Sabelko KA, Sunshine MJ et al. A chemokine, SDF-1, reduces the effectiveness of multiple axonal repellents and is required for normal axon pathfinding. J Neurosci 2003; 23(4):1360-1371.
23. Bandtlow CE. Regeneration in the central nervous system. Exp Gerontol 2003; 38(1-2):79-86.
24. Gertler FB, Niebuhr K, Reinhard M et al. Mena, a relative of VASP and Drosophila Enabled, is implicated in the control of microfilament dynamics. Cell 1996; 87(2):227-239.
25. Waldmann R, Nieberding M, Walter U. Vasodilator-stimulated protein phosphorylation in platelets is mediated by cAMP- and cGMP-dependent protein kinases. Eur J Biochem 1987; 167(3):441-448.
26. Wen Z, Guirland C, Ming GL et al. A CaMKII/calcineurin switch controls the direction of Ca(2+)-dependent growth cone guidance. Neuron 2004; 43(6):835-846.
27. Hong K, Nishiyama M, Henley J et al. Calcium signalling in the guidance of nerve growth by netrin-1. Nature 2000; 403(6765):93-98.
28. Hopker VH, Shewan D, Tessier-Lavigne M et al. Growth-cone attraction to netrin-1 is converted to repulsion by laminin-1. Nature 1999; 401(6748):69-73.
29. Shewan D, Dwivedy A, Anderson R et al. Age-related changes underlie switch in netrin-1 responsiveness as growth cones advance along visual pathway. Nat Neurosci 2002; 5(10):955-962.
30. Cai D, Shen Y, De Bellard M et al. Prior exposure to neurotrophins blocks inhibition of axonal regeneration by MAG and myelin via a cAMP-dependent mechanism. Neuron 1999; 22(1):89-101.
31. Cai D, Qiu J, Cao Z et al. Neuronal cyclic AMP controls the developmental loss in ability of axons to regenerate. J Neurosci 2001; 21(13):4731-4739.
32. Zheng JQ, Zheng Z, Poo M. Long-range signaling in growing neurons after local elevation of cyclic AMP-dependent activity. J Cell Biol 1994; 127(6 Pt 1):1693-1701.
33. Neumann S, Bradke F, Tessier-Lavigne M et al. Regeneration of sensory axons within the injured spinal cord induced by intraganglionic cAMP elevation. Neuron 2002; 34(6):885-893.
34. Qiu J, Cai D, Dai H et al. Spinal axon regeneration induced by elevation of cyclic AMP. Neuron 2002; 34(6):895-903.

CHAPTER 11

Membrane Lipid Rafts and Their Role in Axon Guidance

Carmine Guirland and James Q. Zheng*

Abstract

The plasma membrane of cells contains a variety of lipid and protein molecules that are often segregated and heterogeneously distributed in microdomains. Lipid rafts represent a generalized concept of membrane microdomains that are enriched in cholesterol and sphingolipids and, characteristically, resistant to cold detergent extraction. Lipid rafts have recently received considerable attention because they are thought to be involved in many cellular functions, in particular, signal transduction for extracellular stimuli. Many of these functions are also intimately related to the processes involved in neural development, including neurotrophic factor signaling and synaptic plasticity. Recent studies from our lab and others have indicated an important role for lipid rafts in axonal growth and guidance. Specifically, our data show that lipid rafts on the plasma membrane provide platforms for spatial and temporal control of guidance signaling by extracellular cues. In addition, lipid rafts may also function in other aspects of axonal growth and guidance, including spatial and temporal regulation of adhesion, cytoskeletal dynamics, and growth cone motility. Further elucidating how membrane rafts are involved in guided axonal growth would provide important insights into the intricate signaling mechanisms underlying neuronal wiring, which is fundamental for normal brain development and functional recovery after injury and diseases.

Introduction

In the fluid mosaic model of the plasma membrane posited by Singer and Nichols, the membrane is a bilayer composed of a relatively continuous and homogenous fluid of amphipathic lipids that is interspersed with a mosaic of proteins.[1] Most eukaryotic cells are mainly composed of lipids belonging to three major lipid classes: glycerophospholipids, sphingolipids, and sterols. Various membrane proteins, including receptors, can associate with the plasma membrane by virtue of hydrophobic and electrostatic forces, covalently attached lipid anchors, and membrane-spanning domains. However, this picture of cell membranes has since been evolving steadily.[2] For example, it is known that the lipid and protein constituents of membranes are distributed asymmetrically. Different lipid classes of the membrane have been found in ratios that vary across each leaflet of the membrane, different cell types, and different cell compartments. The diversity of lipids and their distinct spatial distribution suggest that they may be involved in a variety of cellular functions. Arguably the most significant modification of the original fluid mosaic model is the existence of lipid domains of different lipid composition

*James Q. Zheng, Department of Neuroscience and Cell Biology, UMDNJ-R.W. Johnson Medical School, 675 Hoes Lane, Piscataway, New Jersey 08854, U.S.A.
Email: james.zheng@umdnj.edu.

Axon Growth and Guidance, edited by Dominique Bagnard. ©2007 Landes Bioscience and Springer Science+Business Media.

and physical state from the rest of the lipid bilayer. The initial notion of lipid domains was suggested by studies in model membranes, but it was the observation of caveolae, flask-shaped plasmalemmal invaginations of the cell membrane that led to extensive studies on membrane microdomains.[3] Caveolae exhibit several distinct features including a special lipid composition rich in cholesterol and sphingolipids, a striped coat formed by caveolin proteins on the cytoplasmic surface, and in addition to their characteristic flask shape, they can also have vesicular and tubular morphologies. Caveolae were initially thought to mainly function in clathrin-independent endocytosis.[3] Subsequent biochemical analyses of the molecular composition of caveolae, based on the findings that caveolae are low-density membranes and resistant to cold detergent extraction, suggested other possible functions. Most notably, these studies have found the presence of multiple signaling components in caveolae preparations,[3,4] indicating that caveolae may also play a role in signal transduction.

Later studies have pointed out that membrane domains lacking caveolin proteins are also present on the plasma membrane, suggesting the existence of other types of detergent-resistant microdomains (DRMs) that do not involve caveolins.[5] An increasing number of studies have now established that the plasma membrane contains different types of DRMs and caveolae represent a subset.[5] As such the term "lipid rafts" was later used to describe dynamic membrane domains in a broader sense. Before exploring the functions of rafts, it is helpful to consider some of their characteristics. Lipid rafts are small and dynamic: they can be as little as several nanometers in diameter and their transient existence is in the msec range. Both lipid raft size and half-life are flexible parameters that are altered in live cells, which may be involved in lipid raft functions. Rafts are thought to be a liquid ordered phase of membrane, which consists of saturated sphingolipids. The rafts "float" in a liquid disordered phase, which mainly consists of unsaturated glycerophospholipids. Cholesterol is thought to stabilize the sphingolipids in this liquid ordered phase since cholesterol interacts more favorably with sphingolipids over unsaturated phospholipids.[6] This partitioning of the membrane into laterally heterogeneous domains can therefore provide an organized membrane environment for protein interactions or other cellular functions.

Approaches to Study Lipid Rafts and Their Functions

Many studies have relied on two experimental approaches involving detergent resistance and cholesterol dependence to study lipid rafts and their functions. Lipid rafts are biochemically defined on the basis that they remain resistant to cold nonionic detergent treatment and/or are low-density membranes, thus float to the top of a buoyant density gradient. The so-named detergent resistant membranes (DRMs) are also known as detergent-insoluble glycolipid-enriched complexes (DIGs). Proteins that associate with lipid rafts are defined as those that cofractionate with DRM fractions and typically have some lipid modification such as glycosylphosphatidylinositol (GPI) or acyl anchors. Therefore, cold-detergent extraction and membrane fractionation have been extensively used to identify proteins associated with lipid rafts. Using this approach, numerous proteins, including GPI-anchored proteins, caveolins, src-family kinases, and G-proteins, have been shown to associate with lipid raft fractions.[7,8] Since lipid raft integrity depends on cholesterol, cellular functions that require lipid rafts could be affected by manipulating membrane cholesterol. Using various means to manipulate the synthesis or plasma member distribution of cholesterol has been instrumental in investigating the role of lipid rafts in cell functions beyond protein associations. While such experimental approaches have inherent flaws,[6,9] they have proved to be useful methods for identifying many of the constituents and functions involving lipid raft membrane microdomains.

One challenge in studying lipid rafts is the direct visualization of these dynamic microdomains on the native membrane of living cells.[10] While DRMs have been biochemically isolated and analyzed, the dynamics and spatial properties of lipid rafts remain to be directly examined. Detergent extraction does remove some lipids and proteins from rafts, which in combination with methodological differences, may account for the considerable degree of variability in analyses of raft components. However, it is the lack of visual evidence that is fueling the continuing debate over lipid rafts.[6] Past attempts of visualizing lipid rafts by light microscopy and electron microscopy have also generated conflicting results, particularly regarding the size of the rafts on the cell membrane.[10-14] Perhaps the dynamic nature of these membrane domains and their spatial localization on the cell surface contribute to the difficulty in determining their size and distribution. Therefore, future improvement in spatial resolution of current microscopy techniques and the development of new imaging methods on living cells may finally reveal the spatial and temporal properties of DRMs. Among various promising techniques, high-resolution fluorescence resonance energy transfer (FRET) imaging offers the ability to study dynamic membrane microdomains and protein-protein interactions on the plasma membrane.[11,12,15] Such technical advances would allow the validation of the lipid rafts concept and our understanding of the dynamics and functions of these membrane microdomains.

Functions of Lipid Rafts in the Nervous System

Numerous functions of lipid rafts have been implicated in nonneuronal cells (for reviews, see refs. 7,8 and Fig. 1), many of which are likely involved in nerve cells. For example, the first of many functional roles attributed to membrane microdomains was in protein and lipid sorting in polarized epithelial cells. Considering that neurons are highly polarized cells with axonal and dendritic specifications, precise sorting and selective trafficking of different molecules are clearly required for the development and maintenance of specific structures and functions of the neuron. Moreover, membrane lipids and proteins are distributed with spatial differences at various locations of neurons. For example, dendritic spines have been shown to enrich sphingolipids and many postsynaptic proteins[16] while axonal processes contain specific molecules involved in motility and transmitter release, some of which have been shown to associate with membrane rafts.[17] While the exact mechanisms involved in the generation, regulation, and maintenance of neuronal polarity are still under investigation,[18,19] it is conceivable that lipid rafts may play an important role in sorting and trafficking of different molecules to specific neuronal locations, although further evidence is needed.

The notion that lipid rafts are involved in signal transduction was initially suggested by the cofractionation of many signaling components such as GPI-anchored proteins and src-family kinases with detergent resistant membranes.[9] Subsequent studies show that a variety of other receptors and intracellular signaling components are associated with DRMs. It was proposed that sphinglolipid-cholesterol microdomains act as rafts that can selectively associate with proteins and that such raft platforms are functionally involved in membrane trafficking and intracellular signaling.[20] These dynamic rafts are thought to provide suitable microenvironments, which in addition to enabling selective protein-protein interactions may also be involved in localized initiation of signal transduction.[7,8,21] Many different responses to extracellular signals by numerous cell types are currently thought to involve lipid rafts, including immune responses, growth factor signaling, adhesion, and chemotaxis. Importantly, these experiments indicate that lipid rafts can be involved in specific signaling pathways and/or other cell functions by distinct ways. Several models have been proposed on how rafts are involved in signaling, including resident or recruited signaling mechanisms.[8] Proteins with a high affinity for lipid rafts are generally thought to reside within lipid rafts, whereas other proteins with little or no affinity for lipid rafts can be recruited to rafts. Resident proteins associate with lipid rafts in the absence of a stimulus and typically include GPI-anchored proteins and dually acylated ones for the outer and inner leaflets of the plasma membrane, respectively. Even without such lipid modifications other proteins including transmembrane proteins may still reside in rafts by an unknown mechanism.

One of best examples on the role of lipid rafts in signal transduction is growth factor signaling (for review, see ref. 22). Recent evidence indicates that membrane rafts are involved in signaling induced by neurotrophins and glial cell line-derived neurotrophic factor (GDNF) families. The functions of these growth factor families can influence many different neuronal populations and include effects on cell growth, proliferation, differentiation, and survival. The signaling mechanisms that mediate the functions of these two families are similar in that they involve the activation of receptor tyrosine kinases, which leads to the formation of complexes

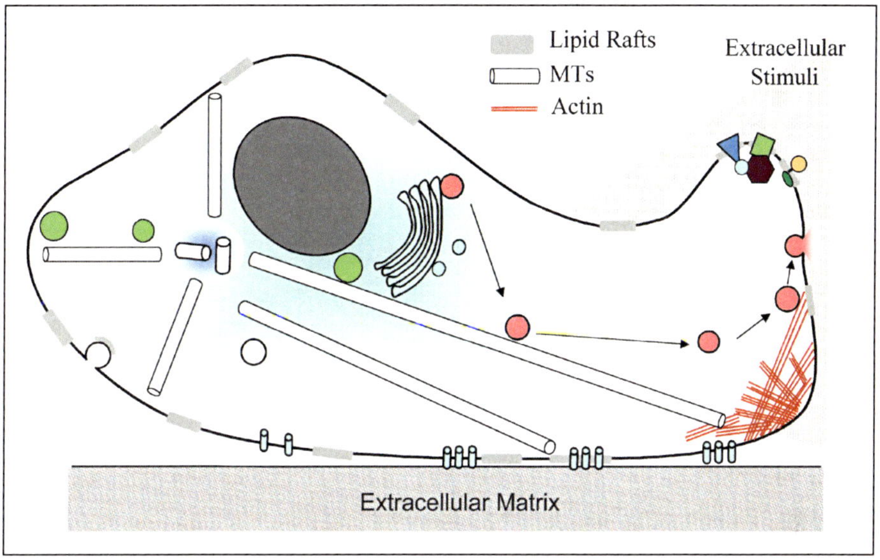

Figure 1. Functions of lipid rafts. This schematic view of a migrating cell (leading edge to the right) depicts the various functions that have been shown to involve lipid rafts. (1) Lipid rafts are involved in the sorting and trafficking of lipids and proteins to the plasma membrane of polarized cells. Vesicles budding from the Golgi are transported along microtubules to the front (red circles with thick border) or rear (green circles with thin border) of the cell. Certain types of endocytosis and exocytosis (unfilled vesicle and omega structure) also involve lipid rafts. It is possible that similar raft-dependent sorting methods are used by migrating cells to localize different sets of signaling components at the leading and trailing edges. (2) Many receptors and intracellular signaling molecules associate with lipid rafts and depend on these membrane microdomains for signal transduction events. Lipid rafts can serve as signaling platforms that enable the coupling of receptors to distinct pathways and can involve different variants of resident or recruited (induced association) mechanisms. (3) Regulation of the cytoskeleton is known to involve various proteins including the RhoGTPases as well as lipids such as certain phosphoinositides. Lipid rafts are thought to play a dynamic role in regulating the efficiency and membrane localization of these important protein and lipid regulators and thus the functions of the cytoskeleton. (4) Cell adhesion involves interactions between specific adhesion molecules on the cell with components of the extracellular matrix or other cells. Certain adhesion molecules associate with lipid rafts and their distribution on the cell may be regulated by lipid rafts and vice versa. It is important to consider that each lipid raft function can operate independent of the others during the overall functioning of the cell. It is also likely that some or all of the lipid raft functions operate in a coordinated fashion.

that are coupled to multiple intracellular signaling events. However, signaling events induced by growth factor stimulation have been shown to involve lipid rafts in different ways.[22] Neurotrophins, such as nerve growth factor (NGF), exert their effects by binding to receptor tyrosine kinase receptors (TrkA, B, and C) or a low-affinity receptor, p75NTR. Trk receptors and p75NTR have been found within lipid raft fractions along with critical intracellular components implicated in downstream signaling of these receptors. Evidence from PC12 cells shows that signaling from TrkA and p75NTR occurs and is enhanced within lipid rafts.[23] Thus neurotrophin binding to TrkA and p75NTR and subsequent signaling occurs within lipid rafts. In contrast to the involvement of rafts in neurotrophin signaling, evidence indicates that GDNF signaling involves recruitment of the receptor tyrosine kinase, c-Ret, to lipid rafts. It is thought that c-Ret recruitment to lipid rafts by GFRalpha enables this receptor tyrosine kinase to selectively associate with its downstream signaling components residing in lipid rafts.[24]

Neurotrophic factors have also been shown to be involved in synaptic plasticity. In particular, brain-derived neurotrophic factor (BDNF) is known to modulate long-term synaptic potentiation. While accumulating evidence has indicated that lipid rafts can influence synaptic transmission through clustering and regulation of neurotransmitter receptors and affect the exocytotic process of transmitter release,[21] recent studies on BDNF effects have revealed some new insights towards how rafts may contribute to BDNF effects on synaptic plasticity.[25] For example, TrkB receptors were recruited to the raft fraction after exposure to BDNF and the translocation depended on tyrosine kinase activity. Furthermore, BDNF recruited TrkB alone into lipid rafts without carrying its associated proteins Shc, Grb2, and PLCγ, which is different from neuregulin-induced recruitment of ErbB4 to lipid rafts. Moreover, the finding that lipid rafts are only involved in BDNF modulation of synaptic activity, but not BDNF enhancement of neuronal survival indicates that these membrane rafts could be involved in signaling specificity of BDNF on developing neurons. Finally, the coreceptor p75 was found to inhibit BDNF-induced TrkB translocation into lipid rafts, suggesting that TrkB and p75 mediate distinct signaling pathways that depend differentially on lipid raft association. Future studies would likely elucidate the intermediate factors that interact with TrkB in the rafts for initiating specific downstream signaling leading to synaptic modification.

Membrane Domains and Growth Cone Motility

Lipid rafts have been implicated in many aspects of cell motility, in particular, cytoskeletal dynamics to substrate adhesion. Significantly, many of the molecular components regulating the actin cytoskeleton, cell motility, and adhesion are associated with rafts, including Rho GTPases, Src-family of tyrosine kinases, phosphoinositides PtdIns(4,5)P2 and PtdIns(3,4,5)P3.[26] In migrating cells, selective adhesion is established by the formation of focal adhesion complexes containing many signaling components and cytoskeletal anchoring. It is well established that cell migration requires dynamic and spatial regulation of focal adhesion complexes: adhesion at the rear end of the cells needs to be removed while the leading front forms new adhesion sites. The findings that distinct raft-associated components are asymmetrically distributed on the leading edge and the uropod suggest that lipid rafts are involved in the spatially-regulated motile activities in migrating cells. Recently, it has been shown that lipid rafts mediate signal transduction events initiated by cell adhesion to the extracellular matrix through integrins. In particular, membrane rafts appear to mediate spatial targeting of Rho GTPases to the plasma membrane for differential association with the downstream effectors for further signaling events, including Rac coupling to focal adhesion kinase and microtubule stabilization by Rho and mDia.[27,28] These findings from nonneuronal cells thus establish that lipid rafts play an important role in cell adhesion and motility by participating in signal transduction and spatial targeting of various signaling components.

In developing neurons, guided elongation of axonal processes depends critically on the motility and pathfinding ability of the tip of the axon, the growth cone for reaching the specific targets. Such directional motility is believed to depend on cytoskeletal dynamics, together with selective adhesion with the substratum, for steering the growth cone in response to a variety of environmental cues. Cell adhesion is mediated by interactions between the growth cone's cell adhesion molecules (CAMs) and the extracellular matrix or other CAMs on neighboring cells. Previous studies have shown that certain CAMs are associated with lipid rafts, indicating that selective adhesion underlying growth cone motility may involve lipid rafts. Recent studies have shown that nerve growth cones exhibit discrete lipid domains on the surface (Fig. 2, see also ref. 29) and raft disruption affected their motility on the adhesion molecule substrates L1, N-cadherin, but not β1 integrin.[29] These results show that growth cone adhesion on selective substrates involves lipid rafts. Furthermore, biochemical analyses have revealed that many other proteins involved in growth cone adhesion and motility are associated with DRMs, including focal adhesion kinases, src family of tyrosine kinases, GAP43, and etc.[17] Therefore, lipid rafts likely play a broader role in growth cone motility during migration. Since directed movement of cells or growth cones requires concerted events among the cytoskeleton, membrane

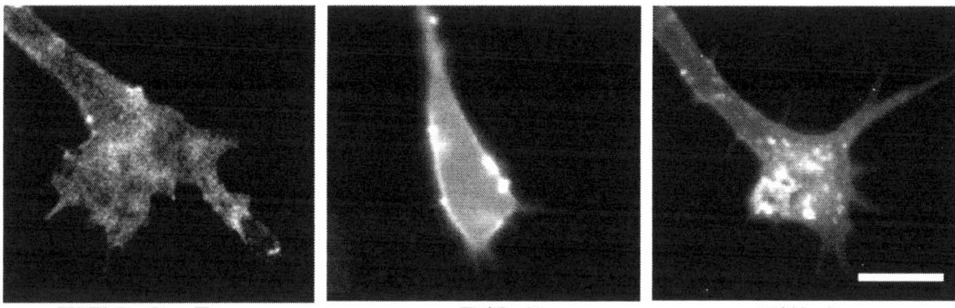

Figure 2. Lipid domains on the growth cone. The plasma membrane of *Xenopus* growth cones was stained with FITC-Cholera Toxin B (CTxB), $DiIC_{18}$, and filipin to examine the distribution of different lipid constituents. Staining of the membrane with the lipophilic dye, $DiIC_{18}$, resulted in a relatively uniform fluorescent signal. On the other hand, labeling of ganglioside G_{M1} by CTxB or cholesterol by filipin showed an apparent heterogeneous distribution, indicating the existence of microdomains. The image of FITC-CTxB staining was processed by applying a digital threshold for better illustration of domains. Scale bar = 10 μm.

anchoring, and cell-substratum adhesion, lipid rafts could serve as the central platforms for spatial and temporal regulation of any of these important events.

Lipid Rafts in Axon Guidance: The Signaling Platforms?

Developing axons are guided to their targets by a variety of extracellular cues that either attract or repel growth cones. Many of these extracellular cues exert their specific actions on developing axons by binding to their surface receptors to initiate complex signaling cascades.[30-32] Therefore, the formation of ligand-receptor complexes on the plasma membrane represents the first step in transduction of guidance signals. In addition, many guidance cues elicit additional steps on and/or within the plasma membrane to generate distinct signaling cascades, including receptor oligomerization and complex formation with coreceptors and/or other membrane-associated components.[33] These protein interactions at/within the plasma membrane are believed to define distinct cellular responses to extracellular stimuli. For example, receptor cross-talk has been shown to specifically enable a particular guidance response while silencing the other.[34] While the membrane components involved in receptor-signaling complexes are being identified, how these receptors, coreceptors, and other membrane-associated components interact on the membrane to generate specific signaling cascades for distinct axonal responses remains elusive. The fact that these important events occur at or within the plasma membrane suggests that the membrane lipid environment could be crucial for the signal transduction of these extracellular guidance cues. On the other hand, specific lipid molecules are known to play an important role in cell signaling. For example, phospholipid phosphatidylinositol-3,4,5-trisphosphate (PIP_3) accumulates at the leading edge of chemotactic cells to recruit signaling proteins containing pleckstrin homology (PH) domains; such localization of PIP_3 at the leading edge is believed to be an essential part of directional responses of chemotactic cells.[35,36] These specific phospholipids (e.g., PIP_2 and PIP_3) may also depend on the lipid environment on the membrane for their localization and functioning.[37] Therefore, the specific lipid composition on the plasma membrane may contribute to not only protein-protein interactions but also lipid signaling in response to extracellular molecules.

So far, only a few guidance molecules are known to have an association with lipid rafts. For example, ephrin ligands and Eph receptor tyrosine kinases are well known molecules involved in axon guidance and topographic mapping of neuronal connections. Ephrin A proteins are GPI-anchored ligands residing in lipid rafts. Interestingly, ephrin B ligands contain a

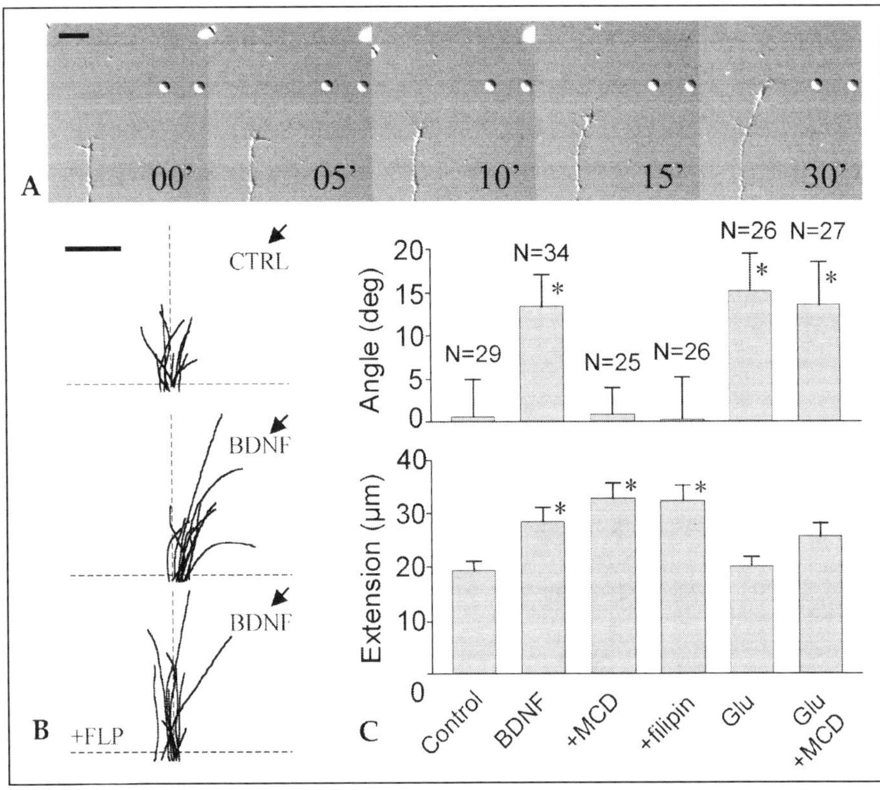

Figure 3. Lipid rafts are involved in functional guidance responses. The requirement of lipid rafts in functional guidance responses was evaluated in vitro by using the growth cone turning assay for the BDNF, netrin-1, Sema3A and glutamate as discussed in the text. We have included some raw data using BDNF as an example. A) Growth cone attraction to a gradient of BDNF is shown in the time lapse DIC images of an individual neurite. The numbers indicate the minutes since the onset of BDNF application. B) Superimposed traces of individual neurite trajectories during the assay period are shown for three experimental groups of growth cones. The origin represents the center of the growth cone that was extending vertical at the beginning of the assay. The arrow indicates the position of the pipette and the "+" sign indicates the addition of a raft disrupting agent. C) Growth cones were incubated with the indicated with two different agents that manipulate membrane cholesterol and thereby disrupting rafts. The growth cones were then exposed to a gradient of BDNF or glutamate to determine whether or not lipid rafts are involved in responses to these ligands. The bar graphs represent the average responses and indicate that lipid raft disruption blocks BDNF, but not glutamate attraction. Interestingly, raft disruption does not block growth promotion by BDNF. These results suggest that lipid rafts are selectively involved in certain functions, namely BDNF attraction, but not BDNF growth promotion, nor glutamate attraction. Scale bars: 20 µm.

transmembrane domain, are also located in lipid rafts, and have been shown to use lipid rafts for signal transduction.[38] Previous studies have also shown that myelin-associated glycoproteins inhibit axonal growth through interactions with specific gangliosides and rafts.[39] Furthermore, NgR and p75NTR, the receptor and coreceptor for Nogo have been shown to reside in lipid rafts.[23,40,41] While these studies indicate the potential involvement of lipid rafts in guidance, they do not directly assess whether or not lipid rafts mediate signal transduction in guidance responses. Using an in vitro guidance assay together with several complementary methods of raft manipulation, we have recently shown that lipid rafts mediate guidance responses of nerve growth cones to BDNF, Netrin-1, and Sema3A gradients[42] (see also Fig. 3). The finding that activation of the MAPK p44/p42 by these guidance molecules could be abolished by raft disruption suggests that lipid rafts are involved in signal transduction of these guidance responses. While the receptors for these ligands were only weakly associated with lipid rafts under control conditions, they increased their affinity with rafts in response to stimulation by the respective ligand. The mechanism responsible for this translocation and downstream events during growth cone guidance are not known. Recent studies on TrkB signaling indicate that activation of TrkB is a preceding requirement for localization into rafts, which specifically mediates BDNF modulation of synaptic transmission.[25] It is possible that association of ligand-receptor complexes within lipid rafts engages distinct signaling pathways from signaling outside rafts.[8,22,25] Growth cone guidance by BDNF and netrin-1 involves phospholipase C (PLC) and PI-3 kinases,[43] both of which associate with lipid rafts.[44] On the other hand, Sema3A signaling involves receptor complexes consisting of neuropillin-1, plexin A, and the adhesion molecule L1.[33,45] It is conceivable that, although all these guidance cues depend on rafts for their effects on growth cones, the specific signaling pathways could differentially rely on distinct raft-dependent mechanisms for generating guidance responses. It will therefore be important to determine whether these relevant signaling components associate with lipid rafts and the relative contributions of signaling in and out of rafts during the growth cone response.

Although we have shown that guidance signaling involves lipid rafts, the contribution of raft-dependent adhesion and/or cytoskeletal regulation in growth cone responses requires further investigation. Many nonreceptor tyrosine kinases involved in adhesion are associated with lipid rafts[8] and active Rho GTPases, which regulate the cytoskeleton, are targeted to lipid rafts for coupling to downstream effectors.[27,28,46] Moreover, cytoskeletal dynamics have been implicated in affecting the position and stability of rafts as well as the associations of certain molecules with rafts.[26] Therefore, lipid rafts may mediate growth cone guidance by providing a critical platform for coupling activated receptors, and/or their downstream effectors with the regulation of adhesion and the cytoskeleton. Rafts have also been implicated in organizing cellular polarity and as such they may be involved in the spatial localization of guidance signaling to mechanisms of adhesion and cytoskeletal regulation.[26] On the other hand, our findings that growth cone attraction induced by glutamate gradients was not affected by raft disruption indicate that lipid rafts were likely involved in signal transduction specific for BDNF, netrin-1, and Sema3A, rather than common steering events. Perhaps, different substrates may contribute to the raft-dependent and -independent adhesion and growth cone motility.[29]

Signal Localization through Lipid Rafts

Similar to chemotactic cells, growth cone turning in responses to guidance gradients involves asymmetric signaling. The recent finding that lipid rafts are functionally involved in such asymmetry provides an exciting avenue for pursuing the subcellular mechanisms of growth cone turning.[42] Some studies on polarization and asymmetric signaling in cell migration suggest that membrane receptors are not spatially redistributed in response to a chemotactic signal and that intracellular gradients are sufficient for encoding spatial information that mediates chemotactic responses.[36] Other studies suggest certain membrane components including receptors and lipid raft markers do exhibit a change of distribution during chemotaxis.[47] Consistent with this latter notion, there is evidence that the TrkB receptor asymmetrically associates with lipid rafts in response to the application of a BDNF gradient. Although the mechanism of this translocation is not known, such asymmetric localization of the receptor is thought to only occur at lipid rafts and would be sufficient to localize subsequent signaling steps required for turning. Furthermore, asymmetric translocation of guidance receptors into lipid rafts after ligand binding could lead to local signal amplification by concentrating signaling molecules and/or excluding unwanted modulatory components,[8] which might be essential for successful

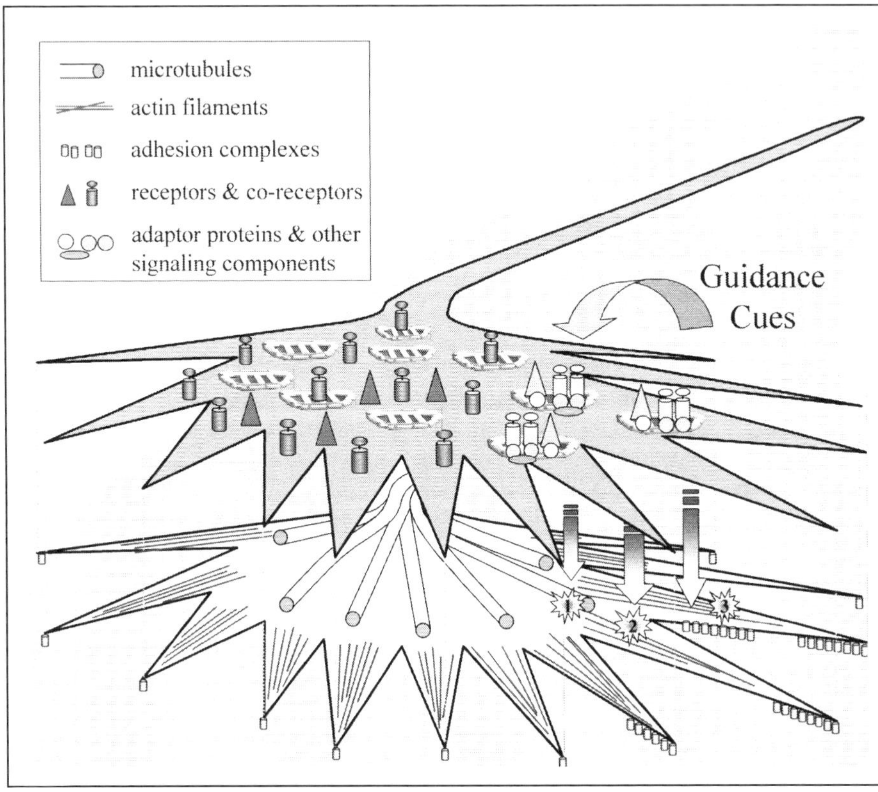

Figure 4. Hypothetical model on the role of lipid rafts in axon guidance and growth cone motility. We propose that lipid rafts provide critical platforms for spatial control of signaling in growth cone guidance to achieve asymmetric signal transduction and growth cone steering. Specifically, membrane receptors and certain raft types are likely to be distributed with relative uniformity on the plasma membrane in the absence of ligand stimulation. In response to stimulation by a guidance cue, the association of the appropriate receptor with lipid rafts increases, which could enable selective interactions with intermediate components residing in rafts for downstream signaling. Such an association may be maintained over time and can serve to amplify the stimulus in an asymmetrically activated signaling complex. Rafts could also be involved in local regulation of both cytoskeletal dynamics and adhesion for directional steering. Importantly, associations with rafts are dynamic events that can be regulated over time and space. For example, modulation of a guidance response may be achieved by regulating the affinity of a signaling component for lipid rafts and thus its ability to interact with other members of the complex. It also stands to reason that any process, which affects the positioning of lipid rafts, would be able to provide an overarching level in a hierarchy of membrane organization.

sensing of extracellular gradients. That raft subtypes asymmetrically redistribute during cell migration, suggests the exciting possibility that a similar mechanism may operate during growth cone guidance.

Based on the evidence discussed above, we propose a model in which lipid rafts provide critical platforms for spatial control of signaling in growth cone guidance to achieve asymmetric signal transduction and growth cone steering (Fig. 4). Lipid rafts can be involved in generating and/or maintaining the asymmetry at several steps along the signal transduction pathway. As the first step, ligand induced translocation of receptors to lipid rafts could enable selective interactions with intermediate components residing in rafts for downstream signaling. Such translocation could also function to shield the activated receptors from nonraft factors that can inactivate the receptors, thus providing a degree of temporal control. Furthermore, membrane microdomains could also provide the platforms for specific targeting and formation of signaling complexes that enable the activation of selective pathways for distinct effects. Finally, rafts could be involved in local control of cytoskeletal dynamics and adhesion for directional steering. Ultimately, different guidance molecules and different environmental settings (e.g., different ECMs) could specifically utilize some of these raft-dependent mechanisms for spatiotemporal regulation of growth cone migration. The challenge would be to dissect the pathways and the specific functions of lipid rafts in each of the many guidance systems.

Future Directions and Concluding Remarks

The findings of a functional role of membrane rafts in axon guidance have also opened new directions for elucidating the molecular mechanisms underlying axon guidance and regeneration. For example, the findings on the ligand-induced translocation of receptors into lipid rafts suggest that coreceptors and/or intermediate signaling components are readily present in lipid rafts, waiting for activated receptors to signal downstream. Therefore, proteomic approaches could be used to analyze membrane raft factions with and without exposure to specific guidance cues, which could potentially lead to the identification of novel intermediate signaling components in the rafts.[48] Moreover, the observation that inhibitory effects of Semaphorin 3A could be abolished by raft manipulation also indicates a potential approach for overcoming inhibition of regenerating axons after nerve injury and diseases. Such an approach could represent a novel strategy for targeting axon inhibition during nerve regeneration and functional recovery. Finally, given the diverse roles potentially played by lipid rafts in various cellular functions, further studies are clearly required to delineate the specific steps of signaling transduction associated with distinct raft functions.

The cell's ability to sense and respond to environmental stimuli is crucial for many functions including neural development, immunity, angiogenesis, wound healing, and embryogenesis. Directed migration of nerve growth cones and chemotactic cells likely requires similar coordinated cellular processes that lead to movement, including sorting and trafficking of specific membrane proteins and lipids, signal amplification and localization, and spatiotemporal regulation of cytoskeletal dynamics and adhesion. Membrane microdomains may be of particular importance for migrating cells as they can serve to spatially and temporally coordinate the component functions required for cell and growth cone movement. Comparative analyses on the various roles and mechanisms related to lipid rafts in cell migration and growth cone guidance may be particularly helpful in understanding the precise functions of lipid rafts in intricate guidance signaling. These studies could in turn provide further mechanistic insights into directed cell migration underlying many important biological responses such as leukocyte chemotaxis during inflammatory response.

References

1. Singer SJ, Nicolson GL. The fluid mosaic model of the structure of cell membranes. Science 1972; 175(23):720-31.
2. Edidin M. Lipids on the frontier: A century of cell-membrane bilayers. Nat Rev Mol Cell Biol 2003; 4(5):414-8.
3. Anderson RG. The caveolae membrane system. Annu Rev Biochem 1998; 67:199-225.
4. Sargiacomo M, Sudol M, Tang Z et al. Signal transducing molecules and glycosyl-phosphatidylinositol-linked proteins form a caveolin-rich insoluble complex in MDCK cells. J Cell Biol 1993; 122(4):789-807.
5. Harder T, Simons K. Caveolae, DIGs, and the dynamics of sphingolipid-cholesterol microdomains. Curr Opin Cell Biol 1997; 9(4):534-42.
6. Munro S. Lipid rafts: Elusive or illusive? Cell 2003; 115(4):377-88.
7. Brown DA, London E. Functions of lipid rafts in biological membranes. Annu Rev Cell Dev Biol 1998; 14:111-36.
8. Simons K, Toomre D. Lipid rafts and signal transduction. Nat Rev Mol Cell Biol 2000; 1(1):31-9.

9. Edidin M. The state of lipid rafts: From model membranes to cells. Annu Rev Biophys Biomol Struct 2003; 32:257-83, (Epub 2003 Jan 16).
10. Lai EC. Lipid rafts make for slippery platforms. J Cell Biol 2003; 162(3):365-70.
11. Kenworthy AK, Petranova N, Edidin M. High-resolution FRET microscopy of cholera toxin B-subunit and GPI- anchored proteins in cell plasma membranes. Mol Biol Cell 2000; 11(5):1645-55.
12. Varma R, Mayor S. GPI-anchored proteins are organized in submicron domains at the cell surface. Nature 1998; 394(6695):798-801.
13. Glebov OO, Nichols BJ. Lipid raft proteins have a random distribution during localized activation of the T-cell receptor. Nat Cell Biol 2004; 6(3):238-43, (Epub 2004 Feb 8).
14. Pitto M, Palestini P, Ferraretto A et al. Dynamics of glycolipid domains in the plasma membrane of living cultured neurons, following protein kinase C activation: A study performed by excimer-formation imaging. Biochem J 1999; 344(Pt 1):177-84.
15. Sekar RB, Periasamy A. Fluorescence resonance energy transfer (FRET) microscopy imaging of live cell protein localizations. J Cell Biol 2003; 160(5):629-33.
16. Hering H, Lin CC, Sheng M. Lipid rafts in the maintenance of synapses, dendritic spines, and surface AMPA receptor stability. J Neurosci 2003; 23(8):3262-71.
17. He Q, Meiri KF. Isolation and characterization of detergent-resistant microdomains responsive to NCAM-mediated signaling from growth cones. Mol Cell Neurosci 2002; 19(1):18-31.
18. Winckler B, Mellman I. Neuronal polarity: Controlling the sorting and diffusion of membrane components. Neuron 1999; 23(4):637-40.
19. Foletti DL, Prekeris R, Scheller RH. Generation and maintenance of neuronal polarity: Mechanisms of transport and targeting. Neuron 1999; 23(4):641-4.
20. Simons K, Ikonen E. Functional rafts in cell membranes. Nature 1997; 387(6633):569-72.
21. Tsui-Pierchala BA, Encinas M, Milbrandt J et al. Lipid rafts in neuronal signaling and function. Trends Neurosci 2002; 25(8):412-7.
22. Paratcha G, Ibanez CF. Lipid rafts and the control of neurotrophic factor signaling in the nervous system: Variations on a theme. Curr Opin Neurobiol 2002; 12(5):542-9.
23. Huang CS, Zhou J, Feng AK et al. Nerve growth factor signaling in caveolae-like domains at the plasma membrane. J Biol Chem 1999; 274(51):36707-14.
24. Paratcha G, Ledda F, Baars L et al. Released GFRalpha1 potentiates downstream signaling, neuronal survival, and differentiation via a novel mechanism of recruitment of c-Ret to lipid rafts. Neuron 2001; 29(1):171-84.
25. Suzuki S, Numakawa T, Shimazu K et al. BDNF-induced recruitment of TrkB receptor into neuronal lipid rafts: Roles in synaptic modulation. J Cell Biol 2004; 167(6):1205-15. (Epub 2004 Dec 13).
26. Golub T, Wacha S, Caroni P. Spatial and temporal control of signaling through lipid rafts. Curr Opin Neurobiol 2004; 14(5):542-50.
27. del Pozo MA, Alderson NB, Kiosses WB et al. Integrins regulate Rac targeting by internalization of membrane domains. Science 2004; 303(5659):839-42.
28. Palazzo AF, Eng CH, Schlaepfer DD et al. Localized stabilization of microtubules by integrin- and FAK-facilitated Rho signaling. Science 2004; 303(5659):836-9.
29. Nakai Y, Kamiguchi H. Migration of nerve growth cones requires detergent-resistant membranes in a spatially defined and substrate-dependent manner. J Cell Biol 2002; 159(6):1097-108.
30. Tessier-Lavigne M, Goodman CS. The Molecular Biology of Axon Guidance. Science 1996; 274:1123-33.
31. Dickson BJ. Molecular mechanisms of axon guidance. Science 2002; 298(5600):1959-64.
32. Pasterkamp RJ, Kolodkin AL. Semaphorin junction: Making tracks toward neural connectivity. Curr Opin Neurobiol 2003; 13(1):79-89.
33. Huber AB, Kolodkin AL, Ginty DD et al. Signaling at the Growth Cone: Ligand-Receptor Complexes and the Control of Axon Growth and Guidance. Annu Rev Neurosci 2003; 28:28.
34. Stein E, Tessier-Lavigne M. Hierarchical organization of guidance receptors: Silencing of netrin attraction by slit through a Robo/DCC receptor complex. Science 2001; 291(5510):1928-38.
35. Dekker LV, Segal AW. Perspectives: Signal transduction. Signals to move cells. Science 2000; 287(5455):982-3, 85.
36. Parent CA, Devreotes PN. A cell's sense of direction. Science 1999; 284(5415):765-70.
37. Laux T, Fukami K, Thelen M et al. GAP43, MARCKS, and CAP23 modulate PI(4,5)P(2) at plasmalemmal rafts, and regulate cell cortex actin dynamics through a common mechanism. J Cell Biol 2000; 149(7):1455-72.
38. Bruckner K, Pablo Labrador J, Scheiffele P et al. EphrinB ligands recruit GRIP family PDZ adaptor proteins into raft membrane microdomains. Neuron 1999; 22(3):511-24.

39. McKerracher L. Ganglioside rafts as MAG receptors that mediate blockade of axon growth. Proc Natl Acad Sci USA 2002; 99(12):7811-3.
40. Yu W, Guo W, Feng L. Segregation of Nogo66 receptors into lipid rafts in rat brain and inhibition of Nogo66 signaling by cholesterol depletion. FEBS Lett 2004; 577(1-2):87-92.
41. Higuchi H, Yamashita T, Yoshikawa H et al. PKA phosphorylates the p75 receptor and regulates its localization to lipid rafts. EMBO J 2003; 22(8):1790-800.
42. Guirland C, Suzuki S, Kojima M et al. Lipid rafts mediate chemotropic guidance of nerve growth cones. Neuron 2004; 42(1):51-62.
43. Ming G, Song H, Berninger B et al. Phospholipase C-gamma and phosphoinositide 3-kinase mediate cytoplasmic signaling in nerve growth cone guidance. Neuron 1999; 23(1):139-48.
44. Inoue H, Miyaji M, Kosugi A et al. Lipid rafts as the signaling scaffold for NK cell activation: Tyrosine phosphorylation and association of LAT with phosphatidylinositol 3- kinase and phospholipase C-gamma following CD2 stimulation. Eur J Immunol 2002; 32(8):2188-98.
45. Castellani V, Rougon G. Control of semaphorin signaling. Curr Opin Neurobiol 2002; 12(5):532-41.
46. Michaely PA, Mineo C, Ying YS et al. Polarized distribution of endogenous Rac1 and RhoA at the cell surface. J Biol Chem 1999; 274(30):21430-6.
47. Gomez-Mouton C, Abad JL, Mira E et al. Segregation of leading-edge and uropod components into specific lipid rafts during T cell polarization. Proc Natl Acad Sci USA 2001; 98(17):9642-7. Epub 2001 Aug 7.
48. Foster LJ, De Hoog CL, Mann M. Unbiased quantitative proteomics of lipid rafts reveals high specificity for signaling factors. Proc Natl Acad Sci USA 2003; 100(10):5813-8. Epub 2003 Apr 30.

CHAPTER 12

The Role of Glial Cells in Axon Guidance, Fasciculation and Targeting

Anabel R. Learte and Alicia Hidalgo*

Abstract

Axons navigate step-wise, from one intermediate target to the next, until they reach their final destination target. In the central nervous system, intermediate targets are often glial cells, and final targeting is also aided by glia. In the peripheral nervous system, however, glial cells most often follow axons, which therefore navigate following other, nonglial clues. Even in the central nervous system, interactions between axons and glia are dynamic and reciprocal, as the neurons regulate migration, survival and proliferation of the glia cells they need for guidance. We review here the experimental evidence investigating roles of glia in axon guidance. Some molecules are known to influence either the neurons or the glia, but the molecular mechanisms underlying axon-glia interactions during pathfinding are only beginning to emerge.

Introduction

In order to wire the nervous system, axons navigate to meet their targets tracing intricate trajectories. In some cases, the target may be a long distance away, for example an interneuron of the spinal cord may have a target in the brain. Furthermore, the trajectories are generally not straight, implying that during navigation growth cones 'make decisions' on whether to continue on a straight trajectory or turn and take a new direction. Thus, to understand what controls axon guidance we need to explain how do neurons 'know' how to get to a target that may be a long distance away and what provokes the changes in the direction of axonal navigation.

In 1976, Bate proposed that axons navigate following 'guidepost cells' located along their trajectories.[1] Accordingly, the trajectory of an axon is covered with guidepost cells and growth cones travel from one guidepost cell to the next, navigating short distances over a territory without 'knowledge' of the ultimate, long distance target. In fact, a growth cone may be within physical reach of the next guidepost cell along its route before it has detached from its prior one. Soon after, Singer proposed the 'blueprint hypothesis', according to which neurites extend following glial patterns and recognise molecular affinities on the surfaces of glial cells.[2] Over the last 30 years, support for the role of glia as guidepost cells came from multiple experiments carried out in which glia are damaged or removed—in earlier days surgically,[3,4] more recently genetically[5,6]—resulting in the disruption of axonal patterns.

Guidepost cells are 'intermediate targets' when an important axon guidance 'decision' is made by growth cones at that cell. Axons extend for some distance in a straight direction, and

*Corresponding Author: Alicia Hidalgo—School of Biosciences, University of Birmingham, Edgbaston, Birmingham B15 2TT, U.K. Email: a.hidalgo@bham.ac.uk

Axon Growth and Guidance, edited by Dominique Bagnard. ©2007 Landes Bioscience and Springer Science+Business Media.

then reach points where axons 'decide' to turn. During their journey, axons may not travel alone, but may adhere to (fasciculate) with other axons taking the same route, forming fascicles. At some points, axons separate (defasciculate). The locations where axons make turning, fasciculation or defasciculation decisions are called 'choice points'. Glial cells often occupy choice point positions and intermediate targets are often glial cells. Thus, axons project to intermediate targets, where they change trajectory, and through subsequent intermediate targets ultimately arrive at their final target, with which they synapse. Thus, the distance to be travelled by a growth cone is broken up into smaller journeys. Glial cells are often located also by the final target. Thus, glial cells are in multiple contexts involved in the decisions of growth cones to turn or not, in the sorting of axons into fascicles and in targeting.

Development presents a further challenge to navigating axons: the territory that an axon travels through may be changing or there may not be a route covered by guidepost cells to follow. For instance, the glial cells the axon uses for navigation may not form a prepatterned 'glial route' before axon extension begins. If the glial cells formed a route all neatly lined up for the axons to follow, then part of the problem in understanding axon guidance would have to be readdressed to the understanding of glial patterning and migration. In some cases, this is in fact what happens. However, in cases in which glial cells may function as guidepost cells, their migration and location may also be influenced by neurons. That is, the neurons influence the environment that they need to follow for navigation.

A functional nervous system not only requires that growth cones meet their targets to establish neural networks, but also that these networks can support neuronal function. Thus, as soon as the networks are established, the axons must be enwrapped by glial cells to provide insulation and homeostasis, supportive of neuronal function. This means that during axon guidance, the number of glial cells must also be adjusted to ensure that by the time the networks are formed, there are enough glial cells as well to ensheath the axons.

Here, we will review the in vivo molecular mechanisms underlying the roles for glial cells during axon guidance, with particular emphasis on evidence from Drosophila (see also refs. 7,8). We will not review the genetic bases of glial cell fate determination or glial differentiation. We will not review either the abundant evidence of glial influences on axonal extension in experimental repair paradigms.

Molecular Mechanisms Involved in Neuron-Glia Interactions during Axon Guidance

Chemoattractants and Chemorepellents Produced by Glial Cells That Influence Axon Guidance

Glial cells secrete important, evolutionary conserved axon guidance molecules such as the chemoattractant Netrin and chemorepellent Slits in *C. elegans*, Drosophila and vertebrates (refs. 9-15). The midline glia of Drosophila glia are functionally equivalent to the vertebrate floorplate, which is also formed by cells with glial features, and produce the same evolutionary conserved axon guidance molecules (Fig. 1A). Interneuron axons cross the midline attracted by Netrin, and after crossing project contralaterally and never cross the midline again as they sense repulsive midline signals such as Slit and Semaphorin. These signalling molecules bind evolutionary conserved receptors on the axons: for instance, Netrins bind Frazzled (Drosophila), Unc (*C. elegans*) and DCC (vertebrates) receptors, Slits bind Robo receptors on axons and Semaphorins bind Plexins and Neuropilins. The floor-plate also secretes the morphogen Sonic Hedgehog, which functions as a chemoattractant for commissural axons in interaction with Netrin-1.[16] Excellent reviews on the mechanisms by which these molecules guide axons, and their signal transduction pathways, are given in other chapters, so they will not be dealt with here any further.

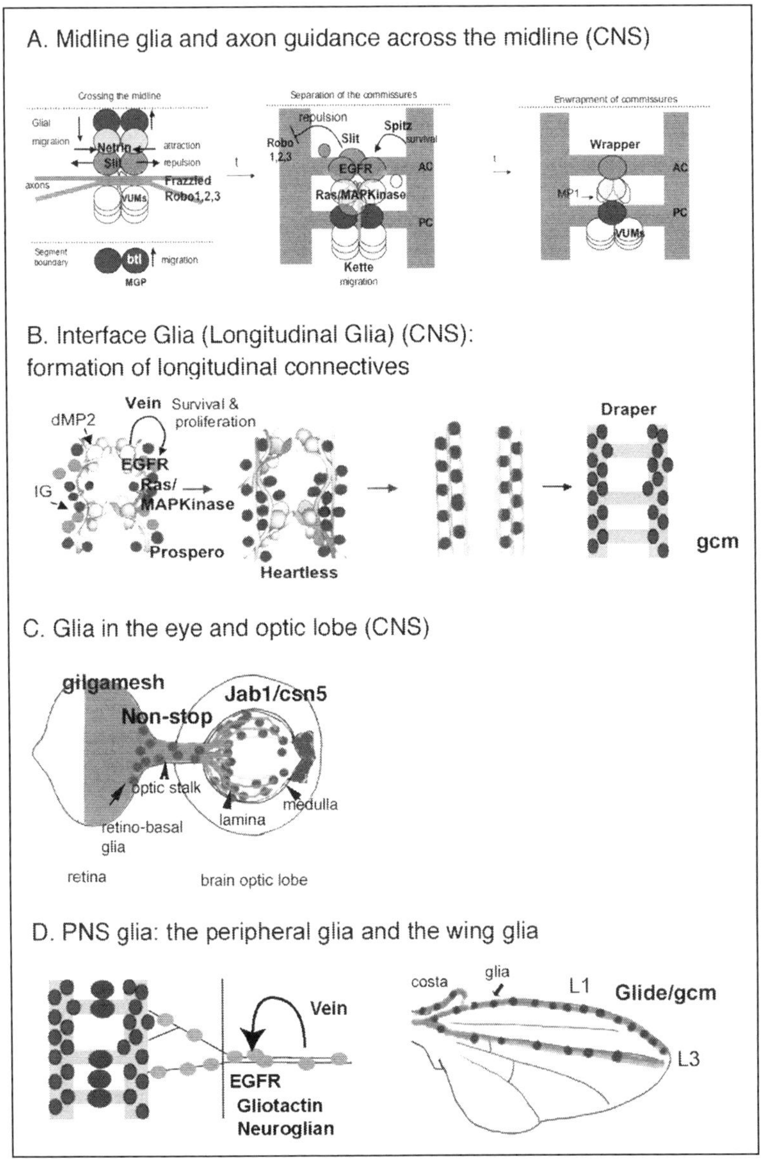

Figure 1. Illustration of different glial cell types in Drosophila and their relationships to axons. A. Midline Glia migrate to trigger the separation of the axonal commissures. The survival of midline glia is regulated through interactions with axons and only about a fourth of the midline glia survive. The Midline Glia produce axon guidance molecules such as Netrin and Slit that control midline crossing. B. The Longitudinal Glia ensheath the longitudinal connectives of the CNS. They are required for axon guidance by pioneer axons of the longitudinal pathways and to trigger axonal defasciculation leading to the formation of the three major longitudinal axonal trajectories. During these events, they are delivered in restricted numbers as their proliferation and survival is regulated by interactions with neurons. C. Glial cells invade the retina by migrating from the optic stalk, glial cells ensheath the optic stalk and are also located at targeting layers in the optic lobe. Mutations that affect glial function in the optic lobe prevent axonal targeting. D. In the PNS, contrary to the CNS, glia migrate following axons.

Molecules Produced by Neurons That Promote Glial Migration, Proliferation or Survival

The main class of neuronal signalling molecules to regulate glial survival are the neuregulins.[14,17] They have been studied in vivo both in vertebrates and in Drosophila. In vertebrates, neuregulin is produced by CNS axons of the optic nerve, and it binds ErbB4 receptors in oligodendrocytes, promoting their survival.[18] Survival of Schwann cells, glia of the PNS, also requires neuregulin, binding to EbrB2 and ErbB3 receptors and neuregulin is involved in multiple other contexts of neuron-glia interactions to promote glial proliferation and differentiation.[17] Signalling through Erb2 and Erb3 receptors also regulates migration of lateral line glia in zebrafish.[19] The Drosophila neuregulin homologue Vein is produced by CNS MP2 pioneer neurons and binds the only EGFReceptor in a subset of the longitudinal glia (the subset of the interface glia derived from the longitudinal glioblast), in the CNS where it activates the Ras/MAP Kinase pathway promoting glial survival and proliferation[20,21] (Fig. 1B). These longitudinal glia are required for growth cone guidance by the longitudinal pioneer axons, including the dMP2s (which produce Vein). Thus, the dMP2 neurons regulate the survival of the glia that they need for guidance. In this way, glial number control is linked to axon guidance. In fact, in *vein* mutants glial apoptosis correlates with axon guidance defects.[21] Vein is also required in the PNS, where it activates EGFR in peripheral glia to promote, at least, glial differentiation.[22] Sensory axons express Vein and use peripheral glia for pathfinding. Another ligand of EGFR, the TGFα Spitz, is also involved in neuron-glia interactions during axon guidance. It binds EGFR in the midline glia, controlling their survival. Spitz binding to EGFR triggers the MAP Kinase pathway which results in the phosphorylation of the pro-apoptotic Hid (now called Wrinkled), causing its inactivation.[23]

The FGF Receptor and PDGF Receptor are also involved in promoting oligodendrocyte survival and proliferation in vertebrates, and at least some of their ligands are produced by neurons.[24,25] However, it is not clear whether they are required in glial cells during axon guidance. In some cases at least FGFs are required in growth cone guidance, but by binding to FGFRs present in the neurons.[26] In grasshopper, FGF coated beads attract CNS glial cells in culture and activate the FGFR which is present at the axon-glia interface, suggesting that glial FGFR responds to secreted FGF.[27] The Drosophila homologue heartless is also expressed in enwrapping CNS glia at the time of axon guidance[28] (Fig. 1B). However, it is not known whether grasshopper or Drosophila FGFRs respond to neuronal ligands. In Drosophila, the FGFR breathless is expressed in midline neurons and mutations in breathless result in midline glia migration defects, but whether this is a primary or a secondary effect is not known[29] (Fig. 1A).

Midline glial cells migrate over the axons of the VUM neurons in a dynamic series of neuron-glia interactions that are necessary for axon guidance across the midline (Fig. 1). Kette is a transmembrane protein produced by VUM axons and mutations in *kette* cause defects in midline glia migration, although this seems to be a secondary effect.[30] In fact, Kette functions with the small GTPases to organise the cytoskeleton during axon guidance. Thus, Kette is necessary in the neuron for pathfinding. It will be interesting to find out whether Kette interacts with glial signals to promote axon pathfinding.

Genes Involved in Glial Development That Enable Axon Guidance, Fasciculation or Targeting

The axons of retinal photoreceptor neurons target to two distinct layers in the Drosophila brain, the lamina and the medulla (Fig. 1C). Targeting at the lamina is enabled by glial cells. Mutations in two genes, nonstop and *Jab5/csn5*, prevent normal glial development, and as a consequence axons that would normally target to the lamina carry on extending to the medulla.[31] This means that lamina glia normally produce a stop signal that enables some of the retinal axons to stop extending and contact their targets at the lamina. Other genes are required in the photoreceptor neurons to respond to the glial stop signal: *Off-trak, misshapen, dreadlocks* and *bifocal*.[32] However, the stop signal produced by the glia is unknown.

A study in the moth, *Manduca sexta*, has shown that EGFR in the neurons is necessary for axonal sorting in the olfactory system.[33] They propose that EGFR may interact with Neuroglian, which, through homophylic interactions, may enable axon-glia interactions and axonal fasciculation. In the PNS, EGFR signaling promotes glial differentiation by regulating Neuroglian and Neurotactin expression.[22] PNS glia in turn promote the expression of Futsch, a microtubule associated protein that is necessary for axon extension.[34] In the absence of PNS glia, sensory axons stall.[22]

Functional Implications

Axon Guidance and Glial Migration in the PNS

One of the first studies on the role of glia in axon guidance was the ablation in grasshopper of the Segment Boundary Cell, a glial cell at the exit of the motoraxons from the CNS.[4] When this glial cell was ablated, the neurons that pioneer the nerves do not turn as normal. In Drosophila, genetic ablation of peripheral glia (which are also placed in the transition zone through which motoraxons exit the CNS) using pro-apoptotic genes causes initial defects in motor-axon guidance, which however are later corrected.[6,35] Sensory axons however have a stronger requirement for glia during pathfinding and in the absence of PNS glia they stall and do not enter the CNS (Fig. 1D).[22] Thus, embryonic peripheral glia are necessary for correct exit of motoraxons from and entry of sensory axons into the CNS.

In fact, there is abundant evidence that sensory axons actively use PNS glia for growth cone guidance. Sensory neurons produce the Drosophila neuregulin Vein, which binds the EGFR in PNS glia, signalling through Ras/MAPKinase to promote glial differentiation by regulating terminal differentiation markers such as Neuroglian and Gliotactin.[22] Upon interference with Ras/MAPkinase signalling in PNS glia, sensory axons stall, just as when PNS glia are ablated. Furthermore, PNS glia regulate the expression of Futsch in sensory neurons, which is necessary for axon extension.[34]

Vertebrate olfactory ensheathing glia are also involved in the transition zone as PNS axons enter the CNS. Axons follow trajectories defined by these glial cells, which produce extracellular and cell adhesion molecules such as N-CAM, β2-laminin, L1, Collagen, etc, permissive for axonal growth.[36] Furthermore, olfactory ensheathing glial cells migrate with extending axons, but are responsible for guidance of the pioneer axons. Finally, these glial cells also produce Nexin, a protease with axon-promoting properties.[36]

In other PNS contexts, however, glia are not required for axon guidance. Instead, glia migrate along axons that follow other navigation cues.[37-39] In the Drosophila wing, glia do not need axons for motility, but they do for direction[37] (Fig. 1D). In the lateral line of zebrafish, time-lapse movies of fluorescently labelled axons and glia in living fish have shown that the glia (Schwann cells) migrate lagging behind the extending axons.[39] Mutations that lead to loss of lateral line glial cells, such as in *Sox10/colourless* mutants, do not affect axon pathfinding. In other mutants that cause abnormal somite patterning, such as *sonic you* and *you too*, the axons take a different route, and glia follow them. Furthermore, in double mutants lacking any of these two genes as well as *Sox10/colourless,* the axons still misroute along the same abnormal pathway in the absence of glia. These findings mean that in the fish lateral line glia are not necessary for correct axonal navigation. Glial cells are required later on for correct fasciculation. Migrating lateral line glia express ErbB receptors and mutations in *erbb2* and *erbb3* receptors and blocking ErbB signalling with pharmacological inhibitors prevents lateral line glial migration.[19] This suggests that neuregulin signalling in zebrafish is responsible for the neuron-glia interactions that control glial migration along the axons. It will be interesting to verify in this context if the neurons express neuregulin.

Interestingly, axons of the lateral line follow a migrating target to trace their trajectory.[40] Hence, although it is not glia that axons follow for guidance in this case, they do not navigate on their own. Thus, understanding axon guidance at the lateral line depends on understanding what controls the migration of this moving target, and how these cells relate to the lateral line axons.

Axon Guidance and Targeting in the CNS

Despite the lack of a role for glia in axon guidance in the PNS, in the CNS of zebrafish, glia are required for axon guidance during the formation of the forebrain commissures.[41] Commissural axons cross the midline over a bridge formed by glial cells. The expression of Slits defines domains of glial exclusion, confining the glia to bands that prefigure the axonal commissures. Radial glia also aid axonal extension in the spinal cord in other vertebrate models.[42]

The role of glia in axon guidance in the CNS has been studied in four glial classes in Drosophila: the midline glia, the longitudinal glia (a subset of the interface glia), the retino-basal of the eye and optic lobe glia of the brain. The CNS glia are required for axon guidance, targeting and fasciculation.

The midline glia produce chemoattractants, such as Netrin, and chemorepellents, such as Slit, that control crossing by many interneuron axons.[9] Thus, interneuron axons project across the midline attracted by Netrin, and after crossing it they respond to the repulsive signal Slit to extend contralaterally. This role of midline glia in axon guidance is well recognised, so it will not be discussed further here.

The formation of the longitudinal pathways proceeds in two steps: first a small number of pioneer axons (four per half-segment, at each side of the midline) traces an axonal scaffold; subsequently, all the remaining follower interneurons (about 350 per half-segment) project, many across the midline, and fasciculate with the pioneer axons along the longitudinal trajectories. Genetic ablation of glia using targeted *ricin* expression and analysis of glial cells missing mutants has revealed that the formation of the pioneer scaffold requires glial cells at choice points for growth cone turning and defasciculation.[5,43] For example, pioneer axons and motoraxons first extend together and glial cells trigger their defasciculation, causing the motoraxon to turn and leave the CNS. After the first longitudinal fascicle is formed, glia trigger defasciculation leading to sorting of axons into three fascicles. The follower neurons also require longitudinal glia for guidance and fasciculation.[43] Further evidence that glia are necessary for guidance of follower neurons comes from looking at axonal trajectories in embryos where the response to the midline repulsive signal Slit has been changed. Ectopic expression of Robo2 causes the displacement of axons that normally project close to the midline to more lateral positions.[44,45] This lateral displacement takes place also in *robo1* mutants, but not in *robo1 glial cells missing* double mutants, in which axons misroute towards the muscle.[46]

Axons of retinal neurons leave the retina to form the optic stalk, reach the optic lobe in the brain and they may target in the lamina or project further to target in the medulla (Fig. 1C). Glial cells are involved at several points along this trajectory. Initially, there are no retinal-derived glial cells, so retinal axons begin to extend without glial cells. But as they leave the retina, retino-basal glia aid their entry into the optic stalk. If glial number is severely reduced, the retinal axons are not able to enter the optic stalk.[47] Furthermore, if glia occupy abnormal positions in the retina, neurons project now towards the ectopic glia rather than exiting the retina into the optic stalk.[47] Thus, retino-basal glia aid guidance of retinal axons into the optic stalk.

When the retinal axons reach the brain, their target fields are covered in glial cells, and glial cells are in fact required for targeting. If glia are abnormal or missing, such as in *nonstop* and in *jab1/csn5* mutants, the neurons that normally would target at the lamina (R1-R6) instead carry on extending past the lamina into the medulla.[31] This means that lamina glia normally produce a 'stop signal' that is received by R1-R6 photoreceptor axons. The nature of this signal is unknown.

Glial Migration in the CNS

Like in the PNS, glial migration also depends on neurons in the CNS. While midline glia attract interneuron axons, their positions are not static. The interneuron axons start crossing the midline along two commissural fascicles that are initially fused at the midline forming what is known as the 'bow-tie'. Midline glia then migrate over the axons of the VUM neurons to trigger defasiculation at the axonal 'bow-tie' and cause the separation of the anterior and posterior commissures[48] (Fig. 1A). The VUMs are a small subset of motorneurons that send axons first along the midline and then away from the CNS and do not project along the commissures. Thus, the midline glia follow cues on the VUMs to migrate, but in turn they send cues to all interneurons to control midline crossing.

Midline glia not only follow clues on the VUMs to migrate, but also, normally, on all commissural axons. In *commissureless* mutants the are no commissures, as all axons run longitudinally without crossing the midline. As a consequence of lack of axonal contact, many of the midline glia die. However, some midline glia do survive and these do not retain their midline positions, but instead migrate to the lateral positions to retain neuronal contact with the only present longitudinal axons.[49] This also implies that the migration preferences of midline glia are in part controlled by trophic factors emanating from the axons.

Longitudinal glia migration also depends on neurons. The longitudinal glia are necessary for guidance of pioneer axons, defasciculation and axonal sorting, but they do not form a prepattern prior to axonal extension. Prior to axon guidance, the longitudinal glia migrate from their site of origin to the edge of the neuroectoderm to the midline, and stop at a certain distance from the midline by sensing midline repulsion from Slit.[50] Subsequently, and as axons begin to extend, the longitudinal glia migrate together with the extending axons, just ahead of the growth cones. Together growth cones and glia explore a territory devoid of either of them, but because glial lamellipodia are so large, a glial network is established across segments before the neurons come into contact. However, ablation of a fraction of the neurons causes the remaining axons to misroute, and the longitudinal glia to migrate following those axons.[21,50] These findings mean that there are reciprocal interactions between neurons and glia during axon guidance.

Similarly circular neuron-glia interactions take place during innervation of the optic lobe by retinal axons. Retinal R1-R6 axons meet their neuronal targets in the lamina of the optic lobe, aided by glial cells. In the absence of Hedgehog signalling from photoreceptor axons, lamina neurons do not develop.[51] Photoreceptor axons induce the formation of an axonal scaffold that optic lobe glia follow to find their locations at the lamina. Thus, the photoreceptor axons induce the migration of the glia so that they occupy the correct positions in which the retinal neurons need them for targeting.[52]

These observations also imply that the molecules controlling neuron-glia interactions that form the pioneer scaffolds are likely to be different from those involving large numbers of neurons. Pioneer neurons and the VUMs belong to a small group of neurons that are highly conserved amongst arthropods.[53] The interactions of glia with these classes of neurons appear to lead to the formation of scaffolds, whereas the interaction of glia with the majority of neurons seems more like paving motorways. Perhaps there is an evolutionary need to separate these two functions, to enable the reliable formation of a scaffold providing robustness to the projections of the majority of neurons. Consistently with this idea, molecules like Vein or Kette are involved in interactions between glia and the pioneer groups but not with the majority of neurons.

The Control of Glial Number during Axon Guidance

Glia are required during axon guidance and fasciculation, but the same glial classes also later on enwrap axons to enable neuronal function. Furthermore, we have seen that the axon influences glial cell migration. It is well known that ultimately the glial mass that ensures appropriate axonal enwrapment is regulated by axons. Glia are overproduced and only those that enable

Table 1. Molecules involved in neuron-glia interactions during axon guidance

	Molecule	Organism	Expressed In	Function	Refs.
Signalling Molecules					
	Netrin	Drosophila Vertebrates *C. elegans*	Midline glia Floorplate	Axonal attraction	9–15
	Slit	Drosophila Vertebrates *C. elegans*	Midline glia Floorplate	Axonal repulsion	9–15
	Sonic hedgehog	Vertebrates	Floorplate	Axonal attraction	13,16
	Semaphorin	Drosophila Vertebrates *C.elegans*	Floorplate	Axonal repulsion	9–15
	Vein/ Neuregulin	Drosophila vertebrates	Neurons	CNS PNS Glial survival proliferation differentiation	17,18, 21,22
	Spitz (TGFα)	Drosophila	unknown	Midline glial survival	23
Receptors					
	Robos/Rig-1	Drosophila Vertebrates *C. elegans*	Neurons	Axonal repulsion	9–15
	Frazzled/ Unc/DCC	Drosophila Vertebrates *C. elegans*	Neurons	Axonal attraction and repulsion	9–15
	Plexins, Neuropilins	Drosophila Vertebrates *C. elegans*	Neurons	Axonal repulsion	9–15
	EGFR/ErbB	Drosophila Vertebrates Manduca	CNS and PNS glia	Glial survival proliferation differentiation migration	19,21,23, 33,22
	FGFR/Heartless	Grasshopper Drosophila Vertebrates	CNS glia	Glial migration Proliferation survival	27,28
	FGFR/Breathless	Drosophila	Midline	Glial migration	29
Other					
Ubiquitin protease	Nonstop	Drosophila	Photoreceptor axons	Targeting	31
Protein degradation	Jab1/csn5	Drosophila	Target glia	Targeting	31
Receptor Tyr Kinase	Off-track	Drosophila	Photoreceptors	Targeting	32
Ser/Thr kinase	Misshapen	Drosophila	Photoreceptors	Targeting	32
Adaptor protein	Dreadlocks (Dock)	Drosophila	Photoreceptors	Targeting	32
Cytoskeletal regulator	Bifocal	Drosophila	Photoreceptors	Targeting	32
Transmembrane	Kette	Drosophila	VUM neurons	Glial migration	30

appropriate enwrapment survive, maintained by trophic factors provided by the axons.[25,54,55] These observations imply that the number of glial cells is regulated during axon guidance and up to enwrapment.

In Drosophila, the neuregulin Vein is expressed by pioneer neurons and it promotes the survival of a subset of longitudinal glial cells, those that the pioneer neurons themselves require for axon guidance[21] (Fig. 1B). Longitudinal glia also migrate together with the longitudinal pioneer axons. Thus, trophic support of the longitudinal glia by the pioneer axons constrains the migration of glia, in turn resulting in correct axonal pathfinding. After the first fascicles form, the axons get reshuffled and sorted into new fascicles. These defasciculation and refasciculation events are also triggered by glial cells.[5] During these events, the longitudinal glioblast lineage produces progeny cells sequentially, delivering longitudinal glia in restricted numbers to enable axonal sorting.[20] This temporal control of glial proliferation is first triggered also by axonal Vein and it requires the transcription factor Prospero and Notch in the glia, which together control their mitotic potential.[20] In this way, glia divide sequentially, delivering restricted numbers of progeny cells that enable the sorting of axons through time.

Conclusions and Perspectives

There is abundant experimental evidence on the involvement of glia and on the complexity of axon-glia interactions during pathfinding. In fact, even in the clearest cases for roles of glia in axon guidance, interactions with axons are reciprocal, as the neurons regulate the survival, proliferation and migration of the glia they need for pathfinding. So far, across species and glial lineages, the neuregulin signalling system is evolutionary conserved. Neuregulin is produced by neurons and it signals through EGFReceptor molecules in glia to regulate the number and differentiation of glia necessary for axonal pathfinding. However, on the whole the molecular mechanisms underlying neuron-glia interactions during axon guidance are far from complete. We have some knowledge of some of the molecules produced by either neurons or glia during pathfinding, but in most cases the partners for these molecules in the reciprocal cell type that would explain axon-glia interactions are not known. It is an exciting time, as these mechanisms begin to emerge.

The principles guiding axonal extension may not be so different in development and in repair. Thus, understanding the role of glia in axon guidance and the transition from axon guidance to enwrapment may enable us to harness the potential of the nervous system to repair itself. Not only during development, but also upon injury adult glia can also promote axonal growth and pathfinding. In fact, transplantation of olfactory ensheathing glia to the site of axonal injury is sufficient to promote axonal regeneration and full recovery of neuronal function.[56] Abundant experimental evidence has shown that axons extend preferentially over glial cells and that glial bridges enable axonal regeneration (e.g., ref. 57). Using model organisms can help us find out which are the key molecules and signalling pathways involved in neuron-glia interactions during axon guidance. We can then extrapolate this knowledge to test the potential of these molecules in promoting repair.

References

1. Bate CM. Pioneer neurones in an insect embryo. Nature 1976; 260:54-55.
2. Singer M, Nordlander RH, Egar M. Axonal guidance during embryogenesis and regeneration in the spinal cord of the newt: The blueprint hypothesis of neuronal patterning. J Comp Neurol 1979; 185:1-22.
3. Silver J, Lorenz SE, Wahlstein D et al. Axonal guidance during development of the great cerebral commissures: Descriptive and experimental studies in vivo on the role of the preformed glial pathways. J Comp Neurology 1982; 210:10-29.
4. Bastiani MJ, Goodman CS. Guidance of neuronal growth cones in the grasshopper embryo. III. Recognition of specific glial pathways. J Neurosci 1986; 6:3542-3551.
5. Hidalgo A, Booth GE. Glia dictate pioneer axon trajectories in the Drosophila embryonic CNS. Development 2000; 127:393-402.
6. Sepp KJ, Schulte J, Auld VJ. Peripheral glia direct axon guidance across the CNS/PNS transition zone. Developmental Biology 2001; 238:47-63.

7. Freeman M, Doherty J. Glial cell biology in Drosophila and vertebrates. Trends in Neurosciences 2006; 29:82-90.
8. Chotard C, Salecker I. Neurons and glia: Team players in axon guidance. Trends in Neurosciences 2004; 27:655-661.
9. Araujo SJ, Tear G. Axon guidance mechanisms and molecules: Lessons from invetebrates. Nature Reviews Neuroscience 2003; 4:916-922.
10. Long H et al. Conserved roles for Slit and Robo proteins in midline commissural axons. Neuron 2004; 42:213-223.
11. Jen J, al e. Mutations in a human Robo gene disrupt hindbrain axon pathway crossing and morphogenesis. Science 2004; 304:1509-1513.
12. Sabatier C et al. The divergent Robo family protein Rig-1/Robo3 is a negative regulator of Slit responsiveness required for midline crossing by commissural axons. Cell 2004; 117:157-169.
13. Kaprielian Z, Runko E, Imondi R. Axon guidance at the midline choice point. Developmental dynamics 2001; 221:154-181.
14. Lemke G. Glial control of neuronal development. Annu Rev Neurosci 2001; 24:87-105.
15. Zou Y, Stoeckli E, Chen H et al. Squeezing axons out of the gray matter: A role for Slit and Semaphorin proteins from midline and ventral spinal cord. Cell 2000; 102:363-375.
16. Charron F, Stein E, Jeong J et al. The morphogen Sonic Hedgehog is an axonal chemoattractant that collborates with Netrin-1 in midline axon guiidance. Cell 2003; 113:11-23.
17. Garratt AN, Britsch S, Birchmeier C. Neuregulin, a factor with many functions in the life of a Schwann cell. BioEssays 2000; 22:987-996.
18. Fernandez PA et al. Evidence that axon-derived Neuregulin promotes oligodendrocyte survival in the developing rat optic nerve. Neuron 2000; 28:81-90.
19. Lyons DA et al. Erbb3 and erbb2 are essential for Schwann cell migration and myelination in zebrafish. Current Biology 2005; 15:513-524.
20. Griffiths RL, Hidalgo A. Prospero maintains the mitotic potential of glial precursors enabling them to respond to neurons. EMBO J 2004; 23:2440-2450.
21. Hidalgo A, Kinrade EFV, Georgiou M. The Drosophila Neuregulin Vein maintains glial survival during axon guidance in the CNS. Developmental Cell 2001.
22. Sepp KJ, Auld VJ. Reciprocal interactions between neurons and glia are required for Drosophila peripheral nervous system development. The Journal of Neuroscience 2003; 23:8221-8230.
23. Bergmann A, Tugentman M, Shilo BZ et al. Regulation of cell number by MAPK-dependent control of apoptosis: A mechanism for trophic survival signaling. Dev Cell 2002; 2:159-170.
24. Hoch RV, Soriano P. Role of PDGF in animal development. Development 2003; 130:4769-4784.
25. Miller RH. Regulation of oligodendrocyte development in the vertebrate CNS. Progress in Neurobiology 2002; 67:451-467.
26. Bovoloenta P. Morphogen signalling at the vertebrate growth cone: A few cases or a general strategy? J Neurobiol 2005; 64:405-416.
27. Condron B. Spatially discrete FGF-mediated signalling directs glial morphogenesis. Development 1999; 126:4635-4641.
28. Shishido E, Ono N, Kojima T et al. Requirements of DFR1/Heartless, a mesoderm specific Drosophila FGFReceptor for the foromation of heart, visceral and somatic muscles, and ensheathing of longitudinal axon tracts in the CNS. Development 1997; 124:2119-2128.
29. Klämbt C, Glazer L, Shilo BZ. Breathless, a Drosophila FGF receptor homolog, is essenlian for migration of tracheal and specific midline glial cells. Genes Dev 1992; 6:1668-1678.
30. Hummel T, Leifker K, Klämbt C. The Drosophila HEM-2/NAP1 homolog KETTE contorls axonal pathfinding and cytoskeletal organization. Genes Dev 2000; 14:863-873.
31. Poeck B, Fischer S, Gunning D et al. Glial cells mediate target layer selection of retinal axons in the developing visual system of Drosophila. Neuron 2001; 29:99-113.
32. Mast JD, Prakash S, Chen PL et al. The mechanisms and molecules that connect photoreceptor axons to their targets in Drosophila. Seminars in Cell and Developmental Biology 2005.
33. Gibson N, Tolbert L. Activation of epidermal growth factor receptor mediates receptor axon sroting and extension in the developing olfactory system of the moth Manduca sexta. J Comparative Neurology 2006; 495:554-572.
34. Hummel T, Krukkert K, Roos J et al. Drosophila Futsch/22c10 is a MAP1B-like protein required for dendritic and axonal development. Neuron 2000; 26:357-370.
35. Sepp KJ, Schulte J, Auld V. Developmental dynamics of peripheral glia in Drosophila melanogaster. Glia 2000; 30:122-133.
36. Vincent A, West A, Chuah M. Morphological and functional plasticity of olfactory ensheathing cells. Journal of Neurocytology 2005; 34:65-80.

37. Aigouy B, Van de Bor V, Boeglin M et al. Time-lapse and cell ablation reveal the role of cell interactions in fly glial migration and proliferation. Development 2004; 131:5127-5138.
38. Giangrande A. Glia in the fly wing are clonally related to epithelial cells and use the nerve as a pathway for migration. Development 1994; 120:523-534.
39. Gilmour D, al e. Migration and function of a glial subtype in the vertebrate peripheral nervous system. Neuron 2002; 34:577-588.
40. Gilmour D, NKnaut H, Maischein HM et al. Towing of sensory axons by their migrating target cells in vivo. Nature Neuroscience 2004; 7:491-492.
41. Barresi M, Hutson L, Chien CB et al. Hedgehog regulated Slit expression determines commissure and glial cell position in the zebrafish forebrain. Development 2005; 132:3643-3656.
42. McDermott K, Barry D, McMahon S. Role of radial glia in cytogenesis, patterning and boundary formation in the developing spinal cord. J Anat 2005; 207:241-250.
43. Hidalgo A, Urban J, Brand AH. Targeted ablation of glia disrupts axon tract formation in the Drosophila CNS. Development 1995; 121:3703-3712.
44. Rajagopalan S, Vivancos V, Nicolas E et al. Selecting a longitudinal pathway: Robo receptors specify the lateral position of axons in the Drosophila CNS. Cell 2000; 103:1033-1045.
45. Simpson JH, Bland KS, Fetter RD et al. Short-range and long-range guidance by Slit and its Robo receptors: A combinatorial code of Robo receptors control lateral position. 2000; 103.
46. Kinrade E, Hidalgo A. Lateral neuron-glia interactions steer the response of axons to the Robo code. Neuron Glia Biology 2004; 1:101-112.
47. Rangarajan R, Gong R, Gaul U. Migration and function of glia in the developing Drosophila eye. Development 1999; 126:3285-3292.
48. Klämbt C, Jacobs JR, Goodman CS. The midline of the Drosophila central nervous system: A model for the genetic analysis of cell fate, cell migration and growth cone guidance. Cell 1991; 64:801-815.
49. Sonnenfeld MJ, Jacobs JR. Apoptosis of the midline glia during Drosophila embryogenesis: A correlation with axon contact. Development 1995; 121:569-578.
50. Kinrade EFV, Brates T, Tear G et al. Roundabout signalling, cell contact and trophic support confine longitudinal glia and axons in the drosophila CNS. Development 2001; 128:207-216.
51. Huang Z, Kunes S. Hedgehog, transmitted along retinal axons, triggers neurogenesis in the developing visual centers of the Drosophila brain. Cell 1996; 86:411-422.
52. Dearborn R, JKunes J. An axon scaffold induced by retinal axons directs glia to destinations in the Drosophila optic lobe. Development 2004; 131:2291-2303.
53. Arendt D, Nübler-Jung K. Comparison of early nerve cord development in insects and vertebrates. Development 1999; 126:2309-2325.
54. Barres BA, Raff MC. Control of oligodendrocyte number in the developing rat optic nerve. Neuron 1994; 12:935-942.
55. Raff M et al. Programmed cell death and the control of cell survival: Lessons from the nervous system. Science 1993; 262:695-700.
56. Ramon-Cueto A, Cordero M, Santos-Benito F et al. Functional recovery of paraplegic rats and motor axon regeneration in their spinal cords by olfactory ensheathing glia. Neuron 2000; 25:425-435.
57. Li Y, Carlstedt T, Berthold CH et al. Interaction of transplanted olfactory-ensheathing cells and host astrocytic processes provides a bridge for axons to regenrate acress the dorsal root entry zone. Experimental Neurology 2004; 188:300-308.

Index